Biotechnology in Animal Health and Production Management

Biotechnology in Animal Health and Production Management

Gaurav Tyagi

RANDOM

RANDOM PUBLICATIONS
NEW DELHI (INDIA)

Biotechnology in Animal Health and Production Management

ISBN 978-93-5111-786-5

Published in 2016 in India by

RANDOM PUBLICATIONS

4376-A/4B, Gali Murari Lal, Ansari Road
New Delhi-110 002
Phone : +9111-43580356, 011-23289044, 011-43142548
e-mail: sales@randompublications.com,
info@randompublications.com, randomexports@gmail.com

Reprinted 2026

Type Setting by : Friends Media, Delhi-110089

Preface

The Animal Science and Management major brings together studies in animal nutrition, growth and development, behaviour and welfare, genetics and breeding, health, livestock science and reproduction, and biotechnology. These studies allow people to manage animals in a way that integrates animal welfare with modern and demanding production requirements.

Techniques of modern biology such as molecular cloning of genes, gene transfer, genetic manipulation of animal and plant embryo transfer, genetic manipulation of rumen microbes, chemical and biological treatment of low quality animal feeds for improved nutritive value, genetically engineered immunodiagnostic and immunoprophylactic agents as well as veterinary vaccines, inter alia, are a reality today and are finding their ways into research and development programmes of developing countries. Biotechnology is offering unprecedented opportunities for increasing agricultural productivity and for protecting the environment through reduced use of agro-chemicals. The major thrust in biotechnology research is currently directed at solving immediate problems of industrialised countries, with major investments coming from transnational companies.

However, many of the new discoveries and products will find their biggest markets in developing countries where the potential for improvement in agricultural productivity and health is greatest. The importance of biotechnology and its relevance is only slowly being accepted by policy makers in developing countries. In the presence of economic crisis, strong fiscal constraints, rapid social change and constant political instabilities, the difficulties associated with major policy changes in developing countries are enormous. This paper reviews available biotechnologies with potential application in livestock improvement and identifies those which have been or may be applied in developing countries in general, and Africa in particular. The review covers biotechnology applications in the areas of animal genetics and breeding, including conservation of animal genetic resources, animal health, physiology of lactation and growth, and animal nutrition.

Considering the potentials of livestock production to provide unique opportunities for meaningful and lucrative employment to the whole lot of small, marginal and landless farmers spreading across more than half of the globe, the book on Animal Health and Production encompasses views of many learned contributors pertaining to issues related to livestock sector's progress and prospects.

– Author

Contents

1

Introduction

In animal biotechnology, embryo transfer, use of monoclonal antibodies for diagnostics and production of efficient vaccines, hormonal manipulation for growth and productivity, and production of transgenics have been the main areas of work in the region.

Nucleus herd improvement using multiple ovulation and embryo transfer (MOET) in Open Nucleus Breeding Systems (ONBS)

Progress through the conventional approaches of genetic improvement of livestock in the region has been slow when compared with the genetic gains recorded in livestock of North America, Europe and Oceania. To overcome this limitation, MOET in ONBS is being advocated in developing countries as the technique may be especially valuable where the use of field progeny testing and artificial insemination (AI) has largely been a failure because of the need for an established infrastructure, both for field recording and AI.

As summarized by Hodges (1988), a genetically superior nucleus flock or herd established under controlled conditions where testing and genetic selection are carried out. The test group is first established by screening the base population for outstanding females. They are then recorded individually in the nucleus herd and the elite females among them are used by MOET with superior sires to contribute embryos which are carried by the recipient cows from the base population. The resulting offspring are reared and recorded and the males among them are evaluated genetically using the performance of their sibs and paternal half-sibs and their own performance where they exhibit the trait.

From these an elite group of males with high breeding value is selected which is used in the base population for genetic improvement by natural service or AI. ONBS yields higher rates of genetic progress per annum than established progeny testing systems mainly because of the shorter generation interval. It is capable of being adapted to different species in which MOET is available. It also permits control of the mix of genes released to the base population. ONBS is flexible for the introduction of exotic germplasm from other populations, for example semen or embryos from temperate breeds and also for the continuous addition of outstanding individuals screened from the base population. China

(Chengdu Fengherangshan Dairy Farm), India (Anand Dairy Cooperatives, Gujrat) and the Republic of Korea (Livestock Station, Suweon) are already practicing MOET in their nucleus herds. Similar work at private dairy farms in the Philippines, Thailand, Pakistan, Indonesia and Malaysia has been established, but it is still slow and needs strengthening.

Embryo transfer research and application of the technique in improved production and genetic upgrading of cattle, buffaloes, swine, goats and sheep, in varying degrees, are being pursued in several of the Asian countries. Notable progress has been made in standardization of superovulation, synchronization and embryo collection techniques, evaluation and splitting of embryos, optimization of superovulation responses by laproscopy and ultrasonography, embryo manipulation, splitting and standardization of techniques for in vitro fertilization and sexing.

Hundreds of calves using these techniques have been produced in China, India and the Republic of Korea and many have also been produced in Pakistan, the Philippines, Indonesia and Thailand, besides the thousands produced in the three developed countries of the region.

In Japan, ever since the standardization of the embryo transfer technique in 1979, the number of calves produced by embryo culture has steadily increased, from 73 in 1980 to 422 in 1984 and to 3 700 in 1988. Artificial production of identical twins (by slicing embryos in two halves) is also commonly practiced In 1988, 268 pairs of such twins were born throughout Japan. The conception rate has been improving, at around 60 percent for fresh embryos and about 40 percent for frozen embryos in the early 1990s. This technique is most commonly applied to the production of high quality beef cattle, such as Japanese Black.

The in vitro fertilization technique in cattle is also being used. In 1987, the first cattle triplet by in vitro fertilization in the world were born at the National Institute of Animal Industry, a research organization of the MAFF. The Institute has also succeeded in nuclear transplantation in cattle.

Since the Asian countries have almost a monopoly on buffalo production, they need special research and development programmes on the species. Embryo transfer technology in buffaloes is much more difficult than in cattle. For instance, on average, five to six transferable embryos per donor are recovered in cattle against only 1.5 in buffaloes. Although some striking results have been obtained, such as the birth of a riverine calf from a swamp recipient in the Philippines, or the birth of the first IVF (in vitro fertilization) buffalo calf in India, the techniques need to be improved considerably to be efficient and commercially viable.

The inception meeting of the regional FAO/UNDP project, RAS/89/001 "Biotechnology Development Network for Animal Production and Health", held in 1988, recommended that a cooperative project be taken up to study the normal hormonal profile of both riverine and swamp buffaloes. The hormonal profile has to be done under varying field conditions. At the same time, the structure of buffalo ovaries needs to be investigated in detail in various breeds. Since the number of primordial oocytes is very low in buffaloes, it is necessary to introduce an entirely different regime for superovulation.

The inception meeting further recommended that once the problem of superovulation in buffaloes had been sorted out, the situation in regard to other items of interest such as embryo splitting in vitro fertilization, etc. could be undertaken. The network should give high priority to the problems of buffalo embryo transfer and countries should work together to produce the necessary data within a defined timeframe on hormonal profiles, structure of the gonad and possible ovulatory regimes that can be taken up for priming the follicles to be followed up by a superovulatory regime.

RECOMBINANT DNA TECHNOLOGY AND ITS APPLICATIONS

Recombinant DNA technology research in Asia is mainly confined to basic research by molecular biologists involving prokaryotes. This entails the discovery of suitable vectors (plasmids and nonplasmids) for the propagation of genes and the expression of these genes in selected hosts. Some biological laboratories are also engaged in DNA fingerprinting or restriction fragment length polymorphism work. Most of these biological laboratories are adequately equipped to undertake major research and train scientists from agricultural disciplines. Only a few animal science/veterinary laboratories in the developing countries were engaged in molecular biology research. There are very few published reports on the work done, hence the information given below is based on information contained in various reports of the regional FAO/UNDP project on animal biotechnology.

In China most of the work is being conducted by the National Laboratory of Agro-biotechnology, Beijing Agriculture University. The following are the highlights of the work in hand:

- The specific gene of equine infectious anemia has been successfully cloned.
- RFLP is used in recharacterizing of some indigenous animal breeds and in classifying some wild animals.
- The gene mapping method is used to identify resistant genes in the MHC B-L subregion of O-type chickens, and to identify other related genes in the vicinity of GH genes in pigs.
- The following gene constructs have been made: bovine genomic growth hormone gene, and procine genomic growth hormone gene, sheep genomic growth hormone gene and procine growth hormone releasing factor gene. Genes for production related hormones (calf renin, bovine pepsin and buffalo renin) have been constructed.

In India major work has been undertaken since the 1980s by the Indian Veterinary Research Institute (IVRI), Izatnagar and Bangalore campus and the National Institute of Immunology (NII), New Delhi, in the following areas.

- Molecular characterization of pox virus, especially caprine virus (both virulent and non-virulent types) is in progress using gene sequencing techniques. Identification of promoter region in the genome is being made in collaboration with the Virus Research Institute, Pirbright, United Kingdom.
- Studies on molecular cloning in Clostridial species for vaccine production have been initiated. Early work involved DNA isolation of non-pathogenic Clostridial species (C. perfringenes, type C and type D), its characterization and isolation of plasmid DNA from host vector.
- Another group of workers has been engaged in the characterization of gene sequences in Clostridium welchi type D and location of gene blocks, responsible for the production of "alfa toxin".
- Molecular characterization of FMD virus type, O, A and C has been attempted. Complementary DNAs for viral RNA type A and O were prepared. A c-DNA probe has been successfully used for identification of virus specific sequences.
- Cloned-DNA for the major antigen of FMD disease has been shown to be expressed in E. coli.
- A partial library of Y chromosome derived DNA sequence of bovine origin in E. cold has been constructed at NII.
- Characterization of extra chromosomal plasmids responsible for desired characters of starter culture and identification of plasmid for lactose utilization in strains of streptococci.

In Japan, research in transgenic animals is highly advanced. In 1989, the National Institute of Animal Industry, following work in the United Kingdom, succeeded in in vitro culture of poultry eggs, achieving a good hatchability of about 34 percent. These results greatly increased the feasibility of transgenic chickens. As regards animal health, a recombinant vaccine of bovine Iymphatic leukaemia virus has been developed by the National Institute. A rabies virus vaccine was also developed. Highly specific and sensitive antibodies have been designed and are being used for the diagnosis of bovine Iymphatic leukaemia, rotavirus infection, mycobacteriosis and paratuberculosis, which are difficult to detect by traditional diagnostic methods.

Japan is harvesting biotechnology for altering silkworms to increase the production and quality of silk filaments and to find new horizons of use other than traditional silk production. Using recombinant DNA technology, the National Institute of Sericultural and Insect Agricultural Technology is transferring the unique glossy yellow greenish silk colour from wild silkworms to domesticated silkworms. A technique of utilizing silkworms as the "factory" for producing useful substances has been established. In this method, for instance, the genes interferon are incorporated into nuclear polyhedrosis virus, and this virus is propagated in the body of silkworms so as to be mass produced. This technique was found to be more efficient than using bacteria as hosts and will be applied to the production of many other useful substances in the years ahead.

The structure and function of hormones controlling the growth and metamorphoses of insects have been understood, and as a result it is now possible to control the growth of insects to some extent. Using this technique to control Silkworm larval cycle and maturity, it is now possible for Japanese scientists to control the thickness of silk filaments from one-tenth to three times that of ordinary ones. In Malaysia, genetic engineering research for animal production and health has been undertaken by the faculty of Biotechnology and Food Sciences and the Veterinary Faculty of University Pertanian Malaysia (UPM) and two Departments of University Malaysia (UM)-the Department of Genetics and Cellular Biology and the Institute for Advanced Studies. Research in progress or completed include:

- Molecular characterization of Newcastle Disease (NCD) virus.
- Collection of baseline data on diversity of rumen bacteria and fungi has been completed and studies on genomic and plasmid DNA in selected microbes and fungi have begun.
- Cloning of xylanase gene for bacteroides succinogens, using E. cold HB 101 as the host.
- Mitochondrial and nuclear RFLP work as markers for the identification of different breeds of buffaloes and chickens is in progress. The cloning of pituitary growth hormone gene using retroviruses as a vector was initiated.
- A study has been completed to relate the transformable antibiotic resistance traits in E. cold strains of bovine, avian and procine sources from slaughterhouses with the frequency of R plasmids, which are extrachromosomal genetic markers, and are suspected to be responsible for drug resistance.

In the Philippines, basic work on several aspects of animal health biotechnology is being conducted at the National Institute of Applied Microbiology and Biotechnology, Los Baños.

The Institute uses protoplast fusion and recombinant DNA technology in the genetic improvement of antibiotic producing microbial strains. Research on local production of tylosin and antibiotics for animal feed is also in progress.

The National Biotechnology Institute, Los Baños, has also undertaken work on the development of low cost biogas systems using crop residues, production of genetically improved lignolytic micro organisms, enzyme engineering techniques for synthesis of ligniases, and the low technology tumbler process of producing microbial proteins using selected strains of fungi and yeast which could be used as a feed ingredient for poultry and pigs.

As regards other Asian countries, the Veterinary Faculty of Kasetsart University, Thailand, the Central Research Institute for Animal Sciences, Bogor, Indonesia, and the Department of Animal Sciences, Agriculture University at Faisalabad, Pakistan, have initiated work on the identification of genetic markers for production and disease traits in livestock.

BIOTECHNOLOGY AND ANIMAL HEALTH

The potential areas of impact for biotechnology on the control of infectious diseases and improved animal health are manifold. They include the identification of susceptible stock, tests for the rapid and early diagnosis of disease, an understanding of the mechanisms of pathogenesis and the development of safe and effective vaccines or prophylactics. The principal methodologies which will enable these targets to be approached are those of monoclonal-antibody production and recombinant-DNA technology. The benefits of selecting stock on the basis of resistance to specific infectious agents would appear at present to be restricted to those areas of the world where the importance of a single disease predominates over all others (*e.g.*, trypanosomiasis). Whilst the possibility of selecting on an antigen non-specific basis is likely to be of more general relevance, at present the development of vaccination strategies that can effectively control infectious agents is likely to be of greatest significance for improved animal health.

To this end monoclonal antibodies can be used in the development of diagnostic sera and reagents, and the characterisation of virulence and adhesion determinants. Once identified such determinants can be isolated in a pure form from crude mixtures with monoclonal immuno-sorbents.

Infectious diseases may be controlled by the prophylactic use of monoclonal antibodies, but this direct effect will be of limited use in animal production, by comparison with their potential use as 'educators' of the immune system in anti-idiotype vaccines. Recombinant-DNA technology can provide a range of potential immunogens including attenuated or non-pathogenic strains and subunit vaccines. Since the majority of infectious agents are met at mucosal surfaces, the problem that remains is how these vaccines may be delivered so as to provide effective presentation and so stimulate a protective response.

MUCOSAL DEFENCE

The level of challenge to, and the degree of sterility maintained differs markedly between the various mucosal surfaces; from the gastrointestinal and upper respiratory tracts to the lower respiratory tract, reproductive tract and mammary gland which at least in health remain sterile. They are presented with a vast array of different antigens ranging from potential pathogens to relatively harmless dietary components. Together with the skin they provide the interface between the host and its environment. However, unlike the skin there are other physiological roles (*e.g.*, gaseous exchange, absorption of nutrients, etc.) which restrict their ability to keep antigens out. Since the mucosal surfaces provide such a primary point of contact it is at these sites that the products of biotechnology might be expected to provide the most relevant and beneficial effects. For convenience the major part of the present review will be restricted to antigens presented to the gut.

The immune system of the gastrointestinal tract is presented with two types of antigens, those capable of causing infection and those, such as dietary antigens, which in themselves impose no direct threat to the animal's health. The type of response stimulated is critical, for whilst failure to respond to potential pathogens may be fatal, an over-reaction to dietary antigens may lead to damaging hypersensitivity responses. Latter may in turn lead to altered handling of potential pathogens. Therefore, although the most obvious area of impact of biotechnology will be directly on infectious agents, regulating responses to food antigens may also be of significance.

THE RESPONSE TO FOOD ANTIGENS

Three responses have been described that may protect an individual from damaging hypersensitivity reactions to dietary antigens. Immune exclusion mediated by locally produced IgA may reduce the absorption of intact proteins across the gut mucosae and in certain species it has been shown that any antigen that is absorbed may be eliminated via bile.

These processes which restrict the access of antigenic material to the immune system are not absolute and the development of an acquired specific hyporesponsive state is probably of greater significance. Thus orally induced tolerance may actively down-regulate the response to dietary proteins. Before the induction of this 'protected state' all animals pass through a transient phase of hypersensitivity, during which time damaging allergic reactions can be demonstrated. In farm species it has been shown that following weaning, piglets are sensitive to antigens in the post-weaning diet and since these changes occur before the proliferation of *Escherichia coli* it has been postulated that response to food proteins may be an important predisposing factor in the aetiology of post-weaning diarrhoea. The importance of gut flora and bacterial lipopolysaccharide in regulating the response to dietary antigens has been clearly shown in rodents, thus by the use of engineered strains of bacteria it may be possible to regulate the response to dietary proteins and so prevent the transient allergic responses that reduce the gut's ability to handle potential pathogens. This brief description of the regulation of the response to food proteins attempts to serve two purposes. First, it may indicate a role for engineered strains in the controls of hypersensitivity reactions and the consequences thereof. Second, the type of response generated to food proteins differs markedly from that required to eliminate infectious agents. It is then perhaps not surprising, as will be discussed later, that dead mucosal microbial vaccines are generally not very effective.

THE RESPONSE TO INFECTION

The implications of biotechnology on animal health directly related to infections are numerous; it is therefore proposed to discuss them sequentially from diagnosis to vaccine development. The conventional methods for the identification of the causal agent have in the past depended heavily on an ability to isolate and grow the organism in pure culture. The use of monoclonal antibodies should aid both the identification of infectious agents and their virulence factors. Unlike conventional anti-serums, which by virtue of the nature of the immune response are very heterogenous, monoclonal proteins have a unique specificity which allows such discrimination to be made. For example, by use of the enzyme-linked immunosorbent assay (ELISA) it is possible to detect subtle antigen differences between strains of an organism as well as to characterise the immune response of the infected animal. Once having identified a particular virus, for example as being important in the pathogenesis of diarrhoea, how then might one proceed towards an effective vaccine? Consider for instance that virus X has been shown to be an important pathogen but is closely related to similar commensals. It can produce disease, and animals that recover from infection show solid protection.

In neutralisation tests it has been possible to show the specificity of X, but it is very difficult to grow! Further progress may then involve the development of a sensitive immunoassay for research and sero-epidemiology. Production of a monoclonal antibody (either conventionally from rodents or preferably from infected farm species) may allow isolation of a strain-specific antigen. Subsequently the antigen should be produced by cloning and expressing the gene in, for example E. *coli,* by recombinant-DNA technology. Such a procedure may form the basis for the development of a vaccine.

CONTROL OF INFECTION

Without attempting to consider ways of eradicating an organism from the environment there are two types of approach to the preventionxontrol of infection within an animal. The first is essentially passive and involves little or no requirement of the host's immune system, whilst the second requires the stimulation of an active vigourous immune response.

PASSIVE PROTECTION

Monoclonal antibodies have been used therapeutically to passively protect neo-natal pigs and calves against challenge with a lethal dose of a K99 positive pathogenic strain of E. *coli.* It has been pointed out that such an approach has a theoretical advantage over conventional vaccination of the dam, in that it does not require anticipation of the problem. Further, in this case where the disease is restricted to the first 24 h of life and the effects of the bacteria are contained within the intestine, it is directly treated by oral administration of the monoclonal antibody.

A similar approach has been used to protect mice from intranasal infection with Respiratory Syncital Virus, but generally it would appear that the therapeutic use of monoclonal antibodies will be of benefit in specific, defined individual areas, notably in human medicine, rather than on a herd basis in veterinary medicine. The rationale for this argument is the amount of antibody required, particularly if it is necessary that it is given orally. Monoclonal antibodies to infectious agents may, however, be of benefit in the 'education' of the immune system and this active process will be discussed later.

Another passive form of treatment involves the use of anti-microbials, and in this context biotechnology has already made a significant impact. Human leucocyte interferon has been isolated by monoclonal anti-immunosorbents and on a commercial level recombinant interferon and interleukin 2 are now generally available.

ACTIVE PROTECTION

Before considering the potential impact of biotechnology on the development of safe, effective vaccines, it is necessary to consider what is required of such products in order that they may stimulate a protective response. As has already been mentioned, the vast majority of infectious agents are met at a mucosal surface and therefore vaccines should be directed towards stimulating protection at these sites.

To this end it is clear that by far the most effective method of eliciting such a response involves presentation of the antigen at the mucosal surface. In general parenteral vaccines have not been effective, and where exceptions exist they almost always involve some mucosal component (*e.g.*, oral + parenteral). The other exception to this principle involves immunisation of the dam to achieve passive protection of the sucking neo-nate, by virtue of the presence of antibodies in colostrum. In the case of domestic species, antibodies in colostrum are derived from serum and these may be effectively stimulated by parenteral immunisation. Unlike colostrum, milk antibodies are locally produced and in this case, like other true mucosal secretions, mucosal immunisation is required to stimulate production of antibodies.

Having established that mucosal vaccines are those of choice for both theoretical and practical reasons, it is, however, disappointing that it is surprisingly difficult to stimulate a vigourous, protective response via this route. As has already been pointed out, this is of benefit with regard to the 'harmless antigens' (*e.g.*, dietary antigens) also presented to the mucosae. To date the majority of effective mucosal vaccines are living attenuated organisms reflecting the now generally held axiom that either replication or at least adherence are essential prerequisites of mucosal immunogenicity. Whilst there is now clear evidence of memory in the secretory immune system, vital if any vaccination regimen is to be effective, it has also been shown that the response is highly dose-dependent. Thus where dead oral vaccines have been used, very large numbers of bacteria are required and even then the level of immunity is sub-optimal by comparison with that following the use of live organisms. Another problem associated with mucosal immunisation, unique to the gastrointestinal tract, relates to an ability to pass through the stomach in a form that remains antigenic.

In this context the use of capsules or liposomes may be of benefit. Alternatively the observed immunological link between mucosal sites, such that oral immunisation leads to the appearance of antibodies in milk, may be of benefit.

The New Vaccines

Vaccines derived as products of biotechnology may take a variety of forms and rather than discussing them on a specific disease basis it is proposed to briefly describe the general feature of each type. The genes coding for protein or peptide antigens derived from a number of bacteria and viruses have been cloned and expressed in appropriate bacteria. For example, with the foot and mouth disease virus one of the structural capsid proteins, WI, has been identified as carrying the antigenic determinants necessary to raise neutralising antibodies in experimental animals.

The viral RNA encoding for WI has been copied into double-stranded cDNA, inserted in a bacterial vector and expressed in *E. coli*. Subsequently a range of oligopeptides from VPI have been isolated or synthesised and a major antigenic determinant comprising nineteen amino acids identified. The question that remains is how these proteins or peptides may be presented to the immune system in order to stimulate a protective response.

The isolated VPI protein, for instance, is not as antigenic as the virus and a 100 to 1000-fold excess on a molar basis is required in order to obtain satisfactory protection. With the oligopeptide it is necessary to couple to a carrier protein (*e.g.*, KLH) and although a single injection of 200 pg has been shown to protect guinea-pigs from virus challenge, to date no satisfactory method has been found to stimulate a protective response in cattle. Similar results have been obtained with *E. coli* antigens (K88ab, ac, ad and K99 have all been cloned and expressed) but it still remains to be determined how a protective mucosal response may be stimulated. Whilst to date this synthetic peptide vaccine approach has been disappointing, its theoretical attraction over conventional dead vaccines remains since it avoids the requirement for handling large volumes of infectious materials, or vaccinal strains that grow poorly in bulk-suspension cultures, and the risk associated with failure to completely inactivate the virus before incorporation into a vaccine. An alternative approach involves the use of attenuated strains and these may take a variety of forms, and have been used against a number of agents. Deletion mutants of Aujeszky's disease virus have been prepared and although shown to have greatly reduced virulence are still highly immunogenic in pigs.

Similarly an attenuated strain of *Salmonella typhi,* Ty2la, which is unable to utilise galactose, has been derived. The organism is able to survive for a limited number of divisions, but antigenicity remains unaltered. For another enteric pathogen, cholera, an appropriate mutant is also available. In this case it produces only the B subunit of the toxin and thus, whilst being able to bind to the epithelial cell and give good protection, does not generally produce diarrhoea. Where attenuated strains of a particular organism are not available or appropriate the use of live recombinant vaccine may be of benefit.

Intranasal inoculation with recombinant *Vaccinia* virus has been shown to protect mice against upper and lower respiratory tract infections with influenza. Alternative vectors to *Vaccinia* are also available, for example yeasts or engineered non-pathogenic strains of E. *coli* that produce only the B subunit of the labile toxin, and so bind to enterocytes but do not produce diarrhoea. The general acceptability of live attenuated vaccines remains to be determined and the possibility of developing techniques whereby peptide vaccines can be delivered in a way that stimulates a protective mucosal response remains attractive.

In this context the use of peptide conjugates coupled to molecules such as cholera toxin or lectins that bind avidly to epithelial cells may be of benefit.

Alternatively the potential role of antiidiotypic antibodies (monoclonal) may be of interest, it having been recently shown that such antibodies may prime for protection against E. *coli*. The capacity of biotechnology to produce a range of antigenic moieties derived from infectious agents would appear to be unlimited. Further, it should not be difficult to identify those antigens that are important in the pathogenesis of disease. The major problem that remains is how best these products might be presented to the immune system in order to stimulate a protective immune response. It is therefore to this target of understanding how antigens are presented to and processed by the mucosal immune system that intense future research should be directed.

USES OF BIOTECHNOLOGY

Biotechnology-the sum of the technologies deriving from molecular and cell biology-will have utility in plant breeding beyond the prebreeding stages. Genetic transformation—non-sexual insertion of alien genes and/or gene regulating systems—is envisaged as eventually being applied, quickly and easily, to finished cultivars, to endow them with a series of useful new traits.

However, experience is beginning to show that genetic background affects expression of genes and gene systems that are transferred *in vitro,* as is also the case when they are transferred via regular sexual mechanisms.

Long periods of trial and error are needed to bring transformed lines into desired levels of expression. Thus, for some years to come, genetic transformation perhaps will be used only as a prebreeding technique, one in which alien genes are put into breeding pools of elite germplasm, following which standard plant breeding methods are used to extract good new varieties containing the alien genes. Cell culture techniques—particularly protoplast fusion techniques—will have utility for making new combinations of nuclear and cytoplasmic genes.

Immediate utility is seen for manufacturing new male sterility systems to make F1 hybrid cultivars, or to develop new kinds of herbicide resistance. With time and experience it may be possible to increase primary productive potential as well, by selecting and fixing optimum mitochondrial and chloroplast genotypes, balanced with proper nuclear genotypes. Such success seems far in the future, however, since at present little is known about optimization of nuclear, mitochondrial, and chloroplastic combinations. DNA technology will help plant breeding in a more traditional way, when information is further developed on maps of restriction fragment length polymorphisms (RFLPs) and their linkages with useful traits in crop plants.

In vitro tests, using RFLP markers, will allow near perfect accuracy in selecting for such useful linked traits. RFLP technology also will be used for precisely characterizing cultivar lineages and relationships, for positive identification of miss-identified or mislabeled cultivars, for characterizing gene frequencies of populations, and even to test hypotheses of landrace evolution, and global movement of crop genotypes. Perhaps the greatest contribution of molecular biology to plant breeding will be a deeper and more precise understanding of gene action and inheritance, of cellular physiology and biochemistry, and perhaps of whole plant physiology. With such deepened understanding, breeding and selection via standard sexual methods will be planned and executed with greater vision and economy. Such knowledge and execution will be, in part, a consequence of efforts to make genetic transformnation a practical reality. The fall-out of knowledge from genetic engineering efforts, wisely utilized by observant biologists, will be of great benefit to plant breeding, and to plant culture in general.

DEVELOPMENT OF NEW CROPS

In addition to broadening the genetic base of established crops, genetic enhancement-and plant breeding in the sense of final cultivar development-can be used in two other ways: to develop new crops from heretofore uncultivated species, and to change old crops into new crops. Breeding of the first type (creating new crops where none existed before) usually is interpreted also to include development of advanced cultivars from landraces.

Usually the advanced cultivars are made for industrial countries and the landraces come from developing countries. Soybeans and amaranth thus are spoken of as "new crops" in the United States, even though they have been cultivated for centuries as landraces in China, and in Central and South America, respectively. Challenges are great and are about the same, for changing either wild species or landraces into advanced cultivars, able to withstand monocropping and intensive culture at high production levels.

Selection for productivity inevitably requires narrowing the germplasm base, in the course of selecting plant architecture and physiology able to withstand crowding and the stress of high yields, and amenable to machine planting, weeding and harvesting. Planting the narrowly selected genotypes in large areas gives excellent access to specially adapted disease and insect pests, as already noted, and so the need for tolerance or resistance always is next to be added to the list of breeding priorities for new crops.

Thus, genetic enhancement first narrows and then later must broaden the genetic base, in the course of new crop development. And as soon as users of the new product get used to larger quantities of it, they will want to standardize and then to change the quality of the product-as, in oil percentage, grain storability and hardness, or flavour and texture. The rule is simple: the more successful the crop, economically, the greater the demands on (and complaints about) the plant breeders. Breeding of the second type-changing old crops into new ones-has been considered as a broad-scale possibility only since concepts of genetic engineering were developed. Ability to produce human insulin in bacteria gave rise to thoughts of producing insulin, or other useful primary gene products, in crop plants. Plants were envisaged as now able to turn sunshine, air and water (and soil nutrients) into high value specialty chemicals, instead of commodity chemicals like corn starch or soybean oil. Most of these ideas still are only ideas, but workers are proceeding to put some of the least complicated ones into action. Modification, via genetic engineering, of the ratios of fatty acid components of rapeseed oil is contemplated. Standard breeding to modify chemical components has been used in Canada already, to make rapeseed into a new crop: "canola."

Canola oil, unlike standard rapeseed oil, is essentially free of erucic acid and glucosinolates and therefore is safe for human consumption when used as a cooking oil. Another possibility derives from recent success in cloning the R gene (first identified by Mendel) in peas which provides the opportunity to identify and transfer, *in vitro*, genes giving new starch branching properties and thus, potentially, useful new kinds of specialty starches in old starch-producing crops such as maize or wheat.

A word of caution is needed for those looking to replace major portions of commodity crop production areas with specialty-product new versions of the old crops. High value specialty products, by definition, require only small amounts of product to saturate the market. They won't replace very much of the production area now used for production of commodity crops. On the other hand, they will make intensive use of crop breeding and crop management skills.

They are "brain-power intensive" crops, and as such, must command higher preunit prices—premiums—or they won't be grown.

PHOTOLYSIS AND CARBON FIXATION

PHOTOSYNTHESIS

Photosynthesis is the means that primary producers (mostly plants) can obtain energy via light energy. The energy gained from light can be used in various processes mentioned below for the creation of energy that the plant will need to survive and grow. Photosynthesis is a reduction process, where hydrogen is reduced by a coenzyme. This is in contrast to respiration where glucose is oxidised. The process is split into two distinct areas, *photolysis* (the photochemical stage) and the *Calvin Cycle*. The diagram below gives a summary of the reaction, where light energy is used to initiate the reaction in its presence:

$$CO_2 + H_2O \rightarrow \text{glucose} + \text{oxygen}$$

PHOTOLYSIS

This part of photosynthesis occurs in the *granum* of *a chloroplast* where light is absorbed by *chlorophyll*; a type of photosynthetic pigment that converts the light to chemical energy. This reacts with water (H_2O) and splits the oxygen and hydrogen molecules apart. From this dissection of water, the oxygen is released as a by-product while the reduced hydrogen acceptor makes its way to the second stage of photosynthesis, the Calvin cycle. Overall, since the water is oxidised (hydrogen is removed) and energy is gained in photolysis which is required in the Calvin cycle

The Calvin Cycle

Also known as the carbon fixation stage, this part of the photosynthetic process occurs in the *stroma* of chloroplasts. The carbon made available from breathing in carbon dioxide enters this cycle, which is illustrated below:

Just like the *Kreb's Cycle* in respiration, a substrate is manipulated into various carbon compounds to produce energy. In the case of photosynthesis, the following steps occur, which create glucose for respiration from the carbon dioxide introduced into the cycle;

- Carbon from CO_2 enters the cycle combining with Ribulose Biphosphate (RuBP)

- A compound formed is unstable and breaks down FROM its 6 carbon nature to a 3 carbon compound called glycerate phosphate (GP)
- Energy is used to break down GP into triose phosphate, while a hydrogen acceptor reduces the compound therefore requiring energy
- Triose Phosphate is the end product of this, a 3 carbon compound which can double up to form glucose, which can be used in respiration.
- The cycle is completed when the leftover GP molecules are met with a carbon acceptor and then turned into RuBP, which is to be joined with the carbon dioxide molecules to re-begin the process.

The energy that is used up in the Calvin cycle is the energy that is made available during photolysis. The glucose that is made via GP can be used in respiration or a building block in forming *starch* and *cellulose*, materials that are commonly in demand in plants.

Limiting Factors in Photosynthesis

Some factors affect the rate of photosynthesis in plants, as follows

- *Temperature* plays a role in affecting the rate of photosynthesis. Enzymes involved in the photosynthetic process are directly affected by the temperature of the organism and its environment
- Light Intensity is also a *limiting factor*, if there is no sunlight, then the photolysis of water cannot occur without the light energy required.
- Carbon Dioxide concentration also plays a factor, due to the supplies of carbon dioxide required in the Calvin cycle stage.

Overall, this is how a plant produces energy which supplies a rich source of glucose for respiration and the building blocks for more complex materials. While animals get their energy FROM food, plants get their energy FROM the sun.

CONTROL OF BIOTECHNOLOGY DEVELOPMENT

The role of the public sector to undertake research and development of agricultural technology as a public service to firms which might profit from translating that R & D into a commercial product or process, to farmers who might profit from adoption of the technology, and perhaps most importantly to consumers at home and abroad who benefited from the lower prices resulting from greater abundance. This was particularly true of the phase of change dominated by improvements in plant and animal breeding, the 'old' biotechnology.

While the public sector had a role in basic research for chemical and mechanical technologies, the benefits of research expenditure in these areas were easier to capture by private companies investing in R & D, so that increasingly the public sector has taken a smaller role in these areas although maintaining a strong regulatory role in regard to agricultural chemicals in particular. Where it is impossible to prevent others from escaping payment for the research costs, either because the product is easily copied (seeds which can be regenerated by farmers or other firms) or because proposals for new methods can be readily implemented, private firms are understandably unwilling to invest.

Machines, insecticides, fungicides and other manufactured inputs do lend themselves more readily to private exploitation. Nevertheless the returns to R & D in these products do depend upon the degree of difficulty potential competitors would have in copying the product. In some cases there are inherent technical difficulties in copying the process, or it would be prohibitively expensive, but in other cases it is the ability to obtain patent protection which creates legal barriers to potential competitors' ability to become 'free-riders' and which protects incentives to private investment in R & D. Traditionally, however, it has been impossible for plant and animal breeders to obtain patent rights for their products, which is one reason why public-sector R & D has remained so important in this area.

For, as under the European Patent Convention (EPC) of 1973, it has been judged impossible for plant varieties to satisfy one of the key criteria to qualify for a patent, namely proof of 'an inventive step'. The application of standard breeding practices to generate new varieties by crossing existing plants or animal strains has not been deemed to be invention. Thus under the EPC one set of exclusions from patentability is 'Plant or animal varieties or essentially biological processes for microbiological processes or the products thereof.' In the absence of patent rights plant-breeding firms in particular have worked hard to obtain other means of protection and royalties for their products.

The history of the development of Plant Breeders' Rights (PER) is presented by Mooney, and is of particular interest because of the concerns he expresses about the consequences of allowing the basic genetic stock, which underpins agriculture and hence our whole society, from becoming private rather than public property. While Mooney's concerns are expressed in relation to 'old' biotechnology varieties of plants, they are of particular importance with respect to the products of the 'new' biotechnology. For bio-engineered products are capable of meeting the inventive-step criterion of patentability, and both micro-organisms and transgenic animals have now been patented in the USA. Before examining developments in the seed industry it is worth touching upon the implications of changes in research policy which stress increasing reliance upon private R & D to develop and exploit the new biotechnology and other technologies.

In the UK government has decided that it should withdraw from funding what it terms 'near-market research', that is, research beyond the basic phase and which is preparation for commercial exploitation. This policy is based in part upon the arguments that industry should be investing more heavily in R & D, and that research strategies at the near-market stage should be driven by assessments of likely commercial success which can best be made by the firms involved, and will therefore lead to greater efficiency in allocating research funds. This has led to the closure of a number of government-financed agricultural research institutes, the scaling down of others and the sale of the National Seeds Organization by auction to Unilever; Unilever won against competition with BP and other major public companies. This deliberate attempt to switch an increasing proportion of agricultural R & D expenditure from the public to the private sector carries with it a number of risks. In the first place there is a controversy about whether there is under-investment in R & D so that the returns to extra expenditure are high, or whether the converse is the case. If there is under-investment, then withdrawal of public support for applied research will exacerbate this, as the private sector will only undertake R & D expenditure on those products and processes from which exclusive benefits can be captured by the investor.

Thus in relation to biotechnology in the USA, where again public support for applied biotechnology research is relatively small, Stallman and Schmid argue that there will be emphasis on technologies which are applied in factory conditions where secrecy and control can be maintained. For technologies which cannot be confined to factories, such as seeds, Stallman and Schmid state that 'Firms are also considering mechanisms which "scramble" the genome of a plant in the second generation', in order to prevent farmers from reproducing seed with the enhanced, engineered characteristics. Clearly such actions are designed to frustrate the maximum spread of benefits from the technology and to maximize private profit for companies investing in research.

Another facet of commercially orientated biotechnology research is that it will aim at the most important crops and livestock products, those produced by the largest farming units, and those which are most heavily subsidized. In the latter case this will worsen the budgetary problems of adjusting agricultural policy in OECD countries. The probable neglect of minor crops, difficult habitats and small farmers means that the public sector will have a defined role in agricultural technology research, but one in which it will be relegated to the second division and where it is unlikely to prove successful in terms of the commercial yardstick of rates of return which is increasingly emphasized by public-research policy.

NEED FOR BIOTECHNOLOGY CAPACITY IN DEVELOPING COUNTRIES

For the developed countries, the need for biotechnology may be to increase the level of affluence, but for developing countries, it could reduce hunger and starvation. The use of bovine somatotropin to increase milk production in economies where farmers are being paid to produce less milk illustrates this point. However, not all available biotechnologies are appropriate or relevant to all countries. Developing countries may need to adapt some of these technologies before they can use them. It is therefore important that developing countries develop capacity to maintain a strong base of applied and adaptive research and some level of training to keep abreast with new developments. Biotechnologies which could have application in developing countries are summarised in Table. While the applications of some of the available technologies are relatively simple, it is unlikely that developing countries will be able to retain those trained to support implementation of biotechnology programmes if adequate financing is not available to provide required equipment, chemicals etc to expand or even maintain such programmes.

Table. Possible Applications of Biotechnology to the Solution of Problems of Livestock Production in Developing Countries.

Problem	Possible solution	Scale of economic impact	Probable time to commercial use, years
Animal diseases	New vaccines and new diagnostics	Large	<5
Poor quality forages	Microbial treatment of forages	Moderate	5-10
	Modification of rumen microflora	Moderate/large	>10
	Genetic improvement of forages and their symbionts	Moderate	5-10
Difficulty of implementing selection programmes	Selection in nucleus herds, using Al, ET, embryo sexing	Large	5-10
	Marker-assisted selection	Moderate	5-10
Difficulty of maintaining dairy cattle performance after F_1 cross	Use of IVF, ET and embryo sexing	Large	>10
	Selection among local breeds using AI, MOET	Medium/large	>10
	Development of composite breeds	Large	>10
Cost and environmental challenge to imported cattle	Use of ET to import embryos	Small	<5
Need for increased efficiency in intensive systems	Use of rBST and rPST in dairy and pig production	Large	<5

The development of biotechnology capacity in developing countries is also justified in order to support research which is only of interest to them or that which, because of restrictions in developed countries, can only be done in developing countries. For example, veterinary regulations in most developed countries generally prohibit the introduction of biological material (*e.g.* infectious agents required for vaccine development) or live animals for research purposes. Therefore developed countries must develop the capacity to undertake such research in the relevant countries. The disparity between industrialised countries and developing regions of the world in terms of veterinary biotechnology has been summarised by Lefevre: Out of 152 laboratories involved in veterinary biotechnology in 1991, only 26 are located in 17 countries in Asia, Africa and South America. Moreover, some of these laboratories may not be actively involved in biotechnology but are merely interested in it. In sub-Saharan Africa (excluding the Republic of South Africa), only ILRAD (based in Kenya) is actively involved in biotechnology research. However, ILRAD's biotechnology work is focussed on only two diseases, trypanosomiasis and theileriosis.

Because the well-equipped and adequately funded laboratories doing research in molecular biology are found almost exclusively in developed countries, the gap between the industrialised and developing countries in technical expertise and relevant research capacity is getting wider and motivated scientists from developing countries with the expertise to carry out sophisticated research are opting to work in laboratories in industrialised countries.Another justification for local capacity in biotechnology research is to provide a home for orphan commodities.

From biotechnology research standpoint, an orphan commodity may be defined as a commodity in which there is or is likely to be little or no investment in modern biotechnology in industrialised countries either because the commodity is not important in temperate areas or because there are no likely profits for transnational companies. Thus an orphan commodity is not necessarily a small commodity. Banana, plantain, cassava, coconut and tropical fruits are examples of orphan crops.

Most African indigenous animals, especially chickens, pigs and goats probably fall in this category although some basic research on these species in the developed countries may benefit the African populations as well. There is thus a need for biotechnology research capacity for problems that may be unique to developing countries. Persley suggested the establishment of a special funding mechanism to provide support for research on orphan commodities by public and private sector institutions in industrialised and developing countries.

BIOTECHNOLOGY ADOPTION AND ADAPTATION

The Case of Artificial Insemination

One of the reasons why technologies developed in the industrialised countries tend not to be implemented with much success in developing countries is the failure to recognise the importance of adapting technology to local conditions. AI is a good example. Where AI technology has been adopted in Africa, not much consideration has been given to adapting it to the circumstances in which it is to be applied to ensure sustainability. Instead, AI uses have been based on sophisticated models intended for countries with good communication and transport systems and with adequate and reliable operating budgets. Such AI use programmes have often collapsed and have had to go through several phases of foreign-aid-supported "rehabilitations" and, in some cases, have eventually collapsed. Government subsidies of the AI system are considered to be another main cause of failure of AI in developing countries.

Privatisation of the services may change the situation. For example, in some of these countries, the concept of farmer co-operatives is well developed and applied for specific cash crops or milk marketing. AI could easily be run by co-operatives organised in schemes in which farmers are "grouped" on the basis of such factors as transport requirements and similarity of systems of production. Semen can easily be delivered over short distances by motorbikes, bicycles, horses, donkeys etc. Under such circumstances, use of fresh semen collected from bulls belonging to the co-operative could be considered. This would eliminate the cost of freezing semen. Such schemes would make it easier to match genotypes to production systems within a country. Indeed, such a scheme could form the basis for the genetic improvement of localised indigenous breeds.

The Case of Embryo Transfer

Embryo transfer (ET) could have a major impact on cattle breeding in developing countries especially as part of a nucleus breeding scheme. However, successful ET requires highly motivated, experienced staff and a high capital investment in facilities, equipment and drugs.

In general, the inappropriateness of ET for developing countries is ascribed to lack of infrastructure. However, in some instances, ET represents a solution to a lack of infrastructure. Thus, establishment of multiple ovulation embryo transfer (MOET) is considered an attractive means of genetic improvement where infrastructure for progeny testing is not available.

Most developing countries have limited financial resources. In addition, equipment and supplies tend to be more expensive than they are in developed countries due to transportation costs, import tariffs, lack of hard currency etc. These make ET technology prohibitive in these countries. However, it is possible to adapt ET techniques to local conditions thereby reducing the cost. Seidel and Seidel pointed out that a lot of the fancy equipment associated with ET are not essential for successful utilisation of the technology. For example, fancy freezing equipment are no more effective than dry ice/alcohol baths for freezing embryos as these save labour and are more convenient. Similarly, filters are not essential for isolation of embryos, neither are disposable sterile syringes and fancy plastic dishes. Thus, by combining good imagination with knowledge of basic principles, the technology can be successfully adapted to local conditions. Therefore research, especially of an applied nature, on such technologies by institutions in developing countries is always justified and often essential. Moreover, researchers need to be exposed to new technologies or procedures to appreciate the power and limitations of such technologies.

Environmental Concerns about Biotechnology

There is much euphoria about developments in biotechnology and potential benefits, but little is said about the risks associated with biotechnology. For example genetically modified organisms could create ecological disaster if released into the environment.

Biosafety is, therefore, an issue of great concern for many developing countries. In a recent meeting of the Intergovernmental Committee on the Convention on Biological Diversity, representatives of developing countries pointed out that biotechnology was evolving more rapidly than the capacity of their countries to install effective safety procedures for the handling and use of living modified organisms and that there was need for adequate and transparent safety procedures to manage and control the risks associated with the use and release of such organisms. To deal with the basic ethical questions and the risks associated with genetic engineering, regulatory mechanisms should be created and internationally acceptable guidelines or regulations put in place.

2

Animals Health Nutrition and Feeding Behaviour

FEEDING BEHAVIOUR

Buffaloes are strict grazers and only browse when feed is utterly scarce. Normally, buffaloes graze during the day. In case of extremely high ambient temperatures, grazing takes place in the morning and afternoon and sometimes during night time. Buffaloes graze more and better than cattle. Thereby they consume more feed and nutrients per kg of body weight than cattle do.

Newborn calves suckle their mothers within two hours of birth. Normal suckling frequency is approximately 6 to 8 times per day. The calves start to nibble grass at 3 to 4 weeks of age, although they are not actually grazing until after a few weeks more. When the calves have reached two months of age, forage starts to become more important than previously and soon most of the nutrient intake comes from forage rather than milk. Natural weaning of calves is usually within a year or before its mother's next parturition.

NUTRITION FOR HORSE

Equine nutrition refers to the feeding of horses, ponies, mules, donkeys, and other equines. Correct and balanced nutrition is a critical component of proper horse care. Horses are herbivores, a type of non-ruminant known as a "hind-gut fermentor." What this means is that horses have only one stomach, similar to humans. However, unlike humans, they also have to digest plant fibre that comes from grass and hay. Therefore, unlike ruminants, who digest fibre in plant matter by use of a multichambered stomach, horses use bacterial fermentation that occurs in the organ known as the *cecum* (or *caecum*) to break down cellulose.

In practical terms, horses prefer to eat small amounts of food steadily throughout the day, as they do in nature when grazing on pasture. Although this is not always possible with modern stabling practices and human schedules that favour feeding horses twice a day, it is important to remember the underlying biology of the animal when determining what to feed, how often, and in what quantities. The digestive system of the horse is somewhat delicate.

Fig. Grass is a Natural Source of Nutrition for a Horse

Because horses are unable to regurgitate food, except from the esophagus, if they overeat or eat something poisonous, vomiting is not an option. They also have a long, complex large intestine and a balance of beneficial bacteria in their cecum that can be upset by rapid changes in feed. Because of these factors, they are very susceptible to colic, which is a leading cause of death in horses. Therefore, horses require clean, high-quality feed, provided at regular intervals, and may become ill if subjected to abrupt changes in their diets.

Horses are also sensitive to molds and toxins. For this reason, they must never be fed contaminated fermentable materials such as lawn clippings. Fermented silage, sometimes called "haylage" is fed to horses in some places; however, contamination or failure of the fermentation process that allows any mold or spoilage may be toxic. Horses and other members of the *Equidae* family are adapted by evolutionary biology to eating small amounts of the same kind of food all day long.

In the wild, horses ate prairie grasses in semi-arid regions and traveled significant distances each day in order to obtain adequate nutrition. Therefore, their digestive system was made to work best with a small but steady flow of food that does not change much from day to day. Digestion begins in the mouth. First, the animal selects pieces of forage and picks up finer foods, such as grain, with sensitive, prehensile, lips.

The front teeth of the horse, called incisors, clip forage, and food is ground up for swallowing by the premolars and molars. The esophagus carries food to the stomach. The esophagus enters the stomach at an acute angle, creating a one-way valve, with a powerful spincter mechanism at the gastroesophageal junction, which is why horses cannot vomit. The esophagus is also the area of the digestive tract where horses may suffer from choke.

Horses have a relatively small stomach for their size, which limits the amount of feed that can be taken in at one time. The average sized horse has a stomach with a capacity of only four gallons, and works best when it contains about two gallons. One reason continuous foraging or several small feedings per day are better than one or two large meals is because the stomach begins to empty when it is 2/3 full, whether the food in the stomach is processed or not.

The small intestine is 50 to 70 feet long and holds 10 to 12 gallons. This is the major digestive organ where most nutrients are absorbed. Bile from the liver acts here, combined with enzymes from the pancreas and small intestine itself. Equids do not have a gall bladder, so bile flows constantly, another reason for a slow but steady supply of food.

Most nutrition is absorbed into the bloodstream from the small intestine. The cecum is the first section of the large intestine. It is also known as the "water gut" or "hind gut." It is a cul-de-sac pouch, about 4 feet long that holds 7 to 8 gallons. It digests cellulose plant fibre through bacterial fermentation. The reason horses must have their diets changed slowly is so the bacteria in the cecum are able to modify and adapt to the different chemical structure of new feedstuffs. Too abrupt a change in diet can cause colic, as the new food is not properly digested.

The large colon, small colon, and rectum make up the remainder of the large intestine. The large colon is 10-12 feet long and holds up to 20 gallons of semi-liquid matter. Its main purpose it to absorb carbohydrates which were broken down from cellulose in the cecum. Due to its many twists and turns, it is a common place for a type of horse colic called an impaction. The small colon is 10-12 feet long, holds about 5 gallons, is the area where the majority of water is absorbed, and where fecal balls are formed. The rectum is about one foot long, and acts as a holding chamber for waste, which is then expelled from the body via the anus.

Nutrients

Like all animals, equines require five main classes of nutrients to survive: water, energy (primarily in the form of fats and carbohydrates), proteins, vitamins, and minerals.

Water makes up between 62-68 per cent of a horse's body weight and is essential for life. Horses can only live a few days without water, becoming dangerously dehydrated if they lose 8-10 per cent of their natural body water. Therefore, it is critically important to provide access to a fresh, clean, and adequate supply of water. An average 450kg horse drinks 10 to 12 gallons of water per day, more in hot weather, when eating dry forage such as hay, or when consuming high levels of salt, potassium, and magnesium. Horses drink less water in cool weather or when on lush pasture, which has a relatively higher water content. When under hard work, or if a mare is lactating, water requirements may be as much as four times greater than normal. Though they need a great deal of water, horses spend very little time drinking; usually 1-8 minutes a day, spread out in 2-8 episodes. Water plays an important part in digestion. The forages and grains horses eat are mixed with saliva in the mouth to make a moist bolus that can be easily swallowed. Therefore, horses produce up to 10 gallons of saliva per day.

Nutritional sources of energy are fat and carbohydrates. Protein is a critical building block for muscles and other tissues. Horses that are heavily exercised, growing, pregnant or lactating need increased energy and protein in their diet. However, if a horse has too much energy in its diet and not enough exercise, it can become too high-spirited and difficult to handle.Fat exists in low levels in plants and can be added to increase the energy density of the diet.

Fat has 9 Mcal/kg of energy, which is 2.25 times that of any carbohydrate source. Because equids have no gall bladder to store large quantities of bile, which flows continuously from the liver directly into the small intestine, fat, though a necessary nutrient, is difficult for them to digest and utilize in large quantities. However, they are able digest a greater amount of fat than can cattle. Horses benefit from up to 8 per cent fat in their diets, but more does not always provide a visible benefit. Horses can only have 15 per cent-20 per cent fat in their diet without the risk of developing diarrhea.

Carbohydrates, the main energy source in most rations, are usually fed in the form of hay, grass and grain. Soluble carbohydrates such as starches and sugars are readily broken down to glucose in the small intestine and absorbed. Insoluble carbohydrates, such as fibre (cellulose), are not digested by enzymes, but are fermented by microbes in the cecum and large colon to break down and release their energy sources, the volatile fatty acids.

Soluble carbohydrates are found in nearly every feed source; corn has the highest amount, then barley and oats. Forages normally have only 6 to 8 per cent soluble carbohydrate, but under certain conditions can have up to 30 per cent. Sudden ingestion of large amounts of starch or high sugar feeds can cause colic or laminitis.Protein is used in all parts of the body, especially muscle, blood, hormones, hooves, and hair cells. The main building blocks of protein are amino acids. Alfalfa and other legumes in hay are good sources of protein that can be easily added to the diet. Most adult horses only require 8 to 10 per cent protein in their diet; however, higher protein is important for lactating mares and young growing foals.

Horses that are not subjected to hard work or extreme conditions usually have more than adequate amounts of vitamins in their diet if they are receiving fresh, green, leafy forages. Sometimes a vitamin supplement is needed when feeding low-quality hay, if a horse is under stress (illness, traveling, showing, racing, and so on), or not eating well. Grain has a different balance of nutrients than forage, and so requires specialized supplementation to prevent an imbalance of vitamins and minerals.

Minerals are required for maintenance and function of the skeleton, nerves and muscles. These include calcium, phosphorus, sodium, potassium, and chloride, and are commonly found in most good-quality feeds. Horses also need trace minerals such as magnesium, selenium, copper, zinc and iodine. Normally, if adult animals at maintenance levels are consuming fresh hay or are on pasture, they will receive adequate amounts of minerals in their diet, with the exception of sodium chloride (salt), which needs to be provided, preferably free choice. Some pastures are deficient in certain minerals, especially selenium or copper, and in such situations, deficiency diseases may occur if the horses' feed is not supplemented.

Calcium and phosphorus are needed in a specific ratio of between 1:1 and 2:1. Adult horses can tolerate up to a 5:1 ratio, foals no more than 3:1. A total ration with a higher ratio of phosphorus than calcium is to be avoided. Over time, the imbalance will ultimately lead to bone problems such as osteoporosis.

Foals and young growing horses through their first three to four years have special nutritional needs and require feeds that are balanced with a proper calcium:phosphorus ratio and other trace minerals. A number of skeletal problems may occur in young animals with an unbalanced diet. Hard work increases the need for minerals; sweating depletes sodium, potassium, and chloride from the horse's system. Therefore, supplementation with electrolytes may be required for horses in intense training, especially in hot weather.

Types of Feed

Equids can consume approximately 2 per cent to 2.5 per cent of their body weight in dry feed each day. Therefore, a 1000 lb adult horse could eat up to 25 pounds of food. Foals less than six months of age eat 2 to 4 per cent of their weight each day.Solid feeds are placed into three categories: forages, concentrates (including grain or pelleted rations), and supplements.

Equine nutritionists recommend that 50 per cent or more of the animal's diet by weight should be forages. If a horse is working hard and requires more energy, the use of grain is increased and the percentage of forage decreased so that the horse obtains the energy content it needs for the work it is performing. However, forage amount should never go below 1 per cent of the horse's body weight per day.

Forages, also known as "roughage," are plant materials classified as legumes or grasses, found in pastures or in hay. Often, pastures and hayfields will contain a bland of both grasses and legumes. Nutrients available in forage vary greatly with maturity of the grasses, fertilization, management, and environmental conditions. Grasses are tolerant of a wide range of conditions and contain most necessary nutrients. Some commonly used grasses include timothy, brome, fescue, coastal bermuda, orchard grass, and Kentucky bluegrass.

Legumes such as clover or alfalfa are usually higher in protein, calcium, and energy than grasses. However, they require warm weather and good soil to produce the best nutrients. Legume hays are generally higher in protein than the grass hays. They are also higher in minerals, particularly calcium, but have an incorrect ratio of calcium to phosphorus. Because they are high in protein, they are very desirable for growing horses or those subjected to very hard work, but the calcium:phosphorus ratio must be balanced by other feeds to prevent bone abnormalities.

Hay is a dried mixture of grasses and legumes. It cut in the field and then dried and baled for storage. Hay is most nutritious when it is cut early on, before the seed heads are fully mature and before the stems of the plants become tough and thick. Hay that is very green can be a good indicator of the amount of nutrients in the hay; however, colour should not be used as sole indicator - smell and texture are also important. Hay can be analyzed by many laboratories and that is the most reliable way to tell the nutritional values it contains. Moldy or dusty hay should not be fed to horses, as it is the most common cause of Recurrent airway obstruction, also known as COPD or "heaves."

Hay, particularly alfalfa, is sometimes compressed into pellets or cubes. Processed hay can be of more consistent quality and is more convenient to ship and to store. It is also easily obtained in areas that may be suffering localized hay shortages. However, these more concentrated forms can be overfed and horses are somewhat more prone to choke on them. On the other hand, hay pellets and cubes can be soaked until they break apart into a pulp or thick slurry, and in this state are a very useful source of food for horses with tooth problems such as dental disease, tooth loss due to age, or structural anomalies.

Haylage, also known as *Round bale silage* is a term for grass sealed in airtight plastic bags, a form of forage that is frequently fed in the United Kingdom and continental Europe, but is not often seen in the United States. Because haylage is a type of silage, hay stored in this fashion must remain completely sealed in plastic, as any holes or tears can stop the preservation properties of fermentation and lead to mold or spoilage. Rodents chewing through the plastic can also spoil the hay introducing contamination to the bale.If a rodent dies inside the plastic, the subsequent botulism toxins released can contaminate the entire bale.

Another type of forage sometimes provided to horses is beet pulp, a byproduct left over from the processing of sugar beets, which is high in energy as well as fibre.Sometimes, straw or chaff from cereal grains is fed to animals. However, this is roughage with little nutritional value. It is not recommended by nutritionists as a horse feed, though it is sometimes used as a filler; it can slow down horses who eat their grain too fast, or it can provide additional fibre when the horse must meet most nutritional needs via concentrated feeds. Straw is more often used as a bedding in stalls to absorb wastes.

Whole or crushed grains are the most common form of concentrated feed, sometimes referred to generically as "oats" or "corn" even if those grains are not present, also sometimes called "straights" in the UK.

Oats are the most popular grain for horses. Oats have a lower digestible energy value and higher fibre content than most other grains. They form a loose mass in the stomach that is well suited to the equine digestive system. They are also more palatable and digestible other grains.

Corn, referred to as Maize in the UK, is the second most palatable grain. It provides twice as much digestible energy as an equal volume of oats and is low in fibre. Because of these characteristics, is easy to over-feed corn, causing obesity, so horses are seldom fed corn all by itself. Nutritionists caution horse owners that moldy corn should never be fed because it is poisonous to horses.

Barley is also fed to horses, but needs to be processed to crack the seed hull and allow easier digestibility. It is frequently fed in combination with oats and corn, a mix informally referred to by the acronym "COB" (for Corn, Oats and Barley).

Wheat is generally not used as a concentrate. However, wheat bran is sometimes added to the diet of a horse for supplemental nutrition, usually moistened and in the form of a bran mash. Wheat bran is high in phosphorus, so must be fed carefully so that it does not cause an imbalance in the Ca:P ratio of a ration. Once touted for a laxative effect, this use of bran is now considered unnecessary, as horses, unlike humans, obtain sufficient fibre in their diets from other sources.

Many feed manufacturers combine various grains and add additional vitamin and mineral supplements to create a complete premixed feed that is easy for owners to feed and of predictable nutritional quality. Some of these prepared feeds are manufactured in pelleted form, others retain the grains in their original form. In many cases molasses as a binder to keep down dust and for increased palatability. Grain mixes with added molasses are usually called "Sweet feed" in the USA and "Coarse mix" in the United Kingdom. Pelleted or *extruded* feeds (sometimes referred to as "nuts" in the UK) may be easier to chew and result in less wasted feed. Horses generally eat pellets as easily as grain. However, pellets are also more expensive, and even "complete" rations do not eliminate the necessity for forage.

The average modern horse on good hay or pasture with light work usually does not need supplements; however, horses subjected to stress due to age, intensive athletic work, or reproduction may need additional nutrition. Extra fat and protein are sometimes added to the horse's diet, along with vitamin and mineral supplements. Soybean meal is a common protein supplement, and averages about 44 per cent crude protein. The protein in soybean meal is high-quality, with the proper ratio of dietary essential amino acids for equids. Cottonseed meal, Linseed meal, and peanut meal are also used, but are not as common.

Vegetable oil is a common fat source added to a ration. Corn oil is particularly popular, but other oils are used as well. Rice bran is a very good fat supplement that contains 20 per cent fat as well as fibre and other nutrients. Flax seed is another good source of fat, though it must ground up for horses to digest it. Some commercial feed manufacturers now make products containing both flaxseed and rice bran.

There are hundreds, if not thousands of commercially prepared vitamin and mineral supplements on the market, many tailored to horses with specialized needs.

Most horses only need quality forage, water and a salt or mineral block. Grain or other concentrates are often not necessary. But, when grain or other concentrates are fed, quantities must be carefully monitored. To do so, horse feed is measured by weight, not volume. For example, 1 lb. of oats has a different volume than 1 lb. of corn. When continuous access to feed is not possible, it is more consistent with natural feeding behaviour to provide three small feedings per day instead of one or two large ones.

However, even two daily feedings is preferable to only one. To gauge the amount to feed, a weight tape, available at most feed stores, can be used to provide a reasonably accurate estimate of a horse's weight. The tape measures the circumference of the horse's barrel, just behind the withers and elbows, and the tape is calibrated to convert inches or centimeters into approximate pounds or kilograms.

Actual amounts fed vary by the size of the horse, the age of the horse, the climate, and the work to which the animal is put. In addition, genetic factors play a role. Some animals are naturally easy keepers, which means that they can thrive on relatively small amounts of food and are prone to obesity and other health problems if overfed. Others are hard keepers, meaning that they are prone to be thin and require considerably more food to maintain a healthy weight.

Veterinarians are usually a good source for recommendations on appropriate types and amounts of feed for a specific horse. There are also numerous books written on the topic. Feed manufacturers usually offer very specific guidelines for how to select and properly feed products from their company, and in the United States, the local office of the Cooperative Extension Service can provide educational materials and expert recommendations.

Equids always require forage. When possible, nutritionists recommend it be available at all times, at least when doing so does not overfeed the animal and lead to obesity. It is safe to feed a ration that is 100 per cent forage (along with water and supplemental salt), and any feed ration should be at least 50 per cent forage. Hay with alfalfa or other legumes has more concentrated nutrition and so is fed in smaller amounts than grass hay, though many hays have a mixture of both types of plant.

When beet pulp is fed, a ration of 2 to 5 pounds is usually soaked in water for 3 to 4 hours prior to feeding in order to make it more palatable, and to minimize the risk of choke and other problems. It is usually soaked in a proportion of one part beet pulp to two parts water. Beet pulp is usually fed in addition to hay, but occasionally is a replacement for hay when fed to very old horses who can no longer chew properly.

Some pelleted rations are designed to be a "complete" feed that contains both hay and grain, meeting all the horse's nutritional needs. However, even these rations should have some hay or pasture provided, a minimum of a half-pound of forage for every 100 pounds of horse, in order to keep the digestive system functioning properly and to meet the horse's urge to graze.

Recent studies address the level of various non-structural carbohydrates (NSC), such as fructan, in forages. Too high an NSC level causes difficulties for animals prone to laminitis or equine polysaccharide storage myopathy (EPSM). NSC cannot be determined by looking at forage, but hay and pasture grasses can be tested for NSC levels.

Concentrates, when fed, are recommended to be provided in quantities no greater than 1 per cent of a horse's body weight per day, and preferably in two or more feedings. If a ration needs to contain a higher percent of concentrates, such as that of a race horse, bulky grains such as oats should be used as much as possible; a loose mass of feed helps prevent impaction colic. Peptic ulcers are linked to a too-high concentration of grain in the diet, particularly noticed modern racehorses, where some studies show such ulcers affecting up to 90 per cent of all race horses.

In general, the portion of the ration that should be grain or other concentrated feed is 0 to 10 per cent grain for mature idle horses; between 20 to 70 per cent for horses at work, depending on age, intensity of activity and energy requirements. Concentrates should not be fed to horses within one hour before or after a heavy workout. Concentrates also need to be adjusted to level of performance. Not only can excess grain and inadequate exercise lead to behaviour problems, it may also trigger Equine Exertional Rhabdomyolysis, or "tying up," in horses prone to the condition.

Horses normally require free access to all the fresh, clean water they want, and to avoid dehydration, should not be kept from water longer than four hours at any one time. However, water may need to be temporarily limited in quantity when a horse is very hot after a heavy workout. As long as a hot horse continues to work, it can drink its fill at periodic intervals, provided that common sense is used and that an overheated horse is not forced to drink from extremely cold water sources. But when the workout is over, a horse needs to be cooled out and walked for 30—90 minutes before it can be allowed all the water it wants at one time. However, dehydration is also a concern, so some water needs to be offered during the cooling off process. A hot horse will properly rehydrate while cooling off if offered a few swallows of water every three to five minutes while being walked. Sometimes the thirst mechanism does not immediately kick in following a heavy workout, which is another reason to offer periodic refills of water throughout the cooling down period.

Even a slightly dehydrated horse is at higher risk of developing impaction colic. Additionally, dehydration can lead to weight loss because the horse cannot produce adequate amounts of saliva, thus decreasing the amount of feed and dry forage consumed. Thus, it is especially important for horse owners to encourage their horses to drink when there is a risk of dehydration; when horses are losing a great deal of water in hot weather due to strenuous work, or in cold weather due to horses' natural tendency to drink less when in a cold environment. To encourage drinking, owners may add electrolytes to the feed, additives to make the water especially palatable (such as apple juice), or, when it is cold, to warm the water so that it is not at a near-freezing temperature.

SPECIAL FEEDING ISSUES FOR PONIES

Ponies and miniature horses are usually easy keepers and need less feed than full-sized horses. This is not only because they are smaller, but also, because they evolved under harsher living conditions than horses, they use feed more efficiently. Ponies easily become obese from overfeeding and therefore are at high risk for colic, Equine Metabolic Syndrome, and, especially, laminitis. Fresh grass is a particular danger to ponies; they can develop laminitis in as little as one hour of grazing on lush pasture.

It is important to track the weight of a pony carefully, by use of a weight tape. Forages may be fed based on weight, at a rate of about 1 pound of forage for every 100 pounds. Forage, along with water and a salt and mineral block, is all most ponies require. If a hard-working pony needs concentrates, a ratio of no more than 30 per cent concentrates to 70 per cent forage is recommended. Concentrates designed for horses, with added vitamins and minerals, will often provide insufficient nutrients at the small serving sizes needed for ponies. Therefore, if a pony requires concentrates, feed and supplements designed specially for ponies should be used. In the UK, extruded pellets designed for ponies are sometimes called "pony nuts.".

Special Feeding Issues for Mules and Donkeys

Like ponies, mules and donkeys are also very hardy and generally need less concentrated feed than horses. Mules need less protein than horses and do best on grass hay with a vitamin and mineral supplement. If mules are fed concentrates, they only need about half of what a horse requires. Like horses, mules require fresh, clean water, but are less likely to over-drink when hot.

Fig. Grass hay for Mules and Donkey

Donkeys, like mules, need less protein and more fibre than horses. They do best when allowed to consume small amounts of food over long periods, as is natural for them in an arid climate. They can meet their nutritional needs on 6 to 7 hours of grazing per day on average dryland pasture that is not stressed by drought. If they are worked long hours or do not have access to pasture, they require hay or a similar dried forage, with no more than a 1:4 ratio of legumes to grass. They also require salt and mineral supplements, and access to clean, fresh water. Like ponies and mules, in a lush climate, donkeys are prone to obesity and are at risk of laminitis.

Many people like to feed horses special treats such as carrots, sugar cubes, peppermint candies or specially manufactured horse "cookies." Horses do not need treats, and due to the risk of colic or choke, many horse owners do not allow their horses to be given treats. There are also behavioural issues that some horses may develop if given too many treats, particularly a tendency to bite if hand-fed, and for this reason many horse trainers and riding instructors discourage the practice.

However, if treats are allowed, carrots and compressed hay pellets are common, nutritious, and generally not harmful. Apples are also acceptable, though it is best if they are first cut into slices. Horse "cookies" are often specially manufactured out of ordinary grains and some added molasses. They generally will not cause nutritional problems when fed in small quantities. However, many types of human foods are potentially dangerous to a horse and should not be fed. This includes bread products, meat products, candy, and carbonated or alcoholic beverages. It was once a common practice to give horses a weekly *bran mash* of wheat bran mixed with warm water and other ingredients.

It is still done regularly in some places. While a warm, soft meal is a treat many horses enjoy, and was once considered helpful for its laxative effect, it is not nutritionally necessary. An old horse with poor teeth may benefit from food softened in water, a mash may help provide extra hydration, and a warm meal may be comforting in cold weather, but horses have far more fibre in their regular diet than do humans, and so any assistance from bran is unnecessary. There is also a risk that too much wheat bran may provide excessive phosphorus, unbalancing the diet, and a feed of unusual contents fed only once a week could trigger a bout of colic.

All hay and concentrated feeds must kept dry and free of mold, rodent feces and other types of contamination that may cause illness in horses. Feed kept outside or otherwise exposed to moisture can develop mold quite quickly. Due to fire hazards, hay is often stored under an open shed or under a tarp, rather than inside a horse barn itself, but should be kept under some kind of cover.

Concentrates take up less storage space, are less of a fire hazard, and are usually kept in a barn or enclosed shed. A secure door or latched gate between the animals and any feed storage area is critical. Horses accidentally getting into stored feed and eating too much at one time is a common but preventable way that horses develop colic or laminitis.

It is also important to never give a horse feed that was contaminated by the remains of a dead animal, it is a potential source of botulism. This is not an uncommon situation. For example, mice and birds can get into poorly stored grain and be trapped; hay bales sometimes accidentally contain snakes, mice, or other small animals that were caught in the baling machinery during the harvesting process.

Horses can become anxious or stressed if there are long periods of time between meals. They also do best when they are fed on a regular schedule, they are creatures of habit and easily upset by changes in routine. When horses are in a herd, their behaviour is hierarchical; the higher-ranked animals in the herd eat and drink first. Low-status animals, who eat last, may not get enough food, and if there is little available feed, higher-ranking horses may keep lower-ranking ones from eating at all.

Therefore, unless a herd is on pasture that meets the nutritional needs of all individuals, it is important to either feed horses separately, or spread feed out in separate areas to be sure all animals get roughly equal amounts of food to eat. In some situations where horses are kept together, they may still be placed into separate herds, depending on nutritional needs; overweight horses are kept separate from thin horses so that rations may be adjusted accordingly.

Horses' teeth continually erupt throughout their life, are worn down as they eat, and can develop uneven wear patterns that can interfere with chewing. For this reason, horses need a dental examination at least once a year, and particular care must be paid to the dental needs of older horses. The process of grinding off uneven wear patterns on a horse's teeth is called *floating* and can be performed by a veterinarian or a specialist in equine dentistry.Diet and Nutrition for Dog

There are a great quantity of commercial foods and treats marketed for dogs, and not all are recommended as part of a balanced, healthy diet.

There is some debate as to whether domestic dogs should be classified as omnivores or carnivores, by diet. The classification in the Order Carnivora does not necessarily mean that a dog's diet must be restricted to meat; unlike an obligate carnivore, such as the cat family with its shorter small intestine, a dog is neither dependent on meat-specific protein nor a very high level of protein in order to fulfill its basic dietary requirements.

Dogs are able to healthily digest a variety of foods including vegetables and grains, and in fact dogs can consume a large proportion of these in their diet. Wild canines not only eat available plants to obtain essential amino acids, but also obtain nutrients from vegetable matter from the stomach and intestinal contents of their herbivorous prey, which they usually consume.

Domestic dogs can survive healthily on a reasonable and carefully designed vegetarian diet, particularly if eggs and milk products are included. Some sources suggest that a dog fed on a strict vegetarian diet without L-carnitine may develop dilated cardiomyopathy, however, L-carnitine is found in many nuts, seeds, beans, vegetables, fruits and whole grains. In the wild, dogs can survive on a vegetarian diet when animal prey is not available.

Observation of extremely stressful conditions such as the Iditarod Trail Sled Dog Race, and scientific studies of similar conditions has shown that high-protein (approximately 40 per cent) diets including meat help prevent damage to muscle tissue in dogs and some other mammals. This level of protein corresponds to the percentage of protein found in the wild dog's diet when prey is abundant; higher levels of protein seem to confer no added benefit.

Dogs frequently eat grass, which is a harmless activity. Explanations abound, but rationales such as that it neutralizes acid, or that dogs eat grass to induce vomiting to remove unwanted substances from their stomachs, are at best educated guesses. Dogs do vomit more readily than humans, as part of their typical feeding behaviour of gulping down food then regurgitating indigestible material such as bones and fur.

This behaviour is typical of pack feeding in the wild, where the most important thing is to get as much of the kill as possible before others consume it all. Individual domestic dogs, however, may be very "picky" eaters, in the absence of this social pressure. Dogs may also appear to eat grass when they are just running the blades through their mouth to gather information. Their sense of smell and taste may act together to detect if other animals have walked through their area or urinated on the grass.

DANGEROUS SUBSTANCES AND HUMAN FOOD

Some foods commonly enjoyed by humans are dangerous to dogs, including chocolate (Theobromine poisoning), onions, grapes and raisins, some types of gum, certain sweeteners and Macadamia nuts. The only known dangerous substance in chocolate is cocoa, so the danger of white chocolate is uncertain. The acute danger from grapes and raisins was discovered around 2000, and has slowly been publicized since then. The cause is not known. Small quantities will induce acute renal failure. Sultanas and currants may also be dangerous. Cooked bones are dangerous for dogs, because the heat of cooking changes their chemical and physical properties so that they cannot be chewed properly. As a result they may splinter into jagged shards that resist digestion. Alcoholic beverages pose comparable hazards to dogs as they do to humans, but due to low body weight and lack of alcohol tolerance they are toxic in much smaller portions.

PLANTS

Plants such as caladium, dieffenbachia and philodendron will cause throat irritations that will burn the throat going down as well as coming up. Hops are particularly dangerous and even small quantities can lead to malignant hyperthermia. Amaryllis, daffodil, english ivy, iris, and tulip (especially the bulbs) cause gastric irritation and sometimes central nervous system excitement followed by coma, and, in severe cases, even death. Ingesting foxglove, lily of the valley, larkspur and oleander can be life threatening because the cardiovascular system is affected. Yew is very dangerous because it affects the nervous system. Immediate veterinary treatment is required for dogs that ingest these.

Household Poisons

Many household cleaners such as ammonia, bleach, disinfectants, drain cleaner, soaps, detergents, and other cleaners, mothballs and matches are dangerous to dogs, as are cosmetics such as deodorants, hair colouring, nail polish and remover, home permanent lotion, and suntan lotion. Dogs find some poisons attractive, such as antifreeze (automotive coolant), slug and snail bait, insect bait, and rodent poisons. Antifreeze is insidious to dogs, either puddled or even partly cleaned residue, because of its sweet taste. A dog may pick up antifreeze on its fur and then lick it off.

Animal feces

Dogs occasionally eat their own feces, or the feces of other dogs and other species if available, such as cats, deer, cows, or horses. This is known as coprophagia. Some dogs develop preferences for one type over another. There is no definitive reason known, although boredom, hunger, and nutritional needs have been suggested. Eating cat feces is common, possibly because of the high protein content of cat food. Dogs eating cat feces from a litter box may lead to to Toxoplasmosis. Dogs seem to have different preferences in relation to eating feces. Some are attracted to the stools of deer, cows, or horses.

Other risks

Human medications may be toxic to dogs, for example paracetamol/acetaminophen (Tylenol). Zinc toxicity, mostly in the form of the ingestion of US pennies minted after 1982, is commonly fatal in dogs where it causes a severe hemolytic anemia.

COW DIET AND STARCH

Corn has been a staple ingredient in lactating dairy cow rations for a hundred years. However, many dairy farms import corn to feed to their dairy cows. The price of this imported corn has been fairly stable for a number of years.

Fig. Feeding Dairy Cow

The government is trying to reduce dependence on foreign oil by supporting efforts to produce fuel (ethanol) from renewable resources like corn or cellulose. There is widespread public support for incorporating ethanol into fuelburning vehicles (gasoline and diesel) because ethanol addition results in greater combustion of these fuels and is nontoxic and biodegradable in the water and the soil. The Congress passed legislation called a renewable fuels standard (RFS) that will at least double the use of ethanol and biodiesel by the year 2012. The number of plants producing ethanol from corn is projected to increase dramatically.

This growth in ethanol production has reduced the amount of corn available for livestock feed and concurrently increased its market price. Increased corn prices have resulted in increased feed costs. While higher milk prices may compensate for ration cost increases, partially replacing corn with other less expensive feedstuffs may be required over the longer term.

ALTERNATE RATION STRATEGIES

Other Grain Sources of Dietary Starch

Fig. Grain Diet for cow

Grains, like corn, are an excellent source of starch which is highly fermentable by ruminal microorganisms. The propionic acid produced by starch-fermenting bacteria is converted to glucose by the cow's liver; this glucose is used to make milk. In addition, the bacteria provide about 50 to 60 per cent of the protein needs of the cow as they are washed out of the rumen and are digested in the abomasum and small intestine. To optimize milk production, starch should make up 24 to 26 per cent of the dietary dry matter. Barley and wheat are rich sources of dietary starch. Shifting cow diets between corn and barley or wheat needs to be done slowly.

The starch in barley and wheat is more rapidly available than that in corn so quick dietary changes can result in digestive upsets and lowered ruminal pH. Wheat flour can sometimes be bought for less than corn, but it is very dusty and potentially explosive so it is often avoided by mills. Field reports suggest that flour might 'paste up" in the rumen, so it is not a popular feed ingredient. Hominy contains less starch but more protein, fibre, and fat compared to corn, however their energy density is similar.

Replacing all of the corn with hominy will reduce the dietary starch concentration about 3 percentage units. Dietary costs would be expected to be reduced by replacing corn with hominy. If no milk is lost, then profit is improved. Experiments comparing performance of cows fed ground corn versus hominy could not be located. However, many Florida dairies feed hominy successfully. Even replacing half the corn with hominy can be a reasonable strategy. Because hominy is higher in fat and phosphorus than corn, care should be taken to avoid overfeeding these nutrients.

Low Starch Feedstuffs as Possible Substitutes for Corn

Because grain sources rich in starch, such as barley or wheat, may not be economically available year-round, other commodity feeds must be considered.

Ingredients with lower prices than corn should be considered as partial replacements for corn. Most common feedstuffs contain quite a bit less starch than corn. The best starch sources after corn include corn silage at 30 per cent, wheat midds at 26 per cent, wheat bran at 23 per cent, rice bran at 19 per cent, and corn gluten feed at 16 per cent (DM basis). All other feedstuffs are less than 10 per cent starch. Replacing some of the corn with these feeds will reduce dietary starch to less than 25 per cent which is a common, US ration target. How far can dietary starch be reduced without significantly affecting milk production?

Wheat Midds

Wheat midds contain 26 per cent starch and 18 per cent protein. Price can be substantially lower than corn at the right time of the year. Wheat midds have not been evaluated as a substitute for corn in many experiments. In a recent study, wheat midds were fed at about 7.5 per cent of the diet, replacing a combination of corn and soybean meal.

Corn was reduced in the diet from 34.1 to 28.9 per cent. Cows fed wheat midds tended to eat less feed dry matter but milk production and milk composition were not affected. This reduced feed intake may have been because wheat midds have a high water-holding capacity, thus increasing gut fill with fibre. Cows fed wheat midds appeared to have looser manure than those fed more corn and soybean meal.

In a second study, wheat midds were fed at 0 or 22.4 per cent of the diet, replacing 35 per cent of the ground corn and 30 per cent of the soybean meal. Intake of feed dry matter, production of milk, and milk composition were not different between the two groups of cows which averaged 150 days in milk at the start of the study. However the digestibility of dry matter and neutral detergent fibre were lower for cows fed wheat midds. Therefore wheat midds do not appear to be an effective feed replacement for ground corn except for lower producing cows.

Corn Feed

Corn gluten feed (CGF) is a byproduct of the manufacture of corn sweeteners, corn starch, corn syrup, and corn oil using the wet milling process. The corn starch is used to make ethanol. The CGF consists of the corn bran and a steep liquor (fermented nutrients extracted from water used to soak corn grain) mixed in approximately a 2:1 ratio. CGF is sold in both wet and dry forms. It contains 16 per cent starch, 36.1 per cent NDF, and 23.5 per cent protein, the protein being highly degradable in the rumen.

Pricing is substantially lower than corn. CGF may serve as a concentrate, replacing only the corn and soybean meal, or as both a concentrate and a fibre source, replacing both concentrate and traditional forages. CGF generally accounted for 10 to 45 per cent of the diet, although some studies went higher. When CGF was fed at about 20 per cent of ration DM, milk production was decreased statistically in 1 study, increased in 1 study, and remained unchanged in 9 other studies. Cows in these studies were milking between 50 and 90 pounds per day.

Increasing the CGF to 30, 40 or even 57 per cent of the diet did not reduce milk in any study with the exception of Staples et al. In two studies, in which CGF was fed between 38 and 40 per cent of the diet replacing both forage and concentrate starting at calving, average milk yield was increased significantly by feeding the corn gluten feed.

As more CGF replaced more corn, the dietary starch concentrations dropped, going as low as 15 per cent in some cases. In spite of lowered starch intake, milk yield did not drop significantly.

The bulk of these studies indicate that dietary starch could be reduced by replacing some corn and protein meals with the digestible fibre found in CGF.

Dropping dietary starch from 26 to 21 per cent may be an acceptable compromise. Sample diets were formulated to include CGF at 0 per cent, 10 per cent, and 20 per cent of the diet; starch was reduced concurrently from 25.9 to 23.5 per cent to 20.9 per cent in these rations. As CGF increases in diet, the proportion of extruded soy meal increased to keep the ruminally undegradable protein constant.

Degradable protein sources had to be reduced to keep the dietary crude protein from exceeding 17 per cent. Whether this approach will maintain milk production is not known. Milk and feed intake will be maintained when CGF is fed at 20 per cent of dietary DM. But feeding CGF at 10 per cent of the diet may be a preferred approach initially to "play it safe." Some signs that may indicate that the dietary starch has gotten too low include; dropped milk production, stiffer manure, increases in milk urea nitrogen, and loss of body condition.

Another concern is that drying CGF can reduce the digestibility of the protein if the drying temperature is too high. This is determined by analyzing CGF for acid detergent insoluble protein (ADIN). Bernard et al.reported that dried CGF had higher concentrations of ADIN than wet CGF. In recent years, high ADIN may not be as big of a problem in CGF based upon analyses reported by Dairy One. The ADIN values were less than 5 per cent of the nitrogen in the majority of samples; ADIN is usually not considered a problem until it reaches 10 per cent of the total nitrogen in the feed.

Commodities need to have some consistent level of nutrient density from delivery to delivery in order to keep the daily ration nutrients consistent for the cows. Dairy One lists on their web site the normal range and standard deviation of the nutrients of most of the feeds they have analysed over the past 6 years.

Corn has an average starch value of 70.6 per cent. The standard deviation is 5.1 per cent. This means that two-thirds of the samples that were analysed ranged between 65.5 per cent and 75.7 per cent which is 5.1 per cent units added to or subtracted from the mean of 70.6 per cent. This also means that one-third of the samples analysed were outside this range. This is a fairly narrow range compared to byproduct feeds. Starch in CGF samples has a typical range of 8.6 per cent to 24.0 per cent with a mean of 16.3 per cent.

This large variation is likely due to the fact that CGF production is not standardized across mills. Likewise, starch in hominy typically ranges between 42.9 per cent and 63.9 per cent with a mean of 53.4 per cent. By feeding these byproducts in smaller amounts, the potential negative effect of nutrient variation on cow performance is reduced. When contracting for these commodities, an acceptable range in variation between loads should be agreed upon ahead of time. Loads delivered outside this agreed-upon-range would then be refused.

Dried Distillers Grains Plus Solubles (DDGS)

Dry corn put through the dry-milling process will produce ethanol, carbon dioxide, and DDGS. Each 56-pound bushel of corn processed through the dry milling process results in 2.85 gallons of ethanol, 18 pounds of carbon dioxide, and 18 pounds of DDGS. Plants using the dry milling process are less expensive to build than the plants using the wet milling process, thus about 75 per cent of the ethanol produced from corn comes from dry milling.

The production of DDGS in the US increased tenfold between the years 1980 and 2000, increasing from 320,000 to 3.5 million metric tons (1 metric ton = 2205 pounds). Production doubled again between 2000 and 2004 to over 7.3 million metric tons. Every dairy cow in the U.S. would need to consume 7.3 pounds per day of DDGS over a 305-day lactation in order to use up this supply, but of course DDGS is also fed to other livestock and is exported.

During the production of ethanol from corn, a syrup product and a cake product are produced. These products are blended together in different proportions to form DDGS. Depending on the plant, the DDGS may be a mixture of syrup to cake ranging in proportion from 35:65 to 55:45. Because of the variation in the ratio of syrup to cake used by different plants to form DDGS, nutrient concentrations will vary from one plant to another.

However the variation looks to be less than that of CGF. Although DDGS is very low in starch, it is an excellent source of protein (30.3 per cent) and fat (13 per cent) and therefore has an energy density very similar to corn. In addition, the ethanol-making process increases the digestibility of the fibre. Unlike CGF, DDGS is a very good source of ruminally undegradable protein

(RUP) but if the temperature of the drying process is too high, the protein may be indigestible. Purchasers of DDGS should analyse for ADIN regularly. Kalscheur et al. reviewed 24 experiments in which 98 comparisons were made between cows fed diets containing DDGS and those not fed DDGS. Cows fed DDGS at up to 30 per cent of the ration produced as much milk as those not fed any DDGS (about 73 lb/day). In spite of decreasing starch content, milk production was maintained. In a study conducted at the University of Florida, whiskey DDGS were fed at 0 per cent or 20 per cent of a 55 per cent concentrate:45 per cent forage diet. The forage was all corn silage in one set of diets and 50 per cent corn silage:50 per cent rhizome peanut silage in another set of diets. The DDGS replaced around 40 per cent of the corn and soybean meal. Milk production was increased from 59.0 to 60.7 pounds per day without changing milk fat test.

The effect of DDGS was the same regardless of whether the forage was totally corn silage or an equal mixture of corn silage and rhizome peanut silage. However, a maximum feeding level of DDGS might be 15 per cent in order to prevent the overfeeding of protein, RUP, unsaturated fat, and phosphorus. Increased corn oil in the DDGS can also lead to decreasing milk fat, so fat content of DDGS should be monitored regularly.

Feeding DDGS at 20 per cent of the diet will increase dietary fat by 2.6 per cent. This will likely be a concern if the diet contains other supplemental fats. Dietary fat should be kept below 6 per cent. The feeding of one pound of DDGS can replace about 0.6 pounds of corn and 0.4 pounds of soybean meal.

Soybean Hulls (SBH)

During the processing of soybeans for oil and meal, the hulls are separated and ground. The resulting soybean hulls consist largely of the outer covering of the soybean so they are high in fibre and contain moderate amounts of protein but very little starch. Protein concentration will depend upon how well the hulls are cleaned. Positive characteristics include a high content of lysine, a low content of phosphorus, and a highly digestible fibre. Fat content can vary widely.

The availability of SBH is likely to increase in coming years as more acres are planted to soybeans in order to use soybean oil for biodiesel production. In 13 separate experiments published between 1976 and 2002, SBH (fed at up to 20 per cent of the dietary DM) successfully replaced ground corn or high moisture corn in rations for lactating cows. In most of these studies, the dietary starch levels were probably greater than what is recommended today, based upon the proportion of corn in the diets fed in these studies. Therefore the replacement of corn with SBH still left enough starch to support adequate intake and milk production. One study fed a more conservative amount of corn in the control diet and deserves further attention here. The control diet was 23 per cent high moisture corn, 26 per cent corn silage, and 26 per cent alfalfa silage.

Adding SBH to the diet at 14 per cent of the dietary DM reduced the corn to 9 per cent of the diet. This dropped the starch from roughly about 25 per cent to 15 per cent, calculated from average values. Diets were fed starting at 8 days fresh. Lactating cows ate more DM when offered the ration containing SBH but milk production was not changed. Lactating heifers fed SBH ate the same amount of feed and produced the same amount of milk as those not fed SBH. Concentrations of milk fat and protein were unchanged. Replacing 60 per cent of the corn with SBH, so that SBH made up 14 per cent of the ration, was effective to support high milk production in early lactation.

Even though dietary NDF increases when SBH replace corn, feed intake has not been decreased. This is in contrast with the well-documented fact that as traditional forage replaces concentrate in the diet, concentration of dietary NDF increases and feed intake decreases. However the NDF in nonforage fibre byproducts like SBH does not have the same physical characteristics as the NDF in traditional, properly chopped forage. The fibre in SBH is highly digestible, very short and has a specific gravity that allows it to move out of the rumen quickly so that SBH fibre does not have the same rumen fill as forage fibre at similar NDF concentrations. In order to use SBH most efficiently, the diet should contain sufficient "effective" fibre from traditional forages.

Corn Silage

Corn silage hybrids differ in the proportion of grain in the total crop. Selecting corn silage hybrids that contain more starch will reduce the amount of ground corn needed. In a test of 55 corn silage hybrids grown in Gainesville in 2006, starch concentration ranged from 22 per cent to 35 per cent of silage dry matter, with an average of 29.3 per cent. Selecting a corn silage hybrid with high levels of starch and digestible fibre, while maintaining high, could be a good strategy to reduce corn costs in the ration. How much could be saved? If a corn silage hybrid had 33 per cent starch instead of 30 per cent, and corn silage was fed at 33 per cent of the ration dry matter, dietary ground corn could be reduced from 18 per cent to 16.5 per cent, which is about 1 lb/day less ground corn. The cost savings would depend upon the relative price of the feedstuff(s) used. Corn silage contains much more starch than other forages fed in Florida. Feeding more corn silage in the ration and less sorghum silage, cottonseed hulls, or bermudagrass will increase the starch in the diet and allow for less ground corn to be fed. If corn silage inventory allows, increasing the corn silage from 33 to 38 per cent of the diet DM by replacing another forage like sorghum silage, will allow ground corn to be reduced by 1.5 percentage units.

Glycerol

Glycerol is a byproduct made as plant and animal fats are processed to make diesel fuels. Diesel fuel produced from fat burns cleaner, as there is no soot or particulate matter produced. Starting with animal fat to make diesel fuel is more profitable than starting with vegetable oil because animal fats are cheaper. However, diesel made from animal fat may not work as well in colder climates because it "clouds up." This process is expanding worldwide and therefore the supply of glycerol in the future is likely to increase. Ten pounds of glycerol result from every 100 pounds of biodiesel produced.

The purity of this byproduct can vary widely. The more impure the product, the more water, methanol, phosphorus, and potassium it contains. A product with these contaminants is labeled glycerin. The glycerin fed in a German study contained 2.2 per cent potassium and up to 2.4 per cent phosphorus. Methanol is present due to its use in the manufacture of biodiesel. Although ruminal bacteria can detoxify methanol, it can be problematic if present in large quantities. According to the FDA, glycerin is considered a substance that is GRAS (Generally Recognized as Safe) for general purpose use in animal feed, unless methanol is present at concentrations exceeding 150 ppm. The energy content of glycerol is similar to corn when fed in high starch diets. Dietary glycerol would be converted mainly to propionate and butyrate by ruminal bacteria. As a feed ingredient, it could substitute for corn or molasses.

Little research has been done regarding the feeding of glycerol to dairy cows. Lactating dairy cows fed glycerol (1.3 per cent methanol) at either 0 per cent, 2.7 per cent, or 5.3 per cent of dietary dry matter ate the same amount of feed and produced the same amount of milk. Milk composition was unchanged with the exception that milk urea nitrogen concentrations were lower in cows fed glycerol.

Efficiency of fat-corrected milk production was improved from 1.46 to 1.59 and 1.60 lb of FCM yield per lb of feed dry matter intake. Interestingly, animals of other species consumed more water when they were fed glycerol. This may have a benefit to dairy cows managed under heat stress conditions. German researchers fed glycerol up to 10 per cent of ration DM successfully. Glycerol appears to be a good pelleting agent when added at 5 per cent.

NUTRITION AND GROWTH OF BACTERIA

Every organism must find in its environment all of the substances required for energy generation and cellular biosynthesis. The chemicals and elements of this environment that are utilized for bacterial growth are referred to as nutrientsor nutritional requirements. Many bacteria can be grown the laboratory inculture media which are designed to provide all the essential nutrients in solution for bacterial growth. Bacteria that are symbionts or obligate intracellular parasites of other cells, usually eucaryotic cells, are (not unexpectedly) difficult to grow outside of their natural host cells. Whether the microbe is a mutualist or parasite, the host cell must ultimately provide the nutritional requirements of its resident. Many bacteria can be identified in the environment by inspection or using genetic techniques, but attempts to isolate and grow them in artificial culture has been unsuccessful. This, in part, is the basis of the estimate that we may know less than one per cent of all procaryotes that exist.

THE MAJOR ELEMENTS

At an elementary level, the nutritional requirements of a bacterium such as E. coli are revealed by the cell's elemental composition, which consists of C, H, O, N, S. P, K, Mg, Fe, Ca, Mn, and traces of Zn, Co, Cu, and Mo. These elements are found in the form of water, inorganic ions, small molecules, and macromolecules which serve either a structural or functional role in the cells. The general physiological functions of the elements are outlined in Table.

Table. Major Elements, their Sources and Functions in Bacterial Cells.

Element	% of dry weight	Source	Function
Carbon	50	Organic compounds or CO_2	Main constituent of cellular material
Oxygen	20	H_2O, organic compunsa, H_2	Constituent of cell material and cell water; O_2 is electron acceptor in aerobic respiration
Nitrogen	14	NH_3, NO_3, organic compounds, N_2	Constituent of amino acids, nucleic acids nucleotides, and coenzymes
Hydrogen	8	H_2O, organic compounds, H_2	Main constituent of organic compounds and cell water
Phosphorus	3	Inorganic phosphates (PO_4)	Constituent of nucleic acids, nucleotides, phospholipids, LPS, teichoic acids
Sulfur	1	SO_4, H_2S, S^o, organic sulfur compounds	Constituent of cysteine, methionine, glutathione, several coenzymes
Potassium	1	Potassium salts	Main cellular inorganic cation and cofactor for certain enzymes
Magnesium	0.5	Magnesium salts	Inorganic cellular cation, cofactor for certain enzymatic reactions
Calcium	0.5	Calcium salts	Inorganic cellular cation, cofactor for certain enzymes and a component of endospores
Iron	0.2	Iron salts	Component of cytochromes and certain nonheme iron-proteins and a cofactor for some enzymatic reactions

TRACE ELEMENTS

Table ignores the occurrence of trace elements in bacterial nutrition. Trace elements are metal ions required by certain cells in such small amounts that it is difficult to detect (measure) them, and it is not necessary to add them to culture media as nutrients. Trace elements are required in such small amounts that they are present as "contaminants" of the water or other media components. As metal ions, the trace elements usually act as cofactors for essential enzymatic reactions in the cell. One organism's trace element may be another's required element andp vice-versa, but the usual cations that qualify as trace elements in bacterial nutrition are Mn, Co, Zn, Cu, and Mo.

CARBON AND ENERGY SOURCES FOR BACTERIAL GROWTH

In order to grow in nature or in the laboratory, a bacterium must have an energy source, a source of carbon and other required nutrients, and a permissive range of physical conditions such as O_2 concentration, temperature, and pH. Sometimes bacteria are referred to as individuals or groups based on their patterns of growth under various chemical (nutritional) or physical conditions. For example, phototrophs are organisms that use light as an energy source; anaerobes are organisms that grow without oxygen; thermophiles are organisms that grow at high temperatures. All living organisms require a source of energy. Organisms that use radiant energy (light) are called phototrophs. Organisms that use (oxidize) an organic form of carbon are called heterotrophs or (chemo)heterotrophs. Organisms that oxidize inorganic compounds are called lithotrophs. The carbon requirements of organisms must be met by organic carbon (a chemical compound with a carbon-hydrogen bond) or by CO_2. Organisms that use organic carbon are heterotrophs and organisms that use CO_2 as a sole source of carbon for growth are called autotrophs. Thus, on the basis of carbon and energy sources for growth four major nutritional types of procaryotes may be defined.

Table. Major Nutritional Types of Procaryotes

Nutritional Type	Energy Source	Carbon Source	Examples
Photoautotroph	Light	CO_2 Green Bacteria	Cyanobacteria, some Purple and
Photoheterotrophs	Light	Organic compounds	Some Purple and Green Bacteria
Chemoautotrophs or Lithotrophs Lithoautotrophs)	Inorganic *e.g.* compounds, H_2, NH_3, NO_2, H_2S	CO_2	A few Bacteria and many Archaea
Chemoheterotrophs Heterotrophs	Organic compounds	Organic compounds	Most Bacteria, some or Archaea

Almost all eucaryotes are either photoautotrophic (*e.g.* plants and algae) or heterotrophic (*e.g.* animals, protozoa, fungi). Lithotrophy is unique to procaryotes and photoheterotrophy, common in the Purple and Green Bacteria, occurs only in a very few eucaryotic algae. Phototrophy has not been found in the Archaea, except for nonphotosynthetic light-driven ATP synthesis in the extreme halophiles.

GROWTH FACTORS

This simplified scheme for use of carbon, either organic carbon or CO_2, ignores the possibility that an organism, whether it is an autotroph or a heterotroph, may require small amounts of certain organic compounds for growth because they are essential substances that the organism is unable to synthesize from available nutrients. Such compounds are called growth factors.

Growth factors are required in small amounts by cells because they fulfill specific roles in biosynthesis.

The need for a growth factor results from either a blocked or missing metabolic pathway in the cells.

Growth factors are organized into three categories:

1. Purines and pyrimidines: required for synthesis of nucleic acids (DNA and RNA)
2. *Amino Acids:* Required for the synthesis of proteins
3. *Vitamins:* Needed as coenzymes and functional groups of certain enzymes

Some bacteria (*e.g.* E. coli) do not require any growth factors: they can synthesize all essential purines, pyrimidines, amino acids and vitamins, starting with their carbon source, as part of their own intermediary metabolism.

Certain other bacteria (*e.g.* Lactobacillus) require purines, pyrimidines, vitamins and several amino acids in order to grow. These compounds must be added in advance to culture media that are used to grow these bacteria.

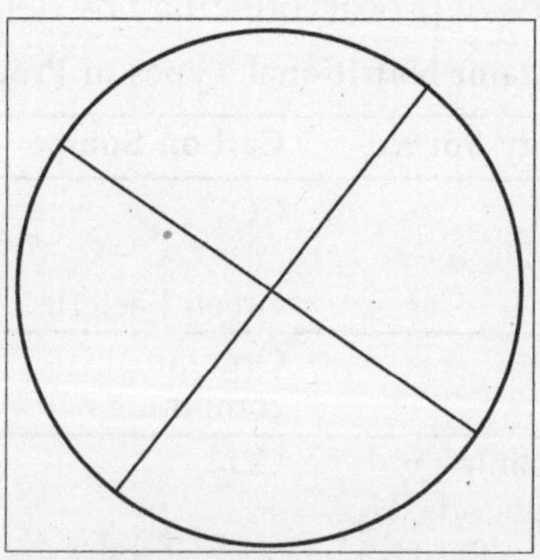

The growth factors are not metabolized directly as sources of carbon or energy, rather they are assimilated by cells to fulfill their specific role in metabolism. Mutant strains of bacteria that require some growth factor not needed by the wild type (parent) strain are referred to as auxotrophs. Thus, a strain of E. coli that requires the amino acid tryptophan in order to grow would be called a tryptophan auxotroph and would be designated E. colitroph.

Some vitamins that are frequently required by certain bacteria as growth factors are listed in Table. The function(s) of these vitamins in essential enzymatic reactions gives a clue why, if the cell cannot make the vitamin, it must be provided exogenously in order for growth to occur.

Table. Common Vitamins Required in the Nutrition of Certain Bacteria.

Vitamin	Coenzyme form	Function
p-Aminobenzoic		Precursor for the biosynthesis-acid (PABA) of folic acid
Folic acid	Tetrahydrofolate	Transfer of one-carbon units and required for synthesis of thymine, purine bases, serine, methionine and pantothenate
Biotin	Biotin	Biosynthetic reactions that require CO_2 fixation
Lipoic acid	Lipoamide	Transfer of acyl groups in oxidation of keto acids
Mercaptoethane -sulfonic acid	Coenzyme M	CH_4 production by methanogens
Nicotinic acid	NAD (nicotinamide adenine dinucleotide) and NADP	Electron carrier in dehydrogenation reactions
Pantothenic acid	Coenzyme A and the Acyl Carrier Protein (ACP)	Oxidation of keto acids and acyl group carriers in metabolism
Pyridoxine (B_6)	Pyridoxal phosphate	Transamination, deamination, decarboxylation and racemation of amino acids
Riboflavin (B_2)	FMN (flavin mononucleotide) an FAD (flavin adenine dinucleotide)	Oxidoreduction reactions
Thiamine (B_1)	Thiamine pyrophosphate (TPP)	Decarboxylation of keto acids and transaminase reactions
Vitamin B_{12}	Cobalamine coupled to adenine nucleoside	Transfer of methyl groups
Vitamin K	Quinones and napthoquinones	Electron transport processes

CULTURE MEDIA FOR THE GROWTH OF BACTERIA

For any bacterium to be propagated for any purpose it is necessary to provide the appropriate biochemical and biophysical environment. The biochemical (nutritional) environment is made available as a culture medium, and depending upon the special needs of particular bacteria (as well as particular investigators) a large variety and types of culture media have been developed with different purposes and uses. Culture media are employed in the isolation and maintenance of pure cultures of bacteria and are also used for identification of bacteria according to their biochemical and physiological properties.

The manner in which bacteria are cultivated, and the purpose of culture media, varies widely. Liquid media are used for growth of pure batch cultures, while solidified media are used widely for the isolation of pure cultures, for estimating viable bacterial populations, and a variety of other purposes.

The usual gelling agent for solid or semisolid medium is agar, a hydrocolloid derived from red algae. Agar is used because of its unique physical properties (it melts at 100°C and remains liquid until cooled to 40°C, the temperature at which it gels) and because it cannot be metabolized by most bacteria. Hence as a medium component it is relatively inert; it simply holds (gels) nutrients that are in aqueous solution.

TYPES OF CULTURE MEDIA

Culture media may be classified into several categories depending on their composition or use. A chemically-defined (synthetic) medium is one in which the exact chemical composition is known. A complex (undefined) medium is one in which the exact chemical constitution of the medium is not known. Defined media are usually composed of pure biochemicals off the shelf; complex media usually contain complex materials of biological origin such as blood or milk or yeast extract or beef extract, the exact chemical composition of which is obviously undetermined. A defined medium is a minimal medium if it provides only the exact nutrients (including any growth factors) needed by the organism for growth.

The use of defined minimal media requires the investigator to know the exact nutritional requirements of the organisms in question. Chemically-defined media are of value in studying the minimal nutritional requirements of microorganisms, for enrichment cultures, and for a wide variety of physiological studies. Complex media usually provide the full range of growth factors that may be required by an organism so they may be more handily used to cultivate unknown bacteria or bacteria whose nutritional requirement are complex (*i.e.*, organisms that require a lot of growth factors, known or unknown). Complex media are usually used for cultivation of bacterial pathogens and other fastidious bacteria.

Most pathogenic bacteria of animals, which have adapted themselves to growth in animal tissues, require complex media for their growth. Blood, serum and tissue extracts are frequently added to culture media for the cultivation of pathogens. Even so, for a few fastidious pathogens such as Treponema pallidum, the agent of syphilis, and Mycobacterium leprae, the cause of leprosy, artificial culture media and conditions have not been established. This fact thwarts the the ability to do basic research on these pathogens and the diseases that they cause.

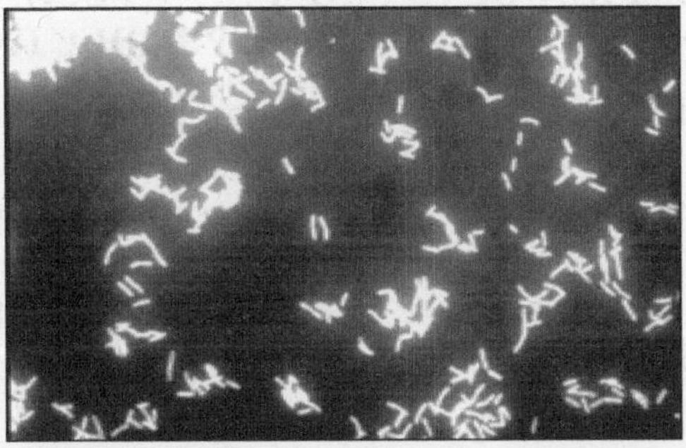

Other concepts employed in the construction of culture media are the principles of selection and enrichment. A selective medium is one which has a component(s) added to it which will inhibit or prevent the growth of certain types or species of bacteria and/or promote the growth of desired species. One can also adjust the physical conditions of a culture medium, such as pH and temperature, to render it selective for organisms that are able to grow under these certain conditions.

A culture medium may also be a differential medium if allows the investigator to distinguish between different types of bacteria based on some observable trait in their pattern of growth on the medium. Thus a selective, differential medium for the isolation of Staphylococcus aureus, the most common bacterial pathogen of humans, contains a very high concentration of salt (which the staph will tolerate) that inhibits most other bacteria, mannitol as a source of fermentable sugar, and a pH indicator dye. From clinical specimens, only staph will grow. S. aureus is differentiated from S. epidermidis (a nonpathogenic component of the normal flora) on the basis of its ability to ferment mannitol. Mannitol-fermenting colonies (S. aureus) produce acid which reacts with the indicator dye forming a coloured halo around the colonies; mannitol non-fermenters (S. epidermidis) use other non-fermentative substrates in the medium for growth and do not form a halo around their colonies.

An enrichment medium employs a slightly different twist. An enrichment medium contains some component that permits the growth of specific types or species of bacteria, usually because they alone can utilize the component from their environment. However, an enrichment medium may have selective features. An enrichment medium for nonsymbiotic nitrogen-fixing bacteria omits a source of added nitrogen to the medium. The medium is inoculated with a potential source of these bacteria (*e.g.* a soil sample) and incubated in the atmosphere wherein the only source of nitrogen available is N_2. A selective enrichment medium for growth of the extreme halophile (Halococcus) contains nearly 25 per cent salt [NaCl], which is required by the extreme halophile and which inhibits the growth of all other procaryotes.

FUTURE BYPRODUCTS FROM NEW TECHNOLOGIES TO IMPROVE ETHANOL PRODUCTION

As the ethanol manufacturing industry works at improving the efficiency of conversion of corn to ethanol, different byproducts will become available. Applegate et al. lists the following potential corn byproducts.

- Called the "quick germ quick fibre method," an enzyme is added to the water used to soak the ground corn, causing the germ and fibre to float before the fermentation process begins. The product from this process contains 28 per cent protein, 5 per cent fat, and 25 per cent NDF.
- Collecting the pericarp fibre and germ prior to fermentation by modifying the dry grinding process (drum degerminator) results in a product that is 24 per cent protein, 8-9 per cent fat, and 28 per cent NDF.
- Removing the fibre by sieving and air aspiration results in a product that is 40+ per cent protein, 15 per cent fat, and 20 per cent NDF.
- Modifying the yeasts used to ferment the sugar to ethanol could allow for an increased concentration of lysine in DDGS, thus giving DDGS a more favourable amino acid profile.

As each mill adopts what they consider the best ethanol-producing technology, the feeding industry will be faced with a variety of byproducts on the market which will need to be properly identified prior to purchase.

With the increased price for corn due to demand for ethanol, farmers will be planting more of their acres to corn. This will reduce the number of acres committed to other crops, such as cottonseed. This shift will likely have a large impact on the market price of a number of feed commodities. The availability and price of each commodity will need to be evaluated for optimal pricing. Some acceptable, alternative feeds may allow some reduction of corn grain in the diet of lactating dairy cows.

FEEDING GOATS

Feeds for goats can be divided into several different categories: forages, grains, protein supplements, minerals and water. Depending on what levels of each category that is fed. It will be found that goats will gain about.25 to.5 a pound each day.

Goats will readily eat a variety of forages. These include pasture, hay, haylage, and brush. Any species of forage is acceptable for goats, although be careful of feeding a lot of alfalfa to wethers and billies because the higher calcium levels can lead to urinary calculi or kidney stones. Goats are very useful in many operations for cleaning up unwanted brushy areas. Most brushy areas will only be able to survive about 2 years with intensive defoliation. Forages are an important part of a ruminant's ration because the high fibre is necessary to keep the digestive system healthy. Grains provide the energy portion of a ration for goats. A variety of grains are available, with corn, oats, wheat and barley being the most widely used. Limit the amount of wheat in the ration to no more than 2 per cent to prevent any digestive problems. When starting goats on a feedlot ration, be sure to gradually introduce them to grain.

Feeding high levels of grain suddenly can cause bloating and overeating disease. Disease can be prevented by vaccinating with type C and D perfringens. This is often available in one vaccine, along with tetanus. Any male goats that are castrated should receive a dose of tetanus vaccine. Castration is a personal consideration as many ethnic groups prefer to purchase intact males.

Protein Supplements

Market goats should be fed a grain ration that runs between 14 per cent and 18 per cent protein. To reach these protein levels, we need to include some sort of protein supplement. Typical supplements will include soybean meal, roasted soybeans, cottonseed, brewers or distiller's grains, or flour mill by-products. Beware of using any dairy protein supplements because of the amount of copper in the supplements. Many nutritionists still disagree over the amount of copper required in a goat ration, but care should still be given to overfeeding this nutrient.

Minerals

One of the best ways to provide minerals to a market goat is through a commercially mixed salt and mineral mix. Be sure to purchase a mix formulated specifically for goats. Allow goat's free access to a mineral feeder that contains the mineral mix at all times. Minerals are especially important for helping to ensure that goats remain healthy. These mineral mixes will often include some of the essential vitamins for goats as well.

Water

Water is often one of the most overlooked aspects of livestock feeding operations. Clean, fresh water should be available at all times for all species of livestock. Animals that have adequate amounts of water are more likely to stay healthy as well as have a faster growth rate. Goats will often be able to supply all of their water needs through dew and lush pasture. However, it is still in their best interest to have a supply of water available to them at all times.

OTHER FEED ADDITIVES

For herds that are prone to urinary calculi, ammonium chloride can be added to prevent the disease. Another common practice is to add a coccidiostat to the ration to prevent as well as control coccidiosis. Common additives used to prevent coccidiosis include Rumensin, Bovatec or Deccox (decoquinate). Lastly, no matter what the ration, be sure to provide enough feeder space for all the goats to eat. Plan for at least 1 foot of feeder space for each goat. And, don't forget to clean feeders and waterers on a regular basis.

Sound feeding practices and good housing facilities result in optimum growth and high milk production and contribute to the good health and comfort of goats.

When feeding goats, keep these objectives must be in mind:

- Feed a young animal enough energy for growth, and feed a mature animal enough energy to maintain a fairly constant body weight;
- Provide enough protein, minerals and vitamins in a balanced feeding programme to maintain a healthy animal; and
- Offer does enough extra food during gestation and lactation for fetus development and milk production.

Optimum growth, good health and high milk production are the results of sound feeding practices. Dairy goats are not unique in their body requirements; they will respond to good nutritional practices. Digestible fibre is especially important in dairy goat diets. Too much grain in relation to forage does not foster good ruminant action and is a costly feeding practice. When feeding, keep minerals and trace mineral salt separate. Always feed them separately.

Feed hay in a rack that will not permit wasting. One recommended type of feeder is a keyhole feeder. Do not overlook forage testing as an economical way to feed the correct ration to herd.

Feeding Kids and Yearlings

Kids are born with no natural protection from disease. The first milk (colostrum) from the mother offers protection and gets the digestive system working. To be most effective, it must be fed before disease-producing organisms enter the mouth and digestive tract. Wash the fresh doe's udders and teats with warm water.

Hand milk half a cup of colostrum and feed it to the kid within 15 minutes of birth. This is the best way to ensure that the newborn receives some milk and to provide it with the most protection from organisms present on the skin of the doe. Complete the first milking and store the colostrum for later feedings if hand feed is elected. Otherwise, permit the kid to nurse at its convenience following the first hand feeding.

Clean the feeding utensils immediately after each use. The same cleaning procedures should be followed for washing milk-handling equipment. Milk replacer may be fed from the fourth day; it should contain at least 20 percent protein, 20 percent fat and be free of vegetable products. A lamb or high-quality calf milk replacer is recommended. Provide hay and grain at one to two weeks of age. Wean from milk when grain intake reaches 1/4-pound daily and kids are readily consuming hay. Make sure the solution is thoroughly mixed. Mix a fresh mixture daily and feed in place of milk. Feed as soon as diarrhea is noticed. Use for 1-1/2 or 2 days, then return to the regular milk diet. After four to six months of age, the kids may be fed a ration similar to that of the milking herd. Good hay and 1/2 pound of grain per day should provide an ample growth rate. Poor hay may require 1 to 1-1/2 pounds of grain daily.

Feeding the Milking Herd

If milk production is important, feed maximum amounts of high quality hay balanced with a grain ration containing enough protein, minerals and vitamins to support production and animal health. Grass or legume hays are equally acceptable. As the percentage of legumes is increased, the need for protein in the grain mix is reduced.

To determine the amount of grain to feed, consider level of milk production, amount and quality of forages consumed, appetite and state of fleshing. Thin, high-producing does should have access to all the hay they can eat plus grain to the limit of their appetite. Does in mid-lactation that are in good flesh should have all the hay they will eat plus 1 pound of grain for each 3 pounds milk produced. Late lactation does may not need more than 1 pound of grain for each 5 pounds of milk.

Feed a grain ration formulated for a milk-producing ruminant (dairy cows). Rolled or cracked grain is more palatable than ground grain. Because of palatability problems, urea is not recommended. Some commercial cow feeds may contain byproduct ingredients unpalatable to goats. Wet molasses is more palatable than dry molasses. Beer or citrus pulp is a valuable source of fibre, especially if the available hay is of low quality.

If the doe is not thin, reduce the amount of grain to 1/2 to 1 pound per day. Feed her all the forage she will eat. Hay fed during the dry period may be of lower quality, but if so, the grain ration should contain additional protein. Browse, leaves and weeds are often useful to recondition the stomach. If the dry ration differs from the milking ration, be sure to change to the milking forage and grain ration two weeks before the doe freshens. Most bucks do not need more than a pound of grain per day plus forages. Don't let them grow fat. Adjust grain upward or downward accordingly. Always feed full forage.

FEEDING AND NUTRITION OF PIG

Feeding guinea pig is a fairly simple manner, though there are a few special needs that we should be aware of. Guinea pigs are strict vegetarians, and in addition to their pellets, they will eat fruits and vegetables, bread, and will even nibble on vegetarian dog biscuits.

Before getting in to the details, it is important to point out that guinea pigs like a routine, predictable life. Sudden changes in their diet can cause stress, so be careful. If the brands of pellets is changed, gradually by adding a few of the new ones to their old pellets, and slowly switching the mixture over. Feeding excessive amounts of fruits and vegetables at one time can cause diarrhea, particularly if new foods are introduced.

All guinea pigs need water to survive. Though they can obtain a large portion of their water from fresh fruits, vegetables and greens, they will still require water from a water bottle to stay healthy. This water should be changed daily, and the water bottles themselves washed periodically to keep them clean. To prevent algae growth, make sure that the sun does not directly shine on the water bottle for long periods of time.

If there is arrangement of tap water bottled water or a water filter system should be considered. Distilled water is not recommended for regular drinking, as it can disturb the osmotic pressure in the body cells, which can potentially lead to osmotic shock.

Be aware that some guinea pigs like to "play" with their water bottle, leaking out large amounts of water by holding the steel ball up in the tube. Others like to try and blow water back up into the bottle, which can contaminate the entire supply. Hay is the mainstay of every guinea pig's diet. Though guinea pig pellets do contain hay, it is not in sufficient quantity to keep guinea pig healthy. Guinea pig must have a fresh supply of hay every day in order to keep its digestive system regular, and as a general rule.

Timothy hay is the best type of hay to feed, though any grass hay will do. If grass hay is not available, a legume hay can be fed, such as alfalfa, though in general this should be avoided as much as possible. Legume hays are high in calcium, and this can lead to bladder stones (uroliths) in some animals.

Most pet stores carry dried timothy hay in large and small bags, but it is far better to purchase hay fresh from a feed and garden store, or directly from a farmer. Buying hay by the bale is especially economical, as a bale of hay will literally last for months, and will stay fresh as long as it is stored properly. Some tips on hay storage can be found here.

Guinea Pig pellets

Guinea pig pellets will provide guinea pig with the proper balance of vitamins (save for Vitamin C), minerals and other nutrients. Although it's not necessary to feed pellets to guinea pigs, it is certainly recommended that they be primary feed, after hay. Guinea pig pellets are, in particular, a prime source of protein; obtaining the necessary amounts of this and other nutrients will require careful dietary planning if pellets are not chosen.

As a general rule of thumb, adult guinea pigs will eat between one and two ounces of pellets each day. More active animals will eat less, and animals without much stimulation will eat more out of boredom. Nursing and pregnant sows may demand a little more than this, and animals that are given plenty of hay and some fresh vegetables each day might eat a little less. It will be easy to tell if they are fed too much or too little: if there are leftover pellets. If the bowl is empty and they seem to be foraging for food, then they aren't fed enough.

Try to feed at the same hour each day, as guinea pigs like a routine and predictable life. Remove any leftover pellets after an hour or so. Guinea pig pellets are a prime source of protein: a diet that is heavy on pellets will lead to a fat guinea pig. Hay is the only food that they should have access to 24 hours a day. When choosing pellets, be sure to *avoid* the guinea pig "mixes" that contain nuts, seeds and dried fruits. These mixes are high in fats and oils, which can lead to excessive weight gain. Additionally, many of these mixes contain sunflower seeds in their shells. Guinea pigs should *never* be fed nuts or seeds that are still in their shells (peanuts, sunflower seeds, etc.): dozens of guinea pigs in the Pacific Northwest alone die each year from choking on these shell fragments.

One thing that should be noted before we move on: guinea pigs should be fed guinea pig pellets, not rabbit pellets or pellets for another species of animal. Guinea pig pellets are nutritionally balanced for guinea pigs, and feeding them pellets for other animals can result in serious health problems.

Though some people may tell that their guinea pigs survive well on rabbit pellets, for instance, it is not trusted that all rabbit food is safe for guinea pigs. Their particular brand may work for guinea pigs, but another brand may lack necessary vitamins or minerals, or have them in insufficient or excessive quantities. There is no regulation in the pet food industry, so pellets and feeds vary considerably between manufacturers. Don't play Russian roulette with cavies: only feed them pellets that are explicitly and *exclusively* for guinea pigs.

It is *extremely important* that guinea pigs receive enough vitamin C each day to prevent scurvy. Guinea pigs cannot manufacture or store vitamin C, so they must obtain it from their diet on a daily basis. Adult guinea pigs will require 10mg of vitamin C each day, and nursing and pregnant sows will require twice that amount (20mg). The caveat: vitamin C, also known as ascorbic acid, breaks down very quickly, so it must be supplied fresh. Although many pellet manufacturers will claim that their feed contains the required amount of vitamin C.

Vitamin C in pellets breaks down during storage, and after about 90 days, there won't be enough vitamin C in the feed to keep guinea pigs healthy. Also note that the food starts breaking down on the shelf, not when the bag of pellets are opened. Unless the manufacturer dates the bag, showing when the feed was packaged, we really have no idea how long pellets have been on the shelf at the local pet store.

Since pellets are generally an inadequate source of vitamin C, most owners will either feed sufficient quantities of fruits and vegetables rich in vitamin C each day, or provide vitamin C in the form of supplements added to the guinea pigs' food or drinking water. Vitamin supplements come in many forms: pre-packaged vitamin supplements for guinea pigs can be bought from pet store, or self grind up vitamin C tablets. In either case, most people prefer to add the vitamins to the animal's drinking water, as that is the only way to guarantee that it will be ingested daily. If the supplement routes are chosen, here are a few pointers:

- Add the appropriate amount of supplement to the amount of water that the guinea pig will drink in a day. Most water bottles work best if they are filled halfway or more, so end up wasting a lot of vitamins. Fortunately, they aren't very expensive. If guinea pig drinks 4 fluid ounces of water a day, and fill water bottle with 8 ounces, then need to add 20mg of vitamin C supplement. This will lead to 4 ounces of wasted water and 10 mg of wasted vitamin C.
- Vitamin C (ascorbic acid) breaks down very quickly in water, so need to add it on a daily basis.
- The chlorine in tap water can actually inactivate ascorbic acid. If vitamin supplements in water is opted, then use tap water that has been standing for at least 24 hours, or filtered tap water from a filter system that reduces chlorine content.

Being vegetarians, guinea pigs will eat many kinds of fruits, vegetables and fresh greens. And, many fruits and vegetables are sources of vitamin C, providing a natural way to meet guinea pig's daily requirement.

Calcium and Phosphorus totals for the given food amounts are presented as well. Although these are necessary minerals, and a part of every guinea pig's diet, they are getting them from their pellets, as well. The nutrients guinea pig is getting along with their vitamin C. Too much calcium can cause bladder sludge, a precursor to bladder stones. The proper balance of phosphorus, calcium and vitamin C prevents guinea pigs from needing vitamin D supplements.

Although most guinea pig pellets contain vitamin D as a precaution, there's no guarantee that particular brand of feed does. In addition to providing vitamin C and variance in cavy's diet, fresh greens and fruits contain a significant amount of water. Feeding "wet foods" will aid nursing sows in lactating, and will reduce the amount of water that the average guinea pig will drink from the water bottle.

It is especially useful to take lettuce, green peppers and other watery greens on trips to provide moisture for guinea pigs while in the car, since water bottles will leak heavily as the bumps in the road rattle the ball in the drinking tube.

To reduce or completely eliminate pellets from guinea pig's diet, they can survive on water, hay and fresh fruits and vegetables. However, this will require very careful dietary planning on part, to make sure that they receive the necessary nutrients in the proper amounts. Also, be aware that many fresh greens are laxative in action, which means that run the risk of giving guinea pigs loose bowels, or even diarrhea, if the are fed too much at one time. If runny droppings are noticed immediately cut fresh greens out of their diet and feed dry foods until the feces returns to normal. Vegetables that are not laxative (such as carrots) may still be fed.

Guinea pigs will not eat what they don't like: some cavies have very discriminating tastes, while others will eat anything that put in front of them. Each pig has its own preferences, and it may take some time to figure out what he or she likes best. Given below is a list of foods that guinea pigs can/will eat. This list is neither comprehensive nor complete.

Apples, bananas, bread (slightly stale & crunchy, but not moldy), broccoli, carrot greens, carrots and baby carrots, celery (cut into small pieces first), cilantro, cucumber, dandelion greens, grass, green & red bell peppers, green leaf & romaine lettuce, kale, kiwi, mustard greens, oats, oranges, parsley, raspberries, spinach, tomatoes.

When feeding "wild" greens, such as grass and dandelion greens, make sure they have not been sprayed with chemicals, or contaminated by droppings or urine from other animals, such as cats, dogs and birds. The dangers of pesticides are obvious, and feces can carry any number of parasites which can be transmitted to cavies eating contaminated greens. Even if dogs and cats are dewormed regularly, their feces can contain protozoa/bacteria which can cause debilitating diarrhea in a guinea pig.

Some foods to avoid are listed below:

- Long celery stalks
- Iceberg lettuce
- Any shelled nuts or seeds
- Raw beans
- Rhubarb

There is some confusion as to whether or not potato peelings are good or bad: some books indicate that they are poisonous to guinea pigs, while others say that they are okay in small amounts. The truth is that potatoes are *okay*, however, any *green* in a potato *is poisoinous*.

If potato peelings are given, make sure there are *no green spots* anywhere in the portions. Also, some people may recommend yogurt in small amounts, for it's bacteria-growing properties (to aid the digestive system). Instead, look into acidophilus powders or liquids: acidophilus is also a bacteria growing culture, and it';s likely to give better results since less of it needs to be fed.

There are several commercially available treats on the market that are aimed at guinea pigs. Berry flavoured "crunchies" and flavoured chew sticks are popular among some owners and guinea pigs, and are generally safe. Some owners also have experimented with vegetarian dog biscuits and dry cereals such as Cheerio's. However, the guinea pig "treat sticks", which are also available commercially. These sticks are essentially seeds and nuts that are held together with honey. Most of the treat sticks on the market contain sunflower seeds that are still in their shells, which is a big no-no for guinea pigs due to the choking hazards.

FISH FEEDING

Providing fish with the right type of food in the right proportions is a very important task. This can be much more of a challenge in a saltwater tank than in a freshwater tank, because most marine fish are still wild-caught, and their natural diet can be difficult to maintain in captivity. With the right information, however, most marine fish will learn to eagerly anticipate their feeding time and will voraciously devour their food.

Fig. Fish Food

A good rule of thumb to follow is that fish should be fed no more than they can consume within a five minute period. Any food that is left behind after feeding time should be removed before it has a chance to begin decomposing. If it is found that fish are consuming the food in far shorter a time than five minutes and seem anxious for more, the amount that are fed must be increased.

On the other hand, if fish are healthy, but they are only consuming about half of what they are fed, probably need to start feeding less. Always keep in mind that much more harm can be done to fish by overfeeding than by underfeeding. In the wild, reef fishes will browse for food throughout the day. Their natural habitat as closely as possible, it is recommended that fish must be fed several small meals every day. At the very least, feed two medium-sized meals per day, one in the morning and one at night.

Types of Fish Food

The kind of food to feed depends on the kind of fish are feeding: herbivore, carnivore, or omnivore.

Herbivores

It is common for fish that graze on algae in the wild to suffer in captivity because they don't receive the appropriate type of food to keep them healthy.

Many of these fish are fed terrestrial greens like spinach and romaine lettuce, but these are not appropriate foods for these fish that are solely algae eaters.

They should be fed either dried or fresh algae. Dried algae is sold in paper-thin sheets that can be broken down into appropriate portion sizes, depending on the number of fish are feeding. Fresh algae can be obtained by placing a glass container, with one rock in aquarium, on a sunny windowsill. Wait for the rock to grow a nice layer of algae, and then place it back in the aquarium for fish to graze on.

Herbivorous fish can also be fed one of the frozen fish foods that is formulated especially for herbivores. This food should contain a variety of marine algae, vegetable matter, and seafood. Whether herbivorous fish are fed dried algae, fresh algae, or a frozen food formula, herbivorous eaters are grazers and should have food available to them at all times, unlike most other fish.

Carnivores

Meat-eating marine fish can enjoy a variety of fresh, frozen, freeze-dried, and live foods. Fresh foods suitable for fish include many seafood items, such as shrimp, scallops, mussels, clams, and some fish. If seafood is fed to fish, cut it into bite-sized pieces (or serve it whole to larger fish). Fresh seafood can be frozen and then thawed as needed.

Frozen foods are available from most fish shops and other sources. These offerings may include brine shrimp, bloodworms, and daphnia. They may be in the form of a slab that can be broken into smaller pieces or small, ready-to-go cubes. To feed frozen foods to carnivorous fish, thaw a small portion in aquarium water, and then drop it in the tank.

Several freeze-dried foods for carnivorous fish are readily available, such as plankton, krill, brine shrimp, and bloodworms. Freeze-dried foods are foods that have been completely dehydrated, yet they maintain most of their nutritional value. They have a relatively long shelf life, particularly when compared to fresh and frozen foods. Freeze-dried foods should be used as supplements for carnivorous fish. Live foods are almost always a favourite of fish. Adult brine shrimp, bloodworms, and Mysis shrimp are among the better choices. Earthworms are also a great live food choice because they come in all sizes, fish love them. Be sure to thoroughly rinse them before feeding them to fish.

Omnivores

The majority of marine fish are omnivorous, which means they need to eat both meat- and plant-based foods. One easy option for omnivorous eaters is commercial fish food, such as flakes or pellets. However, offering a varied diet will give healthier, more colourful fish. Try to feed fish two feedings a day of commercial fish food and one feeding of a meaty item, such as bloodworms or chopped fish. The food offered must be alternated in addition to commercial food to keep the fish from getting bored and to make sure all of their nutritional needs are met.

SHEEP DIET

Sheep are exclusively herbivorous mammals. Like all ruminants, sheep have a complex digestive system composed of four chambers that allows them to break down otherwise inedible cellulose from stems, leaves, and seed hulls in to complex carbohydrates.

When sheep graze, vegetation is chewed in to a mass called a bolus, which is then passed in to the first chamber: the rumen. The rumen is a 5-10 gallon (18.9-37.8 liter) organ which ferments feed via a symbiotic relationship with the millions of bacterium, protozoa, and yeasts which are collectively called flora.

The bolus is periodically regurgitated back in to the mouth as cud for additional chewing and salivation, this process usually takes around six hours. Cud chewing is an evolutionary adaptation that allows ruminants to graze more quickly in the morning, and then fully chew and digest feed later in the day. This is beneficial because grazing, which requires lowering the head, leaves sheep more vulnerable to predators, while cud chewing does not.

During fermentation, the rumen produces excess gas which must be expelled; disturbances of the organ, such as sudden changes in a sheep's diet, can cause potentially fatal conditions such as bloat. After fermentation in the rumen, feed passes in to the reticulum and the omasum; special feeds such as grains may bypass the rumen altogether.

Following the first three chambers, food moves in to the abomasum for final digestion before processing by the intestines. The abomasum is the only one of the three chambers directly equatable with the usual mammalian stomach, and is accordingly sometimes called the "true stomach".

Sheep follow a diurnal pattern of activity, beginning to feed at dawn and throughout the day until dusk, stopping sporadically to rest and chew their cud. Ideal pasture for sheep is not lawn-like grass, but an array of grasses, legumes and forbs. Types of land where sheep are raised vary widely, from pastures that are seeded and improved intentionally to rough, native lands. Common plants toxic to sheep are present in most of the world, and include (but are not limited to) oak and acorns, tomato, yew, rhubarb, potato, and rhododendron. Sheep also may perish from consuming man-made substances.

Sheep are largely grazing herbivores, unlike browsing animals such as goats and deer that prefer taller foliage. With a much narrower face, they can crop plants very close to the ground and can overgraze a pasture much faster than cattle. For this reason, many shepherds utilize some type of managed grazing, whereby a flock is rotated through multiple pastures, giving plants time to recover. Paradoxically, sheep can be both the cause of and solution to one of the major ill effects of overgrazing: the spread of invasive plant species. By disturbing the natural make-up of pasture, sheep and other livestock can pave the way for invasive plants. But sheep also prefer to eat invasives such as cheatgrass, leafy spurge, kudzu and spotted knapweed over native species such as sagebrush—thereby making grazing sheep an effective measure in restoring native pastures.

Other than forage, the only other common staple feed for sheep is hay, most often during the winter months. The ability to survive solely on pasture (even without hay) varies with breed, but all sheep are capable of doing so. Also included in most sheep's diets are minerals, either in a trace mix or in licks.

Naturally, a constant source of potable water is also a fundamental requirement for sheep. The amount of water needed by sheep fluctuates with the season and the type and quality of the food they consume. When sheep feed on large amounts of new growth and there is precipitation (including dew, as sheep are dawn feeders), sheep need less water. When sheep are confined or are eating large amounts of cured hay, more water is typically needed. Sheep also require clean water, and may refuse to drink water that is covered in scum or algae.

COOLING PONDS FOR DAIRY CATTLE

Cattle have used water for cooling probably for as long as there have been cattle. The last survey done in Florida reported that 30 per cent of the dairies had cooling ponds. The reader should be aware that cooling ponds, if not maintained properly, could harbor infectious diseases such as Leptospirosis and many mastitis organisms. The water usage is very high to fill and keep full.

Environmental concerns that must be addressed are the run-off from ponds and ground seepage. Every state seems to have many regulations that must be addressed before building cooling ponds. The benefits of cooling ponds and some suggestions on building and maintaining them. It is the dairymen's responsibility to use them properly and to be aware that ponds can harbor diseases and the pond water must be dealt with according to state and local laws.

What is a cooling pond for dairy cattle? For this discussion they are a man made hole in the ground with a constant inflow of clean fresh water. Usually the ponds drain or overflow from the top. If a series of ponds is used, the overflow water may be run into a retention pond to be irrigated on crops or enter the waste management system. Cooling ponds may also be fed from springs or artesian wells. So cooling ponds for dairy cattle should have constant fresh water in and a constant outflow of water to some place for proper disposal. Cement ponds may also be built. These may drain from the bottom and can be used to flush feed lanes or barns. Bottom draining will help remove some of the solids that have settled to the bottom of the pond. Care must be taken on the entrance and exit to cement ponds.

These may be very slippery if the cement is not cross-grooved. A stagnant pond without fresh water entering, a lagoon, a wallow, and a mud hole do not classify as a cooling pond for dairy cattle.

Site Selection

A site near the feed and drinking water, ideally with some form of shade for cows, is best. The idea is for cows to cool, leave the pond and goes eat and drink, lay down in the shade and then do it all over again. If cows do not have feed and drinking water near, they will stay in the pond and not eat and we have cool wrinkled cows that do not give much milk.

Cooling ponds should be located close to a water supply as fresh water needs to enter the ponds constantly. There must be a way to dispose of runoff water properly. This will vary by areas of the country.

Some slope from the ponds to the waste management area would be helpful, so pumps need not to be used. Cooling ponds should not be too far from the milking parlor as long walks in the sun increase heat stress. Exit lane sprinklers help make the long walk more comfortable.

Concrete Ponds

Concrete ponds will work and have been used but they are expensive to build. However, they are easier to maintain, as the entrances and exits do not have to be rebuilt. The finish of the concrete on exit and entrance slopes should be cross-grooved to prevent cows from falling. A 1:8 slope would be a safe slope if concrete ponds are constructed.

Sizing Cooling Ponds

There is no official listing for cooling pond size. If a holding area requires 15/ft^2 / cow and cows are driven into a pond and taken out as a group then 15/ ft^2 / cow might be adequate. If 50/ft^2 / cow is recommended for most general animal space, then this might be the correct figure for a cooling pond.

This amount of space would allow cows to enter and exit the pond at all times when all cows were not using the pond at the same time. If 50/ft^2 / cow for a pond is used, then 5000/ft^2 is then needed for the 100 cow group.

Pond Shapes and Depths

Using the sample of 5000/ft^2 /100 cows a pond could be 50' wide × 100' long, or a circle 80' in diameter (A= (Pi) r^2; A= 3.14 × 40 = 5024ft^2). If a 50' × 100' pond is used and the sides are fenced so the cows can only enter or exit at each end of the pond, the amount of dirt falling into the pond will be reduced. If a circular pond is made, cow carpet should be used to eliminate the problem of dirt falling in the pond. As cows can swim, pond depth is not a problem. Deeper ponds may allow the settling of organic matter on the bottom of the pond. This organic matter may not be disturbed by cows walking through it as they will swim over the top of it. Preference should be given to depths of 3' or 4' as it seems to cool the cows and let them stand and walk in and out at will. While deep ponds allow cows to submerge, they do not seem to like to float too long so they go to where they can stand and may block cow traffic. A circular pond that is deep in the middle lets cows dunk themselves while others enter and exit the pond at will.

Water Usage

A 100 cow group with 50/ft^2 / cow could have a 50' wide × 100' long × 3' deep cooling pond (it may be 4' in the middle and slope in and out). A cooling pond 50' wide × 100' long × an average 3' deep=15000/ft^3. There are 7.48 gallons/ft^3, thus the pond requires 112,200 gallons of water.

The pond must be filled and then kept fresh with running water at all times while in use. A 1" pipe 300 ft. long and water pressure at 35psi will result in a flow rate of about 23.5 gpm. At this rate it would then take 80 hours to fill the pond initially. The dairy water usage would then be 1,410 gph × 24 hours or 33,840 gallons of water per day. This fresh water must be pumped into the pond and there is going to be close to that amount of run off water that will exit the pond for disposal.

Water Run-off

Disposal of run-off water must be in accordance with whatever local, state, and federal laws that applies. Many dairies run the water into the waste management system, while others add it to retention ponds to be used for irrigation.

Shading over Ponds

Experience has led to discontinued use of shade over ponds as the cows never leave the ponds. Remember, the object of this exercise is to increase dry matter intake and cows will not eat much soaking in the pond.

Pond Maintenance

At least once a year, usually during the winter, the pond should be fenced and the water removed by pumping. The mud and manure on the bottom of the pond should be removed and the pond left empty for the sun to shine on it.

In the spring, the pond should be filled again and the fences removed. Some people just fence the pond in the winter, dig a new pond and fill in the old one with new pond's dirt. Mycoplasma mastitis is often cultured in water from cooling ponds. This is a real problem if pulsators do not work well resulting in damaged teat ends. Many herds with mycoplasma positive ponds, have no mycoplasma in the bulk tank. If the water is filthy, cows should not be permitted in the ponds and the ponds should be pumped dry. The sludge should be cleaned and the ponds filled with water again. Both the sludge and the water shoud be removed from cow areas.

Cooling Ponds to Reduce Body Temperature

In the summer of 1986, an experiment was conducted at a North Florida dairy that was using man-made cooling ponds. Ten early lactation Holstein cows were fitted with radio transmitters that transmitted inner ear temperatures every five minutes to a radio receiver and data logger.

Data were collected over a four-day period in mid-August. The cows' activities were also monitored during this time. Entering and exiting the ponds, eating, drinking and laying were recorded.

The cooling ponds lowered the cows' temperature by 1-2°F depending on the time of day they entered the cooling ponds. The average length per stay in the pond was 18 minutes for events from midnight to noon and 12 minutes per visit from noon to midnight. It was obvious that cooling ponds cooled cows.

Pond Water Quality

In 1987, samples were taken from a dairy's man-made earthen ponds. The ponds were sampled weekly from May 19,1987 to September 28, 1987. Six to ten ponds were sampled each week.

Total bacteria counts (TBC), varied greatly from week to week. This may have been due to no water entering the ponds. There did not seem to be any great increase in total bacteria counts or coliform counts as time progressed. The total bacteria count averaged 3,133,700 CFU/ml for all ponds for the 20 weeks. The coliform counts averaged 14,340 CFU/ml for the same period.

Cooling Ponds and Their Effect on Milk Quality

In a previous study on a dairy in Florida with man-made cooling ponds, it was reported that cows exposed to cooling ponds did not experience more clinical mastitis than cows that did not have access to cooling ponds. In fact, cows exposed to cooling ponds during the trial period were only half as likely to develop a case of clinical mastitis as cows not exposed to cooling ponds.

Clinical Mastitis Organisms

The same dairy had provided its clinical mastitis records for several years. The total number of clinical mastitis cases was high in the first quarter of the year and then declined sharply about the time hot weather came and the cows started using the ponds. The incidence stayed low, even in the last quarter of the year when the ponds were not in use. This can be explained many ways. The first quarter of the year was an extremely wet period.

The cows were quite dirty and the cow wash system was available to one-half of the herd. This may explain some of the variation, even though one-half of the herd had never had a cow wash. Another confounding variable was that this herd switched to Clorox for pre- post-teat dipping in April of that year. It had previously pre- and post-dipped with a Chlorhexidine product. From this it could be concluded that Clorox caused the reduction of clinical mastitis. However, there were no controls in this study, one could conclude nothing except that the ponds did not increase clinical mastitis.

This herd was on a lactation treatment study during the summer of 1987. A total of 40 cows were sampled and treated. There were no unusual organisms treated. These 40 cows were not all the cows treated for clinical mastitis during the period the cows had access to the ponds, but the results were encouraging.

Statewide Effects of Ponds on Milk Quality

Regulatory samples for all the herds in Florida were obtained from the Division of Dairy Industry for the year of 1987 and were analysed for the effects of no ponds, man-made ponds and natural ponds. The numbers presented are least squares estimates of Standard Plate Counts (SPC) and Direct Microscope Somatic Cell Counts (DMSCC) for each month. The data are not biased by different numbers of dairies with no ponds, natural, or man-made ponds. The data were analysed by herds whose milking cows had access to ponds as indicated on the survey.

The data indicated that milking cows with access to man-made cooling ponds had lower SPC and DMSCC counts than other herds in the state with no access to ponds and those herds with natural ponds were higher in both categories. From these data it could be argued that people who built man-made ponds were better managers and thus had higher milk quality.

3

Biotechnology in Medicine for Animal Health

The use of plants as production systems for vaccine antigens has been actively investigated over the last 15 years. The original research focused on the value of this expression system for oral delivery based on the hypothesis that plant-expressed antigens would be more stable within the digestive tract and would allow for the use of the oral route of administration to stimulate a mucosal immune response. However, while first conceived for utility via the oral route, plant-made antigens have also been studied as classical immunogens delivered via a needle to model animal systems.

Antigens have been expressed in a number of whole plant and cell culture systems. Several alternative expression platforms have been developed to increase expression of antigens or to elicit preferred immunological responses.

The biotechnological advances in plant expression and the immunological testing of these antigens will be reviewed in this chapter focusing primarily on diseases of livestock and companion animals.

PLANT-MADE VACCINES

Advances in plant biotechnology have opened the door for the use of transgenic plants as production systems for vaccine antigens. It was over a decade ago when it was first demonstrated that an antigen could be expressed in plants and fed as a vaccine to produce a measurable immune response. Since then, the expression of vaccine antigens in planta has progressed to allow for the expression of bacterial, viral and self-antigens in transgenic whole plants and plant cells.

Many authors have discussed the value of this expression system which includes decreased manufacturing costs and the potential for alternate delivery routes including oral delivery.

Much of the focus of plant-made vaccine studies has been directed towards their utility for oral delivery. The primary advantages of this delivery route are the increased likelihood of eliciting a mucosal immune response and the ease of vaccine administration. The latter is considered key for disease control in third world countries (humans) and for decreased stress and ease of administration in animals. While oral delivery is attractive, the amount of protein required has been estimated to be as large as 1 mg.

This is only a target since the actual amount will depend on antigen presentation, immunogenicity and formulation of the antigen. Nevertheless, the amount of protein required will likely be large for oral delivery and, subsequently, represents a challenge for many plant expression platforms. Vaccines targeted for use by subcutaneous administration will likely require significantly less antigen.

Despite research efforts in this area of vaccinology, to date, there are no commercial products, human or veterinary, that have been approved by the regulatory authorities who license these products, although Dow AgroSciences has entered the regulatory path for licensure of a plant-made vaccine for Newcastle disease in poultry.

In this review, we focus on the current status of in planta expression including whole plant and plant cells and also on the practical application of this technology beyond laboratory phase testing.

METHODOLOGIES FOR EXPRESSION OF VACCINE ANTIGENS IN PLANTA

Expression Platforms

Initial studies of plant-made vaccines utilised easily transformed and manipulatable model crops such as tobacco. Subsequent work has focused on crop plants, most targeted for human use, but some which could overlap for use in animal vaccines including potato, tomato and corn. A considerable amount of work has been done using potato tubers as a platform. The development of an ethanol inducible promoter that allows inducible expression of a transgene in harvested tubers is an interesting improvement that would enhance the utility of tubers as a delivery system for recombinant proteins.

Transgenic, non-cooked tubers have been shown to deliver amounts of antigen necessary to provide some level of protection in disease models. Problems reported have been the slow degradation of protein in harvested and stored tubers and the fact that raw potato is not palatable for humans and has triggered gastrointestinal side effects in volunteer subjects. These and other reasons have led to the more recent shift in opinion that some degree of processing may be necessary or preferred for many of the plant-expressed materials.

A plant species more thoroughly tested for expression of proteins has been corn. Production of proteins in corn seed forms a basis of several studies including those reported by the company Prodigene. The advantages of corn are that the transgenic protein in the dry seed tends to be stable and the seed can be easily stored and transported. In addition, corn crops can be rapidly expanded, depending on demand. Importantly, the matrix of the corn seed extracts has been shown to provide protection against degradation of protein during passage through the gut.

This latter point is of particular significance to this and other orally delivered plant-made materials. The challenge is to thoroughly investigate stability of the expressed protein in the range of gut conditions present among individuals in an animal population and even at different times within an individual. Transgenic corn has been used to obtain protective immune responses against Escherichia coli heat-labile enterotoxin (LT) and an antigen from the swine transmissible gastroenteritis (TGE) virus (TGEV) and LT-B, the B subunit of LT. Tomato has also been a popular plant species for expressing vaccine candidates, providing both containment and ease of processing. The fruit has the advantage over some other expression platforms in that it can be consumed fresh, eliminating the need for post harvest processing. Moreover, existing technologies are available for processing the fruits, formulating with adjuvants and delivering the vaccine. Plant species that are targeted specifically for animal health and have been used for vaccine production include clover, alfalfa and pigeon pea.

The latter plant is used as a possible source of livestock feed in developing countries. Included among livestock-related antigens expressed in alfalfa are those from foot and mouth disease, TGEV, classical swine fever virus and Mannheimia haemolytica. An alfalfa expression system is the basis of the commercial enterprise Medicago, Inc., which has developed alfalfa transformation and protein production strategies to make it an attractive alternative to more common platforms. In particular, Medicago's work has focused on identifying promoter elements that provide high levels of expression using constitutively expressed promoters or inducible promoters. Inducible promoters will be of value for expressing proteins that are detrimental to plant growth and development. Although most researchers developing plant-made vaccines use field-grown plants, some work has been done with cell cultures.

One of the advantages cited for field-grown plants relative to cell cultures grown in bioreactors is the lower cost of goods. However, cell-culture-based platforms do not have to address the issues of containment, manufacturing process controls and chain of custody. Often, the cost of regulatory approval has not been included in the final cost estimates of field-grown products when the above arguments are made. The most widely available cell culture used for testing plant-made vaccine product concepts is the tobacco line BY-2 (or its sibling culture NT-1). This culture was developed as a low-nicotine line that rapidly increases cell mass. Although used extensively for expressing transgenic proteins, its use for producing vaccines has been largely limited to work recently reported. Several antigens have been expressed in tobacco suspension cultures by these groups including those from avian influenza virus (AIV), Newcastle disease virus (NDV), E. coli heat-labile toxin and infectious bursal disease virus (IBDV). All transgenic cultures were shown to produce proteins that retained their biological activity by either *in vitro* tests or direct disease challenge trials.

Another plant species tested for expression of recombinant proteins is safflower. This specialised technology (StratosomeTM, SemBioSys Genetics, Inc) incorporates a unique expression and purification strategy in which the protein of interest is fused to oleosin, the protein that coats oil bodies in safflower seeds. After the seeds are crushed, the oil bodies can be floated away from other seed materials by centrifugation, resulting in a highly purified preparation of transgenic proteins.

Gene Constructs

Most reports in the literature describe early stage product development in which putative protective antigens are expressed in plants and the resultant transgenic material tested for its ability to either elicit an immune response in an animal, or preferably, to provide protection against disease. The majority of these research programmes has been led by academic researchers with little or no consideration of intellectual property ownership and, hence, often use promoters, genetic elements responsible for expressing genes, that are not in the public domain.

The most common among these is the cauliflower mosaic virus 35 S promoter or its variants. This promoter is generally considered a high expressing dicot promoter. Some researchers have used variants of the Agrobacterium mas gene promoter for expression of vaccine candidate proteins because of their reported high expression. The patatin gene promoter was shown to express higher in potato tubers than different versions of the 35 S promoter and the sunflower polyubiquitin gene promoter. Optimisation of promoters for expression of vaccinerelated proteins in plants is an area that needs further exploration. Accumulation of transgenic protein depends not only on the transcription rate of its gene, determined principally by the promoter used, but also on the steady state level of its mRNA, efficiency and rate of translation of the messenger RNA (mRNA) and stability of the protein. Expression of transgenes in heterologous systems has demonstrated the importance of altering codon usage to increase the rate of protein production. The impact of host cell codon bias depends on how closely it matches the codon bias of the source organism of the transgene.

In some cases, altering the codon composition of a transgene is not important, but in others, it can lead to dramatic increases in expression. High mRNA levels require correct and efficient processing of the mRNA, including transcription termination, polyadenylation and splicing. Plants vary from other eukaryotes mechanistically in how they process RNAs and, further, monocots can differ from dicots. Typically, plants tend to be less precise in the sequences that drive these processes. For example, other eukaryotes use a single predominant sequence for directing RNA polyadenylation, whereas plants have many. The challenge for using plants to express antigens is to completely understand the nuances of the plant gene expression machinery and design transgenes accordingly so that signals are not misinterpreted by plant expression machinery.

Most studies testing potential transgene vaccine candidates for animal health ignored plant-specific gene expression issues. How this has impacted expression levels of those proteins, which tend to be low, is worth exploring for the more promising protective antigens. Other gene elements that can affect expression levels are translational enhancing elements that occur in the 50-untranslated region of mRNAs. Most widely incorporated into constructs is the omega element from tobacco mosaic virus (TMV) gene, which increases translation by recruitment of eIF4F translation factor. This element improved expression 11-fold when it was used to express a hepatitis B antigen. Targeting of proteins to different cellular compartments can be critical to the success of potential vaccines expressed in plants. In many cases, targeting the protein to the endoplasmic reticulum (ER) increases accumulation relative to expression in the cytoplasm. There are exceptions to this, which may be due to glycosylation interfering with protein folding or stability. Proposed reasons for the improved protein accumulation in ER include the presence of high levels of chaperones (proteins assisting with protein folding), a reduced level of proteases, a more oxidising environment and presence of protein disulfide isomerase, an enzyme which forms disulfide bonds in proteins. Targeting to the ER requires the presence of a native or heterologous signal peptide on the N-terminus of the protein.

This directs the protein to the endomembrane system, but retention in the ER requires the addition of a retention signal (*e.g.*, KDEL, SEKDEL and HDEL) to the C-terminus of the protein. Addition of these sequences to proteins normally not targeted to the ER may alter protein conformation, a key consideration when designing expression strategies. Absence of the ER retention sequence on a protein steers it into the default secretory pathway, which leads to secretion of proteins that are below the size retained by the cell wall or insertion into the plasma membrane for proteins with membrane anchoring regions.

Transgenic Proteins in Plants

Accumulation and conformation of transgenic proteins are dependent on correct or aberrant post-translational modifications. Most post-translational modifications found in non-plant eukaryotes have also been demonstrated to occur in plants, including glycosylation, S-acylation, N-myristoylation, phosphorylation, and acetylation. One of the most common post-translational modifications is glycosylation, with plant proteins containing both N-linked and O-linked glycosylation. However, plant glycosylation patterns can be significantly different than those found in mammals.

This may be important for *in vivo* use of plant-produced antibodies for passive immunisation or other uses. Among the most noted differences is the presence of b(1,2)-xylose and a(1,3)-fucose residues in plant glycans that are not found in mammals. These different substitutions may lead to rapid clearing of antibodies from the bloodstream when administered to animals. As proteins are transported into the ER, high mannose glycans are added to accessible asparagines in the amino acid sequence context Asn-Xaa-Ser/Thr and more complex glycans are built on the mannose framework in the Golgi apparatus, including addition of xylose and fucose at particular sites in the glycan. Plants can be genetically modified to produce glycans that are more similar to those found in animals.

Expression levels of transgenic proteins are generally from undetectable to 0.4 per cent of total soluble protein (TSP) and, hence, are considered low. While there can be differences between plant species in the level of expression of particular proteins, this difference tends not to be large. For edible vaccines, the expression level typically obtained is often sufficient to elicit some immunological response by feeding, although whether the level would be commercially viable is not reported.

Nevertheless, expression from transgenic inserts located in nuclear genomes and driven by highly expressed promoters clearly will not always be sufficient without some purification or concentration of the antigen. Some alternative production systems for plant-made antigens include oil bodies, viruses and plastids.

Plastids (the most commonly used plastid in plant transformation being the chloroplast) contain a genome that is separate from the nuclear genome. Importantly, there can be multiple copies of the genome per plastid and multiple plastids per cell, resulting in the presence of up to 10,000 copies per cell. This large gene copy number has led to extraordinarily high levels of expression, with typical increases of expression being from 500-to 4000-fold relative to nuclear-located transgenes. The potential for producing large amounts of commercially important proteins led to the development of the commercial enterprise Chlorogen.

An important advantage of plastid transformation is that the transgenic insert is targeted to specific regions of the chloroplast genome by sequences that are homologous to chloroplast DNA that flank the transgenes of interest. Hence, issues of position effects and gene silencing that plague nuclear transformation experiments are not problems with plastid transformation experiments. Key to the success of this expression platform is that chloroplasts have the machinery to fold and process most proteins correctly. Several plants have been transformed using this technology, but transformation of others of potential interest has not been reported.

Daniell *et al.* argue that carrot has advantages for oral delivery since carrots can be easily stored, eaten uncooked and, importantly, the plant is biennial; hence, preventing spread of seed and pollen if harvested after the first generation. A challenge with this technology can be to ensure a high and stable level of plastids carrying the transgene in the crop plant of interest. The primary limitation of the technology is the lack of protein glycosylation. While the activity of many proteins is not affected by glycosylation, it can be critical to protein conformation or function in other cases. The use of this technology for development of vaccines is illustrated by expression of a tetanus toxin gene and a canine parvovirus peptide fused to cholera toxin B subunit (CTB) at 25 and 31 per cent, respectively, of TSP in tobacco.

How antigens are presented to the immune system can drive the type of immune response that occurs. Peptides or protein subdomains can be fused to other proteins or protein domains that stabilise the protein and improve the accumulation of the protein.

The most widely reported protein fusion partners are non-toxic variants of bacterial exotoxins such as heat labile endotoxin (LT) from E. and cholera toxin (CT) from Vitrio cholerae that bind to particular gangliosides on mucosal surfaces. Oral administration of LT or CT fusions target the proteins to M cells in Peyer's patches and subsequent delivery to antigen-presenting cells within this tissue elicits a mucosal response, which can be essential for protection against many enteric pathogens. This has been a widely used strategy for experimental vaccine development. It is not known how an immune response to LT or CT *per se* might alter responsiveness to the target antigen.

A chimeric protein consisting of a protein of interest and a viral coat protein configured in such a way that the coat proteins can assemble into a virus or virus-like particle (VLP) elicits a strong immune response. Key to the success of this approach is that VLPs may be more readily taken up by antigen-presenting dentritic cells.

A limitation of this strategy is the requirement to minimize disruption of critical coat protein structure, which is done, for example, by insertion of peptides into an external loop of the coat protein. Cowpea mosaic virus, tomato bushy stunt virus, TMV and alfalfa mosaic virus (AlMV) have been used successfully to express high levels of peptide sequences in plants.

An example of this approach was the demonstration of disease protection against virulent challenge and IgG and IgA responses in mice against rabies by oral administration of AlMV coat protein fused to a peptide containing antigenic determinants from the rabies virus. There has also been an extensive use of plant virus expression platform in HIV vaccine development.

Immunogenicity of Plant-expressed Antigens

Several viral subunit proteins, expressed in plant systems, have been tested for their potential as oral vaccines. These include mainly viral antigens from pathogens such as Norwalk virus, TGEV, foot and mouth disease virus (FMDV), rotavirus and parvovirus. Not surprising, there are only limited studies involving bacterial pathogens probably due to the complex nature of bacterial pathogenesis and the inability to clearly delineate proteins that are critically associated with protection. While some of these have been characterised immunologically, few studies conclusively elucidate the apparent unique properties of plant-made vaccines or the immune response elicited by using them.

One of the strongest supporting rationales for expression of vaccine antigens in planta is associated with the hypothesis that when delivered in plant material, vaccine antigens will be protected from potential degradation in the gut, leading to increased vaccine efficacy. It has been suggested that plant antigens are 'bioencapsulated' within the plant cell walls and cell membranes and therefore have increased resistance to digestion. This hypothesis was supported in an experiment comparing the immunogenicity of VP2 antigen of canine parvovirus expressed in tobacco. Mice were gavaged with either soluble plant extracts of VP2 antigen of canine parvovirus at two different doses (200 or 1000 mg) or with tobacco leaf extract (600 mg) and the data demonstrated that numerically higher serum IgG titers were obtained in the mice that were gavaged with leaf extract. The authors attributed this increased response to the protective effect of the plant matrix to digestion.

However, since tobacco leaves were unpalatable to mice, gavage was used as the method of delivery of the antigen. It is not known if direct delivery to the stomach limits the potential for antigen presentation along the upper gastrointestinal tract. The ability of orally administered plant-expressed vaccines to stimulate a mucosal response was reported by Wigdorovitz *et al.* Using a mouse model of bovine rotavirus (BRV), these researchers investigated whether females vaccinated orally would transfer lactogenic immunity to their offspring. In this study, the eBRV4 epitope of the VP4 protein of BRV was expressed in transgenic alfalfa and either extracted from the alfalfa for intraperitoneal injection in mice or fed as an oral vaccine to mice.

Both routes of administration stimulated serum antibody titers as determined by ELISA. Female mice receiving these vaccines were bred and 5 days after birth, offspring from these females were challenged orally with a virulent strain of BRV. Offspring of vaccinated females had significantly less clinical disease 24 h post challenge compared to their non-vaccinated cohorts.

The authors asserted that this vaccination regime stimulated lactogenic immunity; however, no data were presented to support this conclusion. Colostrum was not assessed for specific antibody. More importantly, the hemendothelial placentation of mice allows for transfer of large amounts of immunoglobulin across the placenta. Therefore, the protection seen in the offspring of the vaccinated dams may be associated with transplacental transfer of protective IgG and not necessarily lactogenic transfer of rotavirus antibodies as the authors surmised.

Vaccine Testing in Target Animal Species

Although much has been written about the potential for production of vaccines against diseases of livestock, there is scarce information addressing immunogenicity or protective efficacy using relevant challenge models in the species of interest.

Studies in mice or similar model species can show that the plant-expressed antigen retains the ability to drive an immune response but extrapolation to the target species, especially with respect to efficacy, must be done cautiously since model infections in mice frequently have divergent pathogenesis, and by extension, alternative protective immune responses, even when disease in the model appears to physically resemble that in the natural host. There are as yet no published reports of vaccination and challenge studies in ruminants despite activity by several research groups.

Dus Santos and Wigdorovitz (2005) recently reviewed work examining three pathogens of particular significance for the cattle industry of Argentina, FMDV, BRV and bovine viral diarrhea virus (BVDV) expressed in various plants including Arabidopsis thaliana, alfalfa and potato. Vaccines against the FMDV and BRV have demonstrated immunogenicity after intraperitoneal (FMDV and BRV) or oral (BRV) exposure and resistance to challenge with virulent virus in mice.

Researchers in India have Targeted the Highly Contagious and Deadly Disease

Rinderpest, still prevalent in parts of Africa, the Middle East and South Asia, where lack of facilities and extreme heat constrain maintenance of the cold chain necessary to retain viability of the current modified live vaccine. They expressed the protective viral hemagglutinin antigen in pigeon pea (Cajanus cajun), the leaves of which are a common fodder crop readily consumed by livestock, with the intent of producing an easily stored and transported edible vaccine.

In North America, pneumonic pasteurellosis remains an economically important disease of feedlot cattle. Several protective antigens of the principal causative bacterium M. haemolytica have been identified and researchers have succeeded in expressing the critically important leukotoxin (Lkt) in clover and alfalfa, for production of an edible, needle-free vaccine that will be easy to administer, reducing stress on the animals, which can itself precipitate disease, and avoid costly trim losses at injection sites.

The majority of plant-made vaccines have been tested in model animal systems. Vaccination and challenge trials that have been reported in target species are rabbit hemorrhagic disease virus (RHDV) in rabbits; TGEV in pigs; mink enteritis virus in mink; canine parvovirus in dogs; and IBDV, AIV and NDV in chickens. Potato plants expressing VP60 protein of RHDV have been shown to stimulate a protective immune response against challenge when administered four doses (at monthly intervals) of potato leaf extract containing 12 mg of recombinant VP60 with Complete (day 0) and Incomplete (days 30, 60 and 90) Fruend's adjuvant.

Vaccination was by the subcutaneous route for the first vaccination, and intramuscular for the subsequent vaccinations. Rabbits were challenged with a virulent strain of RHDV and vaccinated animals were completely protected from disease while non-vaccinated control rabbits died. Although these data strongly support that this vaccine stimulated a protective response in rabbits, limited animal numbers (n=4) did not allow statistical validation.

Efforts in swine seem to be focused exclusively on development of an oral vaccine for TGE. Although current injectable vaccines are reasonably efficacious in passive protection of piglets by immunisation of sows, they do not induce production of mucosal antibodies and the need for multiple injections poses a management hardship. The TGE spike protein (S) which drives a virus neutralising response was successfully expressed in tobacco plants (Nicotiana tabacam) and three injections of piglets with leaf tissue suspensions induced antibodies in serum detected by ELISA that were weakly virus neutralising. Streatfield *et al.* were able to demonstrate that in swine, plant-made vaccines can stimulate protective immunity. Ten-day-old piglets, fed with seed corn expressing the spike (S) protein of TGEV, were protected against a virulent challenge with TGEV. The protection was similar to that induced by a commercial modified live vaccine (MLV) with none of the risks associated with use of an MLV. The authors noted that the duration of symptoms was shorter in pigs receiving the modified live vaccine.

This may be in part due to differences in the breadth of the immune response stimulated by the modified live vaccine as compared to the non-adjuvanted subunit corn seed vaccine. Since there are limited data generated to demonstrate that plant-made vaccines are protective in host species, these data are a very important advance towards the development of this type of vaccine technology. More recently, Lamphear *et al.* reported an increase in antiviral antibodies in serum, colostrum and milk of sows fed with a corn-based oral TEGV vaccine following priming by a single injection with conventional vaccine. No attempt to assess passive transfer of protection to piglets was reported. There have been relatively more plant-made vaccine vaccination and challenge studies completed in poultry. Vaccination and challenge trials are reported for IBDV and NDV.

For IBDV, live attenuated vaccines are currently employed for control, but highly attenuated strains give only moderate protection, while less attenuated vaccines can produce disease in vaccinates. Wu *et al.* expressed the VP2 protein of IBD virus in A. thaliana and demonstrated a serum antibody response in chickens fed with leaf extracts of the transgenic plants.

Subsequently, the plant-made vaccine was compared to a commercial vaccine of intermediate attenuation given by ocular/nasal drop for protection of specific pathogen-free broiler chickens against challenge infection at 4 weeks of age. The commercial vaccine was administered at 1 and 3 weeks of age, while the plant-made vaccine was given either orally by gavage, as five doses 3 days apart commencing at 1 and 3 weeks, or injected subcutaneously in Freund's complete adjuvant at 1 and 3 weeks. One group of birds received the commercial vaccine at week 1 as a priming dose and the oral vaccine regime at 3 weeks to boost.

Data were pooled from three independent replicates of the experiment, each with ten chickens per group. Chickens in all vaccinated groups produced antibodies by 1 week after the first vaccination that increased following the second dose. Protection against droplet challenge with virulent virus varied from 90 per cent in the orally vaccinated group to 60 per cent in the groups receiving the plant material subcutaneously. The commercial vaccine alone or with the oral plant vaccine boost provided approximately 80 per cent efficacy.

All control chickens vaccinated orally with saline or a non-transformed plant extract succumbed to challenge. These studies are encouraging with respect to the potential to induce protective immunity by feeding transgenic plant material; however, as these researchers acknowledge, current practice in the industry is vaccination in ovo, often using robotics, and further work is needed to determine the relative ease of vaccination and efficacy of oral plant-made vaccines. Protection against IBDV challenge was also demonstrated by Miller *et al.*. These researchers expressed the VP2 protein of IBDV in tobacco cell cultures and produced vaccines using this antigen for demonstration of efficacy by challenge. In the study, birds received three doses of vaccine (days 7, 21 and 35) either via the subcutaneous route of administration, via the oral gavage route of administration (non-adjuvanted) or via the intrabursal route of administration (non-adjuvanted).

All vaccines delivered via the subcutaneous route of administration were protective against challenge except for the VP2 antigen formulated with drakeoil emulsion adjuvant. Those vaccines delivered via oral gavage or intrabursal were not protective. Further application of the plant-made vaccine technology in poultry has been demonstrated by Cardineau *et al.*

In these studies, the hemagglutinin/neuraminidase protein (HN) of NDV was expressed in tobacco cell cultures and chickens were vaccinated with plant-cell-produced HN (adjuvanted or non-adjuvanted) at 2 days of age and 16 days of age via the subcutaneous, intranasal, oral gavage or oral gavage followed by feeding routes of administration. Birds were challenged 14 days after the second vaccination. One hundred per cent of the non-vaccinated control animals died of NDV and 100 per cent of the positive control birds (receiving native NDV subcutaneously) survived. Protection was evident in the birds receiving the plant-cell-produced HN vaccine by the subcutaneous route of administration (non-adjuvanted, 75 per cent; and adjuvanted, 80 per cent), but no protection was evident (100 per cent mortality) by the intranasal, oral gavage or oral feeding routes of administration.

These researchers also reported results from a study where birds received either 20 mg of plant-cell-produced HN protein or 250 mg of plant-cell-produced HN protein with or without MPL–TDM adjuvant (Corixa, Inc). While 100 per cent of the unvaccinated controls died post challenge, 80 per cent of the birds receiving either 20 mg of HN adjuvanted with MPL-TDM or 250 mg of HN non-adjuvanted were protected. Additionally, 100 per cent of the birds receiving adjuvanted 250 mg of HN were protected and 60 per cent of the birds in the non-adjuvanted 20 mg group were protected. The authors concluded that this plant-cell-produced vaccine provided excellent immunity when administered subcutaneously, however, since these antigens are non-invasive and do not have a natural affinity for mucosal surfaces, immunity was not generated by mucosal administration.

Although not of primary veterinary interest, it is worth noting some experiments that have been completed using plant-made vaccines where humans are the target species. Edible plant-made vaccines are of particular interest for protection against enteric diseases since antigen delivered to the intestine has the greatest potential for stimulating relevant mucosal immunity.

Human volunteers ingesting three doses of transgenic potatoes expressing the LT-B gene encoding the E. coli heat-labile enterotoxin (average 0.75 mg dose71) showed antibody-secreting cells in the circulation 7–10 days post ingestion and 10 of 11 had 4-fold rises in serum of LT-specific IgG; six of these also had increased antigen-specific IgA. Similarly, 19 of 20 volunteers who ingested two or three doses of potatoes expressing Norwalk virus capsid protein (NVCP) developed increases in the number of circulating antibody-secreting cells and four developed antigen-specific IgA in feces. The importance of hepatitis B infection globally has spurred considerable interest in development of plant-made vaccines for effective oral delivery in regions where vaccination by injection is problematic.

Efforts over the past several years were recently reviewed by Streatfield. Transgenic lettuce (Lactuca sativa) expressing the surface antigen of hepatitis B virus (HbsAg) induced serum IgG responses in humans fed twice at a 2 month interval. Serum anti-HbsAg antibodies also increased in 10 of 16 humans who ate three doses of transgenic potatoes and in 9 of 17 who ate two doses. Edible plant-made vaccines have also been investigated as a format for vaccination of adults against measles.

CHALLENGES FOR THE TECHNOLOGY

At the Plant Made Pharmaceutical Conference held in Montreal, Canada in 2005, the potential limitation of plant-made vaccines was discussed. While the focus was on orally delivered vaccines targeted for human health, many of the same conclusions are similar for animal health vaccines. There are clear potential benefits of oral delivery of plant-made vaccines; however, there are still questions of expression levels and efficacy in general.

For example, will adjuvants be required to obtain the desired immune response or will VLPs be a major platform due to their expanded immune regulatory potential ? Regulatory approval for a plant-made vaccine has not yet been achieved and, for some production systems, remains uncharted.

In fact, for field-grown materials, multiple requirements will be demanded, both for efficacy of the products and proper stewardship and containment of the plants. To our knowledge, no animal health product derived from field-produced plants has been advanced significantly towards product licensure. Additionally, dependent on the type of system utilised to express the vaccine antigen, the cost of registration and the potential cost of containment can have a significant impact on the commercial viability of the end product. One issue of oral delivery of transgenic proteins that is commonly discussed is the potential for oral tolerance. In fact, oral delivery of auto-antigens may lead to oral tolerance and reversal or reduction of autoimmunity. These reports involve animal models and have yet to be fully evaluated in the target host.

In the research that has focused on oral delivery of plant-expressed antigens as vaccines, there have been no examples of oral tolerance reported. The difference in biological outcome of oral administration of transgenic proteins may be dependent on the specific regimen used for the treatment. For full adoption of a vaccine dependent on oral delivery, optimisation of vaccine delivery and a more thorough understanding of the immune response to vaccination will be necessary.

Since the first published account of the potential of plants as an expression vehicle for antigens, research has focused on antigen expression in plants and models of immunogenicity. Only very limited studies have investigated the efficacy of these plant-made vaccines in the animal host (human or non-human) and the impact of differences in the commonly used model mouse system versus the host animal are yet to be fully understood.

Although many different types of plants have been utilised and many different types of antigens (viral, bacterial and self-antigens) have been expressed, the full applicability of this technology platform for the production of vaccines for animal health will not be fully demonstrated until products are licensed for use in species of importance to veterinary medicine. The future of plant-made vaccines lies in the researcher's ability to develop vaccines against diseases of importance and to test these vaccines for their efficacy in the host. Advantages of the plant expression system over conventional systems will continue to be elucidated as experimental data amass in this new and growing field of vaccinology.

NEW HEALTH PRODUCTS FOR FARM ANIMALS

Animal diseases cost U.S. agriculture $17 billion each year, according to the Congress Office of Technology Assessment. To help prevent these losses, animal scientists are using biotechnology to develop an array of products to diagnose, treat, and prevent disease in farm animals. An area where animal biotechnology has already had a profound impact is diagnostic testing. Many animal diseases are difficult to diagnose. A veterinarian often has to wait hours or days for laboratory results to confirm a diagnosis. In the meantime, the veterinarian must either withhold treatment or risk using an inappropriate therapy.

Using biotechnology, scientists are developing fast, accurate diagnostic tests for many of the most common farm animal diseases. In some cases, a veterinarian can conduct the diagnostic tests right on the farm and immediately start the best therapy. Many of these new tests use monoclonal antibodies. Antibodies are proteins an animal's immune system produces in response to invasion by bacteria, viruses, or parasites. Each kind of antibody binds specifically with a characteristic part of a particular kind of invader. Scientists make monoclonal antibodies by fusing two kinds of cells. One is an immune system cell that produces antibodies, which bind to part of a particular disease-causing microbe; the other is a cancer cell. The cell that results from the fusion inherits from the immune system cell the ability to produce antibodies, and from the cancer cell the ability to reproduce indefinitely.

This new hybrid cell, or hybridoma, acts as an antibody factory. The antibodies are called monoclonal because they are produced by identical "clones" of the original hybridoma-all will bind with the identifying structure on the surface of the microbe in question. One application of monoclonal antibodies is as a test for brucellosis in cattle. This bacterial disease often causes cows to abort pregnancies, and it can be spread to farmers and people who drink milk from infected cows. Vaccines protect pregnant animals from abortion, but vaccinated cows may become carriers of the disease. Conventional diagnostic tests cannot distinguish between the disease-causing microbe and the vaccine, which is made from the microbe. Thus, these tests cannot help the farmer determine which animals are carriers of the disease and which are not. But diagnostic tests using monoclonal antibodies are so specific, they distinguish between cattle that carry the disease-causing bacterium and those that carry only the vaccine. With the aid of these new tests, farmers can isolate carriers to prevent the spread of infection. Diagnostic tests are not just for diseases.

A monoclonal antibody test to detect pregnancy in animals is so simple that a farmer can read the test right on the farm. Monoclonal antibodies are also used in a test to detect a disease called scours in piglets and calves. Scours is a bacterial disease of newborn animals that causes diarrhea. Many of the afflicted animals die of dehydration. By providing quick, accurate diagnosis, the monoclonal antibody test for scours permits immediate administration of therapy. The therapy for scours in calves is also based on monoclonal antibodies. Infected calves are fed monoclonal antibodies that coat the offending bacteria, preventing the microbes from causing diarrhea.

Some other therapies for animal diseases are based on recombinant DNA technology. Genetic engineering allows scientists to insert one or more new genes into an animal, plant, or microbial cell. In developing treatments for an animal disease, for instance, an animal gene is inserted into a microbe or an animal cell that can be grown in a laboratory culture. The inserted gene is one that produces a natural disease-fighting protein. The microbes or animal cells become minute factories, producing large quantities of the therapeutic protein. For example, genetic engineering is used to produce interferons and interleukin-2, natural proteins that fight viruses.

Veterinarians long have used antibiotics against bacterial diseases, but they have had few drugs to treat viral infections. Interferons and interleukin-2 not only kill viruses directly, they also stimulate the animal's immune system, improving the immune system's ability to fight disease naturally. One target of research regarding interferons and interleukin-2 is shipping fever, a disease of cattle caused by several viruses infecting an animal simultaneously.

Although the animal's immune system normally keeps the viruses in check, outbreaks of the disease may result from stress during transport to feedlots or market. Shipping fever costs the beef industry more than $250 million a year. Injections of interferons or interleukin-2 before shipping may help to augment the cattle's natural immunity and prevent shipping fever.

BETTER VACCINES TO PREVENT DISEASE

In addition to producing new diagnostic tests and therapeutic proteins, animal scientists are using biotechnology to develop vaccines to prevent disease. A safe, effective vaccine for swine pseudorabies, a fatal herpes-virus, is already in use. An advantage of the swine pseudorabies vaccine, and other vaccines made through recombinant DNA technology, is that they use only a small portion of the original microbe. Conventional vaccines use killed or weakened forms of the disease-causing microbe.

Sometimes, when the microbes are not completely killed or sufficiently weakened, a conventional vaccine may cause the disease it was supposed to prevent. Disease-causing genes are not included in the genetically engineered vaccines. Therefore, the recombinant vaccines build up the body's immunity without the risk of causing disease. Another advantage of recombinant vaccines is the speed with which they can be developed.

"Conventional vaccine development can take 20 or 30 years, maybe even 100 years," says Karen Jacobsen, D.V.M., University of Georgia College of Veterinary Medicine. "This is one reason why we don't have vaccines for a lot of animal diseases today." Recombinant vaccines are being developed for foot-and-mouth disease, a highly contagious viral disease that infects cattle, sheep, and other animals. Although this disease has been eradicated in North America, it still causes substantial livestock productivity losses elsewhere, particularly in developing countries. Infected herds must be slaughtered, and contaminated ranches must be left idle for months to prevent new outbreaks of the disease. Conventional foot-and-mouth vaccines are made by weakening the virus that causes the disease. These vaccines sometimes revert to the virulent state, and they have caused outbreaks of the disease in Europe. The foot-and-mouth vaccine made with biotechnology cannot cause the disease because, as in other biotech vaccines, the disease-causing genes have been removed.

The biotech vaccine does not need to be kept cold and, therefore, can be used in developing countries. In addition to being used in treating swine pseudorabies and foot-and-mouth disease, recombinant DNA technology is being used to develop vaccines against dozens of livestock and poultry diseases, such as scours and a variety of respiratory disorders. Some vaccines may be injected into eggs so that chickens are immunized before they hatch. Most vaccines protect against viral or bacterial infections, but using genetic engineering, researchers in New Zealand and Australia have developed a vaccine against a parasitic disease. Farmers must destroy the meat from sheep infested with a species of tapeworm common in those two countries. The new vaccine, which protects sheep from the tapeworm, may save farmers millions of dollars annually.

PROTEINS FOR ANIMAL NUTRITION

Natural proteins called somatotropins, or growth hormones, help animals convert feed into muscle or milk. Using recombinant DNA technology, scientists have been able to develop bacteria that produce commercial quantities of somatotropins.

They have found that administering small additional amounts of these proteins to animals helps to increase the animals' efficiency of feed conversion, thereby saving the farmer money. Porcine somatotropin (PST) may help improve human health, as well as lower the farmer's cost of production. PST improves feed efficiency in hogs by 15 to 20 per cent. It also reduces fat deposition, allowing PST-treated hogs to provide consumers with leaner cuts of pork. The ability to produce lean pork has enormous implications for improving human health by reducing dietary fat and cholesterol. Another growth protein, bovine somatotropin (BST), is being administered to make dairy cows more efficient, thereby lowering the dairy farmer's costs. Regular administration of small amounts of BST increases a cow's milk yield by 10 to 25 per cent. Although feed intake also increases, this feed increase is proportionately less than the increase in milk production.

Studies indicate that BST can improve the milk-to-feed ratio by 5 to 15 per cent. Extensive laboratory testing has shown that the meat and milk of treated animals contain no more PST or BST than the meat or milk from untreated animals. Other tests have shown that these proteins do not act on the human body, and therefore, they are safe to eat. Nevertheless, the use of BST remains controversial. Testing is under way to make sure that animals are not adversely affected by treatment with somatotropins. Another protein that can be administered to animals to increase feed efficiency is called growth hormone releasing factor (GHRF). While not a growth protein itself, GHRF causes the animal to increase production of growth proteins.

IMPROVED ANIMAL BREEDS

Long before anyone knew about genes or DNA, people were selectively breeding animals to produce desirable traits, such as strength, feed efficiency, and disease resistance. Selective breeding is a crude technology, because each cross between two animals mixes hundreds of thousands of genes, both desirable and undesirable. Recombinant DNA technology is now making it possible for scientists to breed animals with great precision. One or more new genes can be inserted into an animal embryo without disturbing the rest of the animal's genes. The resulting animal with the new gene or genes is referred to as a transgenic animal. Despite the differences in technique, the scientist using recombinant DNA has the same objectives as the selective breeder: improving efficiency, disease resistance, and other desirable traits. But recombinant DNA provides a wider array of possible new traits, since recombinant techniques allow animal scientists to insert genes not only from the same or closely related species, but also from distantly related animals, plants, and even microbes. One possible application of gene transfer in animals is to develop dairy cows that produce more nutritious milk. Bottle feeding human infants with cow's milk is not an ideal substitute for breast feeding, because cow's milk differs from human milk in some important components. Scientists are looking for ways to transfer genes into dairy cows that will enable the cows to produce milk that more closely resembles human milk.

Advances in Control of Parasitic Diseases Continue to Rely on Biotechnology Parasitic disease continues to be a major health problem worldwide in humans and domestic animals. Applying biotechnology has led to progress in prevention and control in many areas, particularly in regard to diagnosis. Parasites, like most organisms, carry repetitive sequences in their genomes. These are sometimes tandem repeats (back-to-back in a chromosome), or they may be located at random throughout the genome. Often, repetitive sequences are intergenic and non-coding, and are thus far more subject to evolution and change, increasing the likelihood that they will be genus or even species specific. Since detection of these sequences is not dependent upon gene expression, which may vary over time, repetitive sequence-based assays allow detection of the parasite at any life stage.

The effectiveness of repetitive sequence-based polymerase chain reaction (PCR) methods of identification has been demonstrated for parasites of several clinically important genera (1). A PCR assay has also been developed to differentiate two closely related nematode species that are virtually indistinguishable by morphologic examination (2), as well as for diagnosis of Toxoplasma and Cryptosporidium infections (3). Attempts to develop anti-parasite vaccines have met with multiple failures.

The problems often result from lack of post-translational modification by currently used vector-host systems and the inability to regenerate the secondary and tertiary structure of the native immunogens. Antigen presentation in recombinant parasite vaccines has also been thought to be a problem resulting in failed efficacy. There is now some evidence to indicate that cloning of genes for antigens of interest to yield chimeric molecules expressed in the BCG strain of Mycobacterium bovis may provide improved antigen presentation to stimulate both humoural and cellular responses (4); however, it is unlikely that future applications of this technology will be simple. Nonetheless, work is proceeding on the application of biotechnology to the development of vaccines against parasites in most food animals.

A particularly striking exception to the marginal successes reported in other systems has been a recombinant vaccine for a parasitic helminth of sheep (5). This product yielded greater than 94 per cent protection in experimental work and is now available as a commercial product. A further possibility is the application of anti-idiotypic monoclonal antibodies. When monoclonal antibodies are prepared against a monoclonal antibody, it is possible to obtain, in a very small number of cases, an antibody which mimics the original antigen. Results of work with Cryptosporidium perfringens epsilon toxin suggest that animals immunised with anti-idiotype monoclonals can be protected against toxin challenge.

FOOD AND DRUG ADMINISTRATION

In May 1992, the US Food and Drug Administration, responding to corporate pressures to remove the prospect of regulation of genetically altered foods, issued a set of rules which have largely left responsibility for protecting the public health and safety in the hands of the industry, permitting products to be marketed without scrutiny unless the industry indicated to the agency that it believed governmental oversight was justifiable.

However, the proposal as published by the FDA in the Federal Register perversely noted several important problem areas: creating allergens, additions of genes from sources which might violate religious and cultural norms (of vegetarians, Jews or Muslims, etc.), and implications for animal welfare and well-being (for example, the incorporation of human growth hormone-producing gene into a pig's genome produced a highly arthritic animal).

Indeed, the commissioner of the FDA, David A. Kessler, and his colleagues noted the possibility that genetically engineered food might 'contain high levels of unexpected, acutely toxic substances'. In order to begin to address ethical issues presented by the genetic modification of foodstuffs, we need to have some ideas of the purposes for which these modifications are performed. (For example, if we wish to approach an analysis on utilitarian cost-benefit grounds.) Despite a great number of general statements, human nutrition and hunger do not appear to be the actual driving forces behind the development of genetically engineered foods. Other than claiming increased shelf life (which can be indirectly related to nutrition in the sense of reducing spoilage or the consumption of foods that have begun to spoil), there is hardly any indication at all that genetic engineering is being directed to create nutritious substances out of non-nutritious ones; similarly, there are few real instances of genetic engineering reducing the cost of foodstuffs, or increasing the quantities of foods actually available to populations that are hungry because of the non-existence of consumables (as opposed to hungry because they lack money to buy sufficient foods, no matter how produced).

Indeed, early genetic manipulations seemed perversely designed to minimize the achievement of such goals: the creation of 'herbicide tolerant plants' which do not reduce the applications of dangerous chemicals to agricultural fields (which might occur if genetic manipulations of food crops were designed to make them resistant to insects, fungi, or disease) but instead permit higher levels of chemical application, or the introduction of recombinant bovine growth hormone to produce more milk at a time when developed countries suffer from milk gluts and have instituted programmemes to kill cows and physically dump milk (milk as a commodity is price inelastic; quantity increases do not lead to price decreases). Indeed, as the above examples indicate, the goals of genetically engineering foods seem to be more closely tied to increasing the economic gain and power of the corporations involved in food production (for example, the development of herbicide-tolerant farm crops has been led by corporations which manufacture herbicides), making production easier for corporations, reducing or altering packaging and transportation costs for the corporations, etc.

PRODUCTION OF FRUIT ALCOHOL

Alcohol and acids are two primary products of fermentation, both used to good effect in the preservation of foods. Several alcohol-fermented foods are preceded by an acid fermentation and in the presence of oxygen and acetobacter, alcohol can be fermented to produce acetic acid. Most food spoilage organisms cannot survive in either alcoholic or acidic environments.

Therefore, the production of both these end products can prevent a food from undergoing spoilage and extend its shelf life.Primitive wines and beers have been produced, with the aid of yeasts, for thousands of years, although it was not until about four hundred years ago that micro-organisms associated with the fermentation were observed and identified. It was not until the 1850's that Louis Pasteur demonstrated unequivocally the involvement of yeasts in the production of wines and beers. Since then, the knowledge of yeasts and the conditions necessary for fermentation of wine and beer has increased to the point where pure culture fermentations are now used to ensure consistent product quality. Originally, alcoholic fermentations would have been spontaneous events that resulted from the activity of micro-organisms naturally present. These non-scientific methods are still used today for the home preparation of many of the worlds traditional beers and wines.Alcoholic drinks fall into two broad categories: wines and beers. Wines are made from the juice of fruits and beers from cereal grains. The principal carbohydrates in fruit juices are soluble sugars; the principal carbohydrate in grains is starch, an insoluble polysaccharide.

The yeasts that bring about alcoholic fermentation can attack soluble sugars but do not produce starch-splitting enzymes. Wines can therefore be made by the direct fermentation of the raw material, while the production of beer requires the hydrolysis of starch to yield sugars fermentable by yeast, as a preliminary step.Raw fruit juice is usually a strongly acidic solution, containing from 10 to 25 percent soluble sugars. Its acidity and high sugar concentration make it an unfavourable medium for the growth of bacteria but highly suitable for yeasts and moulds. Raw fruit juice naturally contains many yeasts, moulds, and bacteria, derived from the surface of the fruit. Normally the yeast used in alcoholic fermentation is a strain of the species Saccharomyces cerevisiae. The fermentation may be allowed to proceed spontaneously, or can be "started" by inoculation with a must that has been previously successfully fermented by S. cerevisiae var. ellipsoideus. Many modern wineries eliminate the original microbial population of the must by pasteurisation or by treatment with sulphur dioxide.

The must is then inoculated with a starter culture derived from a pure culture of a suitable strain of wine yeast. This procedure eliminates many of the uncertainties and difficulties of older methods. At the start of the fermentation, the must is aerated slightly to build up a large and vigourous yeast population; once fermentation sets in, the rapid production of carbon dioxide maintains anaerobic conditions, which prevent the growth of undesirable aerobic organisms, such as bacteria and moulds. The temperature of fermentation is usually from 25 to 30°C, and the duration of the fermentation process may extend from a few days to two weeks. As soon as the desired degree of sugar disappearance and alcohol production has been attained, the microbiological phase of wine making is over. Thereafter, the quality and stability of the wine depend very largely on preventing further microbial activity, both during the "aging" in wooden casks and after bottling.At all stages during its manufacture, fruit juice alcohol is subject to spoilage by undesirable microorganisms. Pasteur, whose descriptions of the organisms responsible and recommendations for overcoming them are still valid today, first scientifically explored the problem of the "diseases" of wines. The most serious aerobic spoilage processes are brought about by film-forming yeasts and acetic acid bacteria, both of which grow at the expense of the alcohol, converting it to acetic acid or to carbon dioxide and water. The chief danger from these organisms arises when access of air is not carefully regulated during aging.

Much more serious are the diseases caused by fermentative bacteria, particularly rod-shaped lactic acid bacteria, which utilise any residual sugar and impart a mousy taste to the wine. Such wines are known as turned wines. Since oxygen is unnecessary for the growth of lactic acid bacteria, wine spoilage of this kind can occur even after bottling. These risks of spoilage can be minimised by pasteurisation after bottling.

GRAPE WINE

Grape wine is perhaps the most common fruit juice alcohol. Because of the commercialisation of the product for industry, the process has received most research attention and is documented in detail. The production of grape wine involves the following basic steps: crushing the grapes to extract the juice; alcoholic fermentation; maltolactic fermentation if desired; bulk storage and maturation of the wine in a cellar; clarification and packaging. Although the process is fairly simple, quality control demands that the fermentation is carried out under controlled conditions to ensure a high quality product.

The distinctive flavour of grape wine originates from the grapes as raw material and subsequent processing operations. The grapes contribute trace elements of many volatile substances (mainly terpenes) which give the final product the distinctive fruity character. In addition, they contribute non-volatile compounds (tartaric and malic acids) which impact on flavour and tannins which give bitterness and astringency. The latter are more prominent in red wines as the tannin components are located in the grape skins.

Although yeasts are the principal organisms involved, filamentous fungi, lactic acid bacteria, acetic acid bacteria and other bacterial groups all play a role in the production of alcoholic fruit products. Normal grapes harbour a diverse micro-flora, of which the principal yeasts (Saccharomyces cerevisiae) involved in desirable fermentation are in the minority. Lactic acid bacteria and acetic acid bacteria are also present. The proportions of each and total numbers present are dependent upon a number of external environmental factors including the temperature, humidity, stage of maturity, damage at harvest and application of fungicides.

It is essential to ensure proliferation of the desired species at the expense of the non-desired ones. This is achieved through ensuring fermentation conditions are such to encourage Saccharomyces species. The fermentation may be initiated using a starter culture of Saccharomyces cerevisiae - in which case the juice is inoculated with populations of yeast of 106 to 107 cfu/ml juice. This approach produces a wine of generally expected taste and quality. If the fermentation is allowed to proceed naturally, utilising the yeasts present on the surface of the fruits, the end result is less controllable, but produces wines with a range of flavour characteristics. It is likely that natural fermentations are practiced widely around the world, especially for home production of wine.

During alcoholic fermentation, yeasts are the prominent species. The composition of fruit juice - its acid and sugar level and low pH favour the growth of yeasts and production of ethanol that restricts the growth of bacteria and fungi. In natural fermentations, there is a progressive pattern of yeast growth. Several species of yeast, including Kloeckera, Hanseniaspora, Candida and Metschnikowia, are active for the first two to three days of fermentation.

The build up of end products (ethanol) is toxic to these yeasts and they die off, leaving Saccharomyces cerevisiae to continue the fermentation to the end. S. cerevisiae can tolerate much higher levels of ethanol (up to 15 per cent v/v or more) than the other species who only tolerate up to 5 or 8 per cent alcohol. Because of its tolerance of alcohol, S. cerevisiae dominates wine fermentation and is the species that has been commercialised for starter cultures.Traditionally, fermentation was carried out in large wooden barrels or concrete tanks. Modern wineries now use stainless steel tanks as these are more hygienic and provide better temperature control.

White wines are fermented at 10 to 18° C for about seven to fourteen days. The low temperature and slow fermentation favours the retention of volatile compounds. Red wines are fermented at 20 to 30°C for about seven days. This higher temperature is necessary to extract the pigment from the grape skins.

FACTORS AFFECTING WINE FERMENTATION

There are several variables which can affect the fermentation process and final quality of wine. Factors which are most important to control are:

- The clarification and pre-treatment of juice
- Chemical composition of the juice

- Temperature of the fermentation
- The influences of other micro-organisms
- Clarification and pre-Treatment of Juice

Excessive clarification removes many of the natural yeasts and flora. This is beneficial if a tightly controlled induced fermentation is desired, but less so if the fermentation is a natural one. Long periods of settling out however, encourage the growth of natural flora, which can contribute to the fermentation.

Chemical Composition of Juice

The main consituents of grape juice are glucose (75 to 150 g/l), fructose (75 to 150 g/l), tartaric acid (2 to 10 g/l), malic acid (1 to 8 g/l) and free amino acids (0.2 to 2.5 g/l). The main reaction is the fermentation of glucose and fructose to ethanol and carbon dioxide. However, the presence of nitrogenous and sulphurous products also contributes to the fermentation. The addition of sulphur dioxide to the juice delays the growth of yeast, but does not necessarily inhibit growth of the non-Saccharomyces strains. Fruits generally contain sufficient substrates - soluble sugars - for the yeast to ferment and convert into an acceptable concentration of alcohol. Sugar can be added to fruit juices with a low sugar content, to increase the amount of fermentable substrate.

Temperature

Temperature has an impact on the growth and activity of different strains of yeast. At temperatures of 10 to 15 C, the non-Saccharomyces species have an increased tolerance to alcohol and therefore have the potential to contribute to the fermentation.

Influence of other Micro-Organisms

Other micro-organisms have the potential to influence wine production at all stages of the process. Prior to harvest, yeasts grow on the surface of grapes. Fungicides are used in an attempt to control their growth, but these disturb the natural balance of flora, thus making it difficult to carry out a 'natural' fermentation.Overuse of fungicides can lead to the development of resistant strains of yeast which have the potential to produce toxins which destroy the desirable yeast species.

These yeasts are known as 'killer' strains. Other microbes have further chances to influence the fermentation during the clarification process, after fermentation and during maturation and bottling when acetobacter species can oxidise the alcohol and produce acetic acid.About two to three weeks after the alcoholic fermentation is finished wines often undergo a malo-lactic fermentation. This occurs naturally and lasts for about four weeks. It is a lactic acid fermentation, initiated by lactic acid bacteria resident in the wine. Inoculating the fermented wine with cultures of Leuconostoc oenos can start the process if it is desired. The main reaction of these bacteria is the decarboxylation of L-malic acid to L-lactic acid, which decreases the acidity of the wine and increases its pH by about 0.3 to 0.5 units.Wines produced from grapes grown in colder climates tend to have a higher concentration of malic acid and a lower pH (3.0 to 3.5) and the taste benefits from this slight decrease in acidity.

The benefits of this process are that it imparts a more mellow flavour to the wine. The growth of malo-lactic bacteria also contributes to the taste of the wine. Wines that have undergone a malo-lactic fermentation appear to be less susceptible to any further damage from other bacteria. This could be because L. oenos has used up all available substrate, or it may have secreted bacteriocins which prevent the growth of other species.Although the malo-lactic fermentation seems to be a useful process, not all wines benefit from it. Wines produced from grapes in warmer climates tend to be less acidic (pH > 3.5) and a further reduction in acidity may have adverse effects on the quality of the wine.

Decreasing the acidity also increases the pH to values which can allow spoilage organisms to multiply. It is difficult to prevent the malo-lactic fermentation from taking place naturally, especially later on after the wine has been bottled.In low acid wines, the acidity may be adjusted after this fermentation has taken place. The malo-lactic fermentation can be prevented by controlling several factors: the wine pH (< 3.2); ethanol content (> 14 per cent) and levels of sulphur dioxide (>50 mg/l). The bacteriocin nisin can also be used to control the growth of malo-lactic bacteria. However the subtle blend of aromas and flavours that contribute to the final taste may be lost by such stringent control. The conversion of malic acid to lactic acid is one of the main reactions carried out by wine lactic acid bacteria. L oenos needs to be present in significant numbers (greater than 106 cfu/ml) for the reaction to take place at a suitable pace. The bacteria use residual pentose and hexose sugars in the wine as a substrate for growth. The main reaction is the deacidification (or decarboxylation) of malic acid. In addition to this, the by-products of the reaction impart flavours and aromas to the wine. During storage, wines are prone to non-desirable microbial changes. Yeasts, lactic acid bacteria, acetic acid bacteria and fungi can all spoil or taint wines after the fermentation process is completed. The changes that occur are increased acidification through the formation of acetic and other acids from alcohol; increased carbonation through a secondary fermentation of residual sugars and flavour changes through the metabolism of numerous compounds.

THE DIVERSITY OF FERMENTED FOODS

Numerous fermented foods are consumed around the world. Each nation has its own types of fermented food, representing the staple diet and the raw ingredients available in that particular place. Although the products are well know to the individual, they may not be associated with fermentation. Indeed, it is likely that the methods of producing many of the worlds fermented foods are unknown and came about by chance.

Some of the more obvious fermented fruit and vegetable products are the alcoholic beverages - beers and wines. However, several more fermented fruit and vegetable products arise from lactic acid fermentation and are extremely important in meeting the nutritional requirements of a large proportion of the worlds population.

Organisms Responsible for Food Fermentations

The most common groups of micro-organisms involved in food fermentations are:

- Bacteria
- Yeasts
- Moulds

Bacteria

Several bacterial families are present in foods, the majority of which are concerned with food spoilage. As a result, the important role of bacteria in the fermentation of foods is often overlooked. The most important bacteria in desirable food fermentations are the lactobacillaceae which have the ability to produce lactic acid from carbohydrates. Other important bacteria, especially in the fermentation of fruits and vegetables, are the acetic acid producing acetobacter species.

Yeasts

Yeasts and yeast-like fungi are widely distributed in nature. They are present in orchards and vineyards, in the air, the soil and in the intestinal tract of animals. Like bacteria and moulds, yeasts can have beneficial and non-beneficial effects in foods.

Yeasts are unicellular organisms that reproduce asexually by budding. In general, yeasts are larger than most bacteria. Yeasts play an important role in the food industry as they produce enzymes that favour desirable chemical reactions such as the leavening of bread and the production of alcohol and invert sugar. Moulds are also important organisms in the food industry, both as spoilers and preservers of foods. Certain moulds produce undesirable toxins and contribute to the spoilage of foods.

Moulds

The Aspergillus species are often responsible for undesirable changes in foods. These moulds are frequently found in foods and can tolerate high concentrations of salt and sugar. However, others impart characteristic flavours to foods and others produce enzymes, such as amylase for bread making. Moulds from the genus Penicillium are associated with the ripening and flavour of cheeses. Moulds are aerobic and therefore require oxygen for growth.

They also have the greatest array of enzymes, and can colonise and grow on most types of food. Moulds do not play a significant role in the desirable fermentation of fruit and vegetable products. When micro-organisms metabolise and grow they release by-products. In food fermentations the by-products play a beneficial role in preserving and changing the texture and flavour of the food substrate. For example, acetic acid is the by-product of the fermentations of some fruits. This acid not only affects the flavour of the final product, but more importantly has a preservative effect on the food. For food fermentations, micro-organisms are often classified according to these by-products.

The fermentation of milk to yoghurt involves a specific group of bacteria called the lactic acid bacteria (Lactobacillus species). This is a general name attributed to those bacteria which produce lactic acid as they grow. Acidic foods are less susceptible to spoilage than neutral or alkaline foods and hence the acid helps to preserve the product. Fermentations also result in a change in texture. In the case of milk, the acid causes the precipitation of milk protein to a solid curd.

Enzymes

The changes that occur during fermentation of foods are the result of enzymic activity. Enzymes are complex proteins produced by living cells to carry out specific biochemical reactions. They are known as catalysts since their role is to initiate and control reactions, rather than being used in a reaction. Because they are proteinaceous in nature, they are sensitive to fluctuations in temperature, pH, moisture content, ionic strength and concentrations of substrate and inhibitors. Each enzyme has requirements at which it will operate most efficiently.

Extremes of temperature and pH will denature the protein and destroy enzyme activity. Because they are so sensitive, enzymic reactions can easily be controlled by slight adjustments to temperature, pH or other reaction conditions. In the food industry, enzymes have several roles - the liquefaction and saccharification of starch, the conversion of sugars and the modification of proteins. Microbial enzymes play a role in the fermentation of fruits and vegetables. Nearly all food fermentations are the result of more than one micro-organism, either working together or in a sequence.

For example, vinegar production is a joint effort between yeast and acetic acid forming bacteria. The yeast convert sugars to alcohol, which is the substrate required by the acetobacter to produce acetic acid. Bacteria from different species and the various micro-organisms - yeast and moulds -all have their own preferences for growing conditions, which are set within narrow limits. There are very few pure culture fermentations.

An organism that initiates fermentation will grow there until it's by-products inhibit further growth and activity. During this initial growth period, other organisms develop which are ready to take over when the conditions become intolerable for the former ones.In general, growth will be initiated by bacteria, followed by yeasts and then moulds. There are definite reasons for this type of sequence. The smaller micro-organisms are the ones that multiply and take up nutrients from the surrounding area most rapidly. Bacteria are the smallest of micro-organisms, followed by yeasts and moulds. The smaller bacteria, such as Leuconostoc and Streptococcus grow and ferment more rapidly than their close relations and are therefore often the first species to colonise a substrate.

DESIRABLE FERMENTATION

It is essential with any fermentation to ensure that only the desired bacteria, yeasts or moulds start to multiply and grow on the substrate. This has the effect of suppressing other micro-organisms which may be either pathogenic and cause food poisoning or will generally spoil the fermentation process, resulting in an end-product which is neither expected or desired. An everyday example used to illustrate this point is the differences in spoilage between pasteurised and unpasteurised milk. Unpasteurised milk will spoil naturally to produce a sour tasting product which can be used in baking to improve the texture of certain breads.

Pasteurised milk, however, spoils (non-desirable fermentation) to produce an unpleasant product which has to be disposed of. The reason for this difference is that pasteurisation (despite being a very important process to destroy pathogenic micro-organisms) changes the micro-organism environment and if pasteurised milk is kept unrefrigerated for some time, undesirable micro-organisms start to grow and multiply before the desirable ones.

In the case of unpasteurised milk, the non-pathogenic lactic acid bacteria start to grow and multiply at a greater rate that any pathogenic bacteria. Not only do the larger numbers of lactic acid bacteria compete more successfully for the available nutrients, but as they grow they produce lactic acid which increases the acidity of the substrate and further suppresses the bacteria which cannot tolerate an acid environment.Most food spoilage organisms cannot survive in either alcoholic or acidic environments. Therefore, the production of both these end products can prevent a food from spoilage and extend the shelf life. Alcoholic and acidic fermentations generally offer cost effective methods of preserving food for people in developing countries, where more sophisticated means of preservation are unaffordable and therefore not an option.

The principles of microbial action are identical both in the use of micro-organisms in food preservation, such as through desirable fermentations, and also as agents of destruction via food spoilage. The type of organisms present and the environmental conditions will determine the nature of the reaction and the ultimate products. By manipulating the external reaction conditions, microbial reactions can be controlled to produce desirable results. There are several means of altering the reaction environment to encourage the growth of desirable organisms.

MANIPULATION OF MICROBIAL GROWTH AND ACTIVITY

There are six major factors that influence the growth and activity of micro-organisms in foods. These are moisture, oxygen concentration, temperature, nutrients, pH and inhibitors. The food supply available to the micro-organisms depends on the composition of the food on which they grow. All micro-organisms differ in their ability to metabolise proteins, carbohydrates and fats. Obviously, by manipulating any of these six factors, the activity of micro-organisms within foods can be controlled.

Moisture

Water is essential for the growth and metabolism of all cells. If it is reduced or removed, cellular activity is decreased. For example, the removal of water from cells by drying or the change in state of water (from liquid to solid) affected by freezing, reduces the availability of water to cells (including microbial cells) for metabolic activity. The form in which water exists within the food is important as far as microbial activity is concerned.

There are two types of water - free and bound. Bound water is present within the tissue and is vital to all the physiological processes within the cell.Free water exists in and around the tissues and can be removed from cells without seriously interfering with the vital processes. Free water is essential for the survival and activity of micro-organisms. Therefore, by removing free water, the level of microbial activity can be controlled. The amount of water available for micro-organisms is referred to as the water activity (aw). Pure water has a water activity of 1.0.Bacteria require more water than yeasts, which require more water than moulds to carry out their metabolic activities.

Almost all microbial activity is inhibited below aw of 0.6. Most fungi are inhibited below aw of 0.7, most yeasts are inhibited below aw of 0.8 and most bacteria below aw 0.9. Naturally, there are exceptions to these guidelines and several species of micro-organism can exist outside the stated range. The water activity of foods can be changed by altering the amount of free water available. There are several ways to achieve this - drying to remove water; freezing to change the state of water from liquid to solid; increasing or decreasing the concentration of solutes by adding salt or sugar or other hydrophylic compounds (salt and sugar are the two common additives used for food preservation). Addition of salt or sugar to a food will bind free water and so decrease the aw. Alternatively, decreasing the concentration will increase the amount of free water and in turn the aw. Manipulation of the aw in this manner can be used to encourage the growth of favourable micro-organisms and discourage the growth of spoilage ones.

OXIDATION-REDUCTION POTENTIAL .

Oxygen is essential to carry out metabolic activities that support all forms of life. Free atmospheric oxygen is utilised by some groups of micro-organisms, while others are able to metabolise the oxygen which is bound to other compounds such as carbohydrates. This bound oxygen is in a reduced form. Micro-organisms can be broadly classified into two groups - aerobic and anaerobic. Aerobes grow in the presence of atmospheric oxygen while anaerobes grow in the absence of atmospheric oxygen. In the middle of these two extremes are the facultative anaerobes which can adapt to the prevailing conditions and grow in either the absence or presence of atmospheric oxygen.Microaerophilic organisms grow in the presence of reduced amounts of atmospheric oxygen. Thus, controlling the availability of free oxygen is one means of controlling microbial activity within a food. In aerobic fermentations, the amount of oxygen present is one of the limiting factors. It determines the type and amount of biological product obtained, the amount of substrate consumed and the energy released from the reaction. Moulds do not grow well in anaerobic conditions, therefore they are not important in terms of food spoilage or beneficial fermentation, in conditions of low oxygen availability.

Temperature

Temperature affects the growth and activity of all living cells. At high temperatures, organisms are destroyed, while at low temperatures, their rate of activity is decreased or suspended. Micro-organisms can be classified into three distinct categories according to their temperature preference.

Nutritional Requirements

The majority of organisms are dependent on nutrients for both energy and growth. Organisms vary in their specificity towards different substrates and usually only colonise foods which contain the substrates they require. Sources of energy vary from simple sugars to complex carbohydrates and proteins. The energy requirements of micro-organisms are very high. Limiting the amount of substrate available can check their growth.

Hydrogen Ion Concentration (pH)

The pH of a substrate is a measure of the hydrogen ion concentration. A food with a pH of 4.6 or less is termed a high acid or acid food and will not permit the growth of bacterial spores. Foods with a pH above 4.6. are termed low acid and will not inhibit the growth of bacterial spores. By acidifying foods and achieving a final pH of less than 4.6, most foods are resistant to bacterial spoilage.

The optimum pH for most micro-organisms is near the neutral point (pH 7.0). Certain bacteria are acid tolerant and will survive at reduced pH levels. Notable acid-tolerant bacteria include the Lactobacillus and Streptococcus species, which play a role in the fermentation of dairy and vegetable products. Moulds and yeasts are usually acid tolerant and are therefore associated with spoilage of acidic foods.Micro-organisms vary in their optimal pH requirements for growth. Most bacteria favour conditions with a near neutral pH (7). Yeasts can grow in a pH range of 4 to 4.5 and moulds can grow from pH 2 to 8.5, but favour an acid pH. The varied pH requirements of different groups of micro-organisms is used to good effect in fermented foods where successions of micro-organisms take over from each other as the pH of the environment changes. For instance, some groups of micro-organisms ferment sugars so that the pH becomes too low for the survival of those microbes. The acidophilic micro-organisms then take over and continue the reaction. The affinity for different pH can also be used to good effect to occlude spoilage organisms.

Inhibitors

Many chemical compounds can inhibit the growth and activity of micro-organisms. They do so by preventing metabolism, denaturation of the protein or by causing physical damage to the cell. The production of substrates as part of the metabolic reaction also acts to inhibit microbial action.

CONTROLLED FERMENTATION

Controlled fermentations are used to produce a range of fermented foods, including sauerkraut, pickles, olives, vinegar, dairy and other products. Controlled fermentation is a form of food preservation since it generally results in a reduction of acidity of the food, thus preventing the growth of spoilage micro-organisms. The two most common acids produced are lactic acid and acetic acid, although small amounts of other acids such as propionic, fumaric and malic acid are also formed during fermentation.

It is highly probable that the first controlled food fermentations came into existence through trial and error and a need to preserve foods for consumption later in the season. It is possible that the initial attempts at preservation involved the addition of salt or seawater. During the removal of the salt prior to consumption, the foods would pass through stages favourable to acid fermentation. Although the process worked, it is likely that the causative agents were unknown. Today, there are numerous examples of controlled fermentation for the preservation and processing of foods. However, only a few of these have been studied in any detail - these include sauerkraut, pickles, kimchi, beer, wine and vinegar production. Although the general principles and processes for many of the fermented fruit and vegetable products are the same -relying mainly on lactic acid and acetic acid forming bacteria, yeasts and moulds, the reactions have not been quantified for each product. The reactions are usually very complex and involve a series of micro-organisms, either working together or in succession to achieve the final product.

YEAST FERMENTATIONS

What are yeasts?

A yeast is a unicellular fungus which reproduces asexually by budding or division, especially the genus Saccharomyces which is important in food fermentations. Yeasts and yeast-like fungi are widely distributed in nature. They are present in orchards and vineyards, in the air, the soil and the intestinal tract of animals. Like bacteria and moulds, they can have beneficial and non-beneficial effects in foods. Most yeasts are larger than most bacteria. The most well known examples of yeast fermentation are in the production of alcoholic drinks and the leavening of bread. For their participation in these two processes, yeasts are of major importance in the food industry. Some yeasts are chromogenic and produce a variety of pigments, including green, yellow and black. Others are capable of synthesising essential B group vitamins.Although there is a large diversity of yeasts and yeast-like fungi, (about 500 species), only a few are commonly associated with the production of fermented foods.

They are all either ascomycetous yeasts or members of the genus Candida. Varieties of the Saccharomyces cervisiae genus are the most common yeasts in fermented foods and beverages based on fruit and vegetables. All strains of this genus ferment glucose and many ferment other plant derived carbohydrates such as sucrose, maltose and raffinose. In the tropics, Saccharomyces pombe is the dominant yeast in the production of traditional fermented beverages, especially those derived from maize and millet.

CONDITIONS NECESSARY FOR FERMENTATION

Most yeasts require an abundance of oxygen for growth, therefore by controlling the supply of oxygen, their growth can be checked. In addition to oxygen, they require a basic substrate such as sugar. Some yeasts can ferment sugars to alcohol and carbon dioxide in the absence of air but require oxygen for growth. They produce ethyl alcohol and carbon dioxide from simple sugars such as glucose and fructose.

$$C_6H_{12}O_6 \rightarrow 2C_2H_5OH + 2CO_2$$

In conditions of excess oxygen (and in the presence of acetobacter) the alcohol can be oxidised to form acetic acid. This is undesirable if the end product is a fruit alcohol, but is a technique employed for the production of fruit vinegars.Yeasts are active in a very broad temperature range - from 0 to 50 C, with an optimum temperature range of 20 to 30 C. The optimum pH for most micro-organisms is near the neutral point (pH 7.0). Moulds and yeasts are usually acid tolerant and are therefore associated with the spoilage of acidic foods. Yeasts can grow in a pH range of 4 to 4.5 and moulds can grow from pH 2 to 8.5, but favour an acid pH.In terms of water requirements, yeasts are intermediate between bacteria and moulds. Bacteria have the highest demands for water, while moulds have the least need. Normal yeasts require a minimum water activity of 0.85 or a relative humidity of 88 per cent.Yeasts are fairly tolerant of high concentrations of sugar and grow well in solutions containing 40 per cent sugar. At concentrations higher than this, only a certain group of yeasts - the osmophilic type - can survive.

There are only a few yeasts that can tolerate sugar concentrations of 65-70 per cent and these grow very slowly in these conditions. Some yeasts - for example the Debaromyces - can tolerate high salt concentrations. Another group which can tolerate high salt concentrations and low water activity is Zygosaccharomyces rouxii, which is associated with fermentations in which salting is an integral part of the process.

GENETICAL MODIFIED ORGANISM-DERIVED FOODS

The concept of substantial equivalence was originally developed through discussions at the Organisation for Economic Co-operation and Development, though, to a large extent, these discussions built on previous work done by the World Health Organization (WHO) and the Food and Agriculture Organization. In 1991, a Joint FAO/WHO Consultation had concluded that the evaluation of a food derived through modern biotechnology should consider both food safety and nutritional value using similar conventional food products as a standard and taking into account the processing of the food and its intended use. In the 1980s, the OECD established a Group of National Experts on Safety in Biotechnology (GNE), which continued to work through to 1993.

As an intergovernmental organization, OECD's GNE was attended by delegates nominated by the governments of the OECD member countries, who were, for the most part, representatives of those agencies and ministries with a responsibility for safety in biotechnology. As a result of the work of the GNE, the OECD published a number of important documents relevant to the safety assessment of biotechnology-derived products during this period, which deal with a range of issues, including safety considerations for industrial, agricultural as well as environmental applications of organisms derived by recombinant DNA techniques.

It was recognized at this stage that the safety assessment of an organism derived through recombinant DNA techniques would rely heavily on the knowledge of its parental organism, as well as on an analysis of how the new organism appears to differ from the parent. OECD recommended that the considerable data on environmental and human health effects of living organisms that exists should be used to guide the risk assessment.

By the early 1990s, there had already been large numbers of small-scale field trials of new crop varieties derived through recombinant DNA techniques, and it became clear that new foods derived from these varieties would be marketed during the 1990s. In order to proactively address food safety issues related to these novel foods, in 1990, the GNE established a Working Group on Food Safety and Biotechnology comprising experts, mainly from the ministries and agencies responsible for food safety issues in OECD member countries.

The main objective of the Working Group was to elabourate scientific principles and concepts to be used when evaluating the safety of new foods and food components of terrestrial microbial, plant or animal origin. The Working Group did not consider the safety assessment of food additives, contaminants, processing aids or packaging materials. Nor did it consider environmental safety issues which had been (or were being) addressed by other groups of the GNE. Early in the life of the Working Group, an important recognition was that traditionally, the safety of food for human consumption had been based on the reasonable certainty that no harm will result from intended uses under the anticipated conditions of consumption. Foods prepared and used in traditional ways have usually been considered safe on the basis of long-term experience, even though they may have contained natural toxicants or anti-nutritional substances.

Normally, new varieties of foods or crops have not been subjected to traditional toxicological testing. In fact, where toxicological testing has been applied to whole foods, the results have often been difficult to interpret. Although the OECD's Working Group recognized that modern biotechnology might extend the scope of genetic changes that can be made - and might even broaden the range of the possible sources of foods - it was recognized that this would not inherently lead to foods that are less safe than those developed through conventional techniques. The Working Group therefore indicated that the evaluation of foods or food components derived through modern biotechnology does not require a fundamental change in established principles, nor does it require a different standard of safety. Realising the importance of using examples of new foods to identify and demonstrate the applicability of the proposed scientific principles, the Working Group organized a number of meetings and intergovernmental consultations which included case study presentations of novel foods such as enzymes, genetically modified bakers' yeast, mycoprotein and a number of genetically modified varieties of crop species.

Although these case studies were not intended to be formal safety evaluations they were crucial in illustrating important concepts in the safety assessment of novel foods. Therefore, it was through their application in real examples that the concepts and principles were identified. The Working Group proposed a scientific approach to the evaluation of foods derived through modern biotechnology, which is based on a comparison with traditional foods that have a safe history of use.

One of the main concepts developed was that of substantial equivalence, and in describing this concept, the Group stressed that this was a principle that had been used in the past (perhaps intuitively) even if it had not been articulated as such. The concept of substantial equivalence is described as embodying the idea that existing organisms used as food, or as a source of food, can be used as the basis for comparison when assessing the safety of human consumption of a food or food component that has been modified or is new. As previously indicated, in elabourating the concept of substantial equivalence, the OECD Working Group noted that food safety is considered as a reasonable certainty, that no harm will result from intended uses under the anticipated conditions of consumption and that the most practical approach to the determination of safety is to consider whether a genetically modified organism (GMO)-derived food is comparable to an analogous conventional food product.

The concept of substantial equivalence therefore relies on the existing history of safe food use of a conventional food product as a useful consideration in framing the safety assessment of a GMO-derived food by permitting the identification of similarities and differences which can be considered in the assessment. In 1996, participants at an expert FAO/WHO consultation recommended that safety assessment based upon the concept of substantial equivalence be applied in establishing the safety of foods and food components derived from genetically modified organisms.

Assessing substantial equivalence was recognized as not being a safety assessment *per se*, but a process which establishes that the characteristics and composition of the new GMO-derived food are comparable to those of a familiar, conventional food which has a history of safe consumption. A Joint FAO/WHO Expert Consultation on Foods Derived from Biotechnology was convened in 2000 to address food safety and nutritional questions regarding foods derived from GM plants, including a review of the scientific basis, application and limitations of the concept of substantial equivalence. It concluded, based on the current application of substantial equivalence and alternative strategies that this concept contributed to a robust safety assessment framework. Moreover, the Consultation noted that substantial equivalence is a concept used to identify similarities and differences between GM food and a comparator with a history of safe food use which subsequently guides the safety assessment process.

While not a safety assessment *per se*, the substantial equivalence approach allows the structuring of the safety assessment through characterizing the similarities and identifying the differences which can then be the focus of further consideration. This approach has been referred to as a useful safety standard. It therefore permits inference that the new food under consideration will be no less safe than the conventional food under conditions of similar exposure, consumption patterns and processing practices.

Substantial equivalence is therefore clearly not intended to be a measure of absolute safety, but instead recognizes that while demonstrating absolute safety is an impractical goal, demonstrating that there is reasonable assurance that the GMO-derived product under consideration is no less safe than a conventional food product is an achievable goal. Since it's development, the use of the substantial equivalence concept in the safety assessment of GMO-derived foods has been subject to misinterpretation and criticism.

The comparative nature of the substantial equivalence concept without prescribing the extent of the phenotypic and compositional comparisons has led some to criticize the concept as not being measurable, and therefore inappropriate in safety assessment. This and other criticisms relate, in part, to the mistaken perception that the determination of substantial equivalence was the end point of a safety assessment rather than the starting point. After several OECD countries had gained experience with safety assessment of GMO-derived foods, an OECD workshop examined the effectiveness of the application of substantial equivalence in safety assessment. This workshop concluded that the substantial equivalence approach provides equal or increased assurance of the safety of foods derived from genetically modified plants, as compared with foods derived through conventional methods. In 2000, the OECD Task Force for the Safety of Novel Foods and Feeds also reviewed the substantial equivalence concept, its interpretation and its application. While noting that the majority of international guidance documents addressing the safety assessment of genetically modified (GM) plants had interpreted substantial equivalence consistently, this Task Force reported that there were differences in how it was applied and that these differences needed to be resolved.

The Codex *Ad Hoc* Intergovernmental Task Force on Foods Derived from Biotechnology pursued the development of international guidance on the safety assessment of GMO-derived foods. To date, this Task Force has developed proposed draft principles and guidelines for the safety assessment of foods derived from modern biotechnology. These guidelines interpreted the concept of substantial equivalence as a way of structuring the safety assessment, consistent with the FAO/WHO interpretation.

ANIMAL HEALTH AND WELFARE

REPRODUCTIVE BIOTECHNOLOGIES AND TECHNIQUES

All reproductive manipulations, including semen collection, repeated artificial insemination, embryo collection and transfer, or the inducement of superovulation can cause stress to animals during constraint and handling. These methods are not specific to GM animals and few have been studied from the perspective of animal welfare. The intensity of violation during artificial insemination and embryo collection varies according to the species. These two techniques are relatively uninvasive for cattle and are done with local anaesthetic.

On the other hand, for smaller species such as sheep, goats and pigs, they entail invasive surgical intervention (laparotomy or laparoscopy) and thus cause operative and post-operative pain. Cows carrying an embryo are subject to amniocentesis to check the genotype. Cows carrying nontransgenic embryos or male embryos are aborted and have another embryo inserted. This limits the number of recipient females used to give birth to transgenic animals, but it also raises welfare concerns as these animals are regularly exposed to stress and pain. At the same time, they can be prone to slow or difficult calving (dystocia) requiring increasingly frequent caesarean sections. Abnormalities in the offspring can also raise concern for an animal's welfare. Alternatives to such methods causing stress and pain are sometimes suggested, such as the removal of ova in the slaughterhouse, non-invasive embryo collection or techniques of cloning and in vitro culture to reduce the number of recipient animals.

In Vitro Culture

In vitro culture tends to extend the gestation period and to produce higher birthweights, resulting in difficult deliveries and frequent surgical intervention. These animals have more congenital malformations (skeletal malformation, incomplete development of the vascular system or urogenital tract, immune system dysfunction, brain lesions) and high perinatal mortality. The mechanisms responsible for these abnormalities are not known, but chromosomal anomalies, abnormal regulation of gene expression during embryo development and problems of communication between foetus and mother have been implicated. Culture conditions also appear to be directly associated with the level of developmental abnormalities and perinatal mortality.

Efficiency of Biotechnologies

Efficiency varies according to technique. Microinjection is particularly inefficient, with reported successful transgenic offspring of only 1%. Embryo mortality stands at 80% to 90% and there is high perinatal mortality. In addition, few animals possess the inserted gene at birth and many show anatomical, physiological or behavioural abnormalities. The efficiency of methods used also depends on the species modified. Efficiency has a direct bearing on the number of attempts needed for success and the number of animals used. While several embryos can be implanted into mice and pigs to offset embryo mortality and thus considerably reduce the number of recipient females, this does not seem possible with cattle because of the phenomenon of freemartinism. Transgenic animals do not necessarily pass the transgene to their offspring. It has also been reported that offspring can have a limited lifespan, a variety of malformations and general or specific malfunctions. This is attributed to so-called "mosaic" transgenic animals that do not carry the transgene uniformly in all their cells.

Gene Expression

The random insertion of transgene sequences in the genome of a target cell can have major consequences, especially in the expression of certain genes. Numerous abnormalities have been noted among transgenic animals, depending on the function of the transformed gene. Recessive silent abnormalities can be expressed during breeding between transgenic animals, and their homozygous offspring could present a number of abnormalities. Controlling genetic expression is variable and difficult requiring the use of appropriate promoters and regulatory sequences. Many abnormalities are still observed among resulting animals, both mammals and fish (physiological and morphological problems, sensitivity to disease and stress, behavioural problems, etc.). These will undoubtedly lessen as techniques improve.

Benefits to Animal Welfare and Integrity

Biotechnologies are also promoted as improving animal welfare and health, with reduced mortality and morbidity because of greater resistance to disease and parasites; the elimination of phenotype characteristics such as horns; the adaptation of animals to intensive production; a reduced sensitivity to suffering and stress. They also help control the sex ratio among animal populations, especially chickens, thus avoiding the slaughter of male chicks. Others also argue that animals with greater genetic value receive better care which can only enhance their welfare. However, a problem of animal health does arise. The shrinking gene pool and the uniformity of genomes could weaken animal populations and make them more susceptible to disease.

Nuclear Transfer

The method of somatic cell nuclear transfer is relatively new and inefficient. The failure rate is high, as is prenatal and postnatal mortality, and there are many abnormalities causing problems of health and welfare from the immediate postnatal period: respiratory distress, lack of suckling reflex, physiological and morphological abnormalities, behavioural problems. The cause of these problems has not been identified with any certainty (stages of the technique itself, difficulty in genetically reprogramming the inserted nucleus, etc.). Study findings do however vary considerably and many transgenic animals have been successfully produced. Further studies are required.

BIOMEDICAL APPLICATIONS

There are fewer welfare problems with biomedical applications as there are few secondary modifications (although phenotypic problems can still occur). Also, as these animals have high value, their care is undoubtedly better than that of farm animals.

Biopharmaceuticals

The proteins or substances secreted in the physiological products should not have any effects on the organism of the secreting transgenic animal. If the substance is carried in the circulation and is biologically active in the animal's organism, it can cause certain physiological dysfunctions which can cause pain and discomfort. Secretion of the substance should be confined to the expressing tissue, but this is difficult to control. It has been reported that the early expression of a transgene can cause the premature development of the mammary gland and render any subsequent lactation painful.

Xenotransplantation

No phenotypical abnormality has been noted in pigs produced for xenotransplantation, so there are no problems of animal welfare, even though the efficiency of the microinjection technique used to obtain such transgenic pigs is low. As regards animal health, modification of their cell structure could render pigs susceptible to infection from human viral disease. The special conditions of husbandry also raise concerns about the pigs' welfare. The isolation needed for pathogen-free pigs and while awaiting organ donation causes abnormal behaviour among pigs which are extremely sociable animals.

Other

Problems of welfare (and ethics) arise in connection with genetically engineered farm animals that are created to study human diseases or the toxicity of medical drugs or other substances. Standard testing procedures need to be drawn up to avoid all stress and suffering. The impact of genetic engineering on the health and welfare of animals is a major public concern. Perception of animal welfare varies widely because determined by cultural attitude to animals, and involves ethical judgements. Biotechnologies can cause suffering, physical and psychological stress, behavioural and physiological abnormalities and/or problems of health among animals, whatever their intended use or function. On the other hand, they also have the potential to reduce or eliminate such problems, which also arise under intensive animal production practices.

MANIPULATION OF REPRODUCTION IN ANIMALS

One of the natural laws of life is to be many from one which is possible through reproduction. In animals, sexual reproduction occurs through fertilization of female gametes (ova) by male gametes (sperms) followed by fusion of two nuclei coming from two gametes. Thus sexual reproduction maintains the genetic traits of organism. But offspring develop and born in natural way. To meet the demand of animal products due to gradual increasing propulsion there is need of increase in number of animals too which is possible bio technologically because natural process of increase in number is slow.

In nature female mammals produce only one egg in a month except pigs. Secondly, ruminant females can have only one offspring in a year. Therefore, biotechnology can help to meet increasing demand and need of quality products of animals. In this regard principles of artificial breeding system has been discussed that may be useful further in the area of animal biotechnology.

ARTIFICIAL INSEMINATION

A male animal produces millions of sperms daily. Theoretically, it can inseminate females regularity and produce several offspring. This excess capacity of male has been utilized through developing new technologies for artificial insemination which can be said as the first animal biotechnology.

The most effective factor that has increased the productivity of cattle is the artificial insemination. However, the breeder must replace the nature through artificial insemination if he ensures about the ovulation of female in herd together at a time. In contrast, if a breeder awaits for female to ovulate and then inseminate separately, the importance and economic significance of artificial insemination get reduced. Therefore, through artificial insemination technology increased number of females can be inseminated by a male.

SEMEN AND ITS STORAGE

In addition methods have been developed to produce semen from male by ejaculation. Semen ejaculate is collected and diluted (extended). Sperm motility and there number per millilitre are examined under the microscope. About 0.2 ml bull semen contains about 10 million motile sperms. The diluted sperms may be used fresh within a few days or cryopreserved at -196 Degree Celsius by using liquid nitrogen. The cryopreserved semen can be stored for a long time and easily transported across states or countries. Thus cryopreserved semen of a single male is capable of inseminating thousands of females of a country or other countries. For example, one ejaculate semen of a bull is sufficient to inseminate about 500 cows.

Animals are a source of high protein, food, fibre and leather which men have been using since the time immemorial. In India milk and milk products have become a part of our civilization. Even in any small festival, demand of these two gets increased as several milk-based sweets and food items are prepared. Right from God worship to dining table use of milk and milk products is unavoidable. However, milk production could not keep pace with milk demand.

Therefore, to meet milk demand synthetic milk and synthetic khoa are illegally sold in market even after several punishment threats declared by the Government. Due to unaware use of synthetic milk several infants have been reported to lose their lives

Production of food from animals is inexpensive and exhausting, and less efficient than plants. The most common farm hoofed animals are cattle, sheep, goats and pigs but unlike plants their management is expensive. Inspite of expenses rendered on nutrition, shelter and management, animals have been our companion since the dawn of civilization. Moreover, there is regional variation on availability of animal protein to human diet due to cost, culture and religion.

In biological sense, despite several similarities animals differ from plants in their reproduction systems such as number of gametes (ova), completion of life cycle, production of products, etc. Except pigs females of all common livestocks produce one egg at about monthly intervals. Therefore, through biotechnological methods discovered in recent years, improvement in reproductive systems of animals, their milk proteins and milk-products, and several novel products which normally lack in animals can be carried out.

Embryo Transfer

In 1890, the first case of developing pregnancy in rabbits through embryo transfer is known in literature. During 1930s the same method was used in goat and sheep. In catties cases of embryo transfer were reported after 1950 (BIOTOL series, 1992). Embryo transfer method cannot be used widely because of its high cost, technical difficulties and limited supply of embryo from superovulated donors. The embryo develops in foster mother (recipient) which simply acts as incubator and does not make any genetic contribution to the offspring.

Secondly, ruminant female carries one pregnancy at a time as only one egg is produced and fertilized with male's sperm. Thus, there is a chance for increasing the number of egg production at a time and transfer of fertilized eggs *i.e.* embryo into uterus of less important foster mothers other than original female in farm animal.

Multiple Ovulation (Superovulation)

The reproductive cycle of ruminant female is such that the ovarian follicle of a non-pregnant female matures and releases single egg at a time. The time of ovulation differs in different animals, for example 21 days in cows and horse, 16 days in sheep and goats, etc. Normally ovulation occurs as a result of circulation of gonadotropic hormone. But by increasing the concentration of hormone the number of egg production gets increased. In well managed domestic catties 8-10 eggs are superovulated; the number may go to 60. However, this depends on health, nutrition, breed of animals and environment in which they live. Price (1991) has reviewed the current practice of multiple ovulation and superovulation. This technique has spread from cattle and sheep to goats, horses and deers.

He has emphasized that:

- General selection for increased litter size is useful only in sheep,
- The gonadotropic hormone induces superovulation in goats, sheeps and cattle but the response varies so much,
- Immunization against ovarian steroid hormones can increase litter size in sheep. Therefore, different molecular forms of follicle stimulation hormones (FSH) should be characterized in the farm animals.

Misra *et al.* have reported multiple ovulation and embryo transfer in Indian buffalo (Bubalus bubalis). After injecting the gonadotrophin the females are induced for superovulation. During follicular phase (second phase of oestrous cycle) about 20 ovarian follicles are induced. Eventually these grow and filled with fluid. The space of follicle which is filled with fluid is called antrum, and such follicles as antral follicles. In normal course, only one follicle develops and releases one egg after maturation. However, before ovulation the follicles laying against surface of ovary looks large sized (8 mm in sheep and pigs, 15 mm in cattle). Therefore, immature oocytes from follicles of donor females are recovered surgically by using laparoscope. The females to be superovulated are frequently injected with prostaglandin F2a (PGF2a) so that synchronized oestrous could develop in them. After 10 days of oestrous they are injected with hormone FSH up to 4 days followed by PGF2a treatment so that oestrous may be maintained. The FSH treatment induces superovulation. The females are artificially inseminated. The eggs are fertilized. After fertilization embryos undergo developmental stages.

Multiple Ovulation with Embryo Transfer (MOET)

After 6-8 days of fertilization (for sheep and goat) the embryos are recovered. During this stage embryos have come to morula or blastula atage and remain in female's oviduct. In cattle the embryos are recovered without surgery by inserting a catheter into oviduct. A saline solution is drained in oviduct and embryos flushed out through catheter into a storage bottle. The oviduct of sheep is small, therefore, it is exposed surgically and embryos are recovered by syringe adopting the same method for cattle. In storage bottle embryos settle down and solution decanted. Embryos with small volume of saline are poured into a Petri dish. Then they are examined under microscope. The identified embryos are transferred into synchronized recipient females (|as artificial insemination), when they are in 6-8 days of cycle of embryo development, stored and transported or manipulated as desired. In general, one superovulated female results into 5-6 progenies. In cattle 50-60 per cent of pregnancy can be achieved from embryo transfer method.

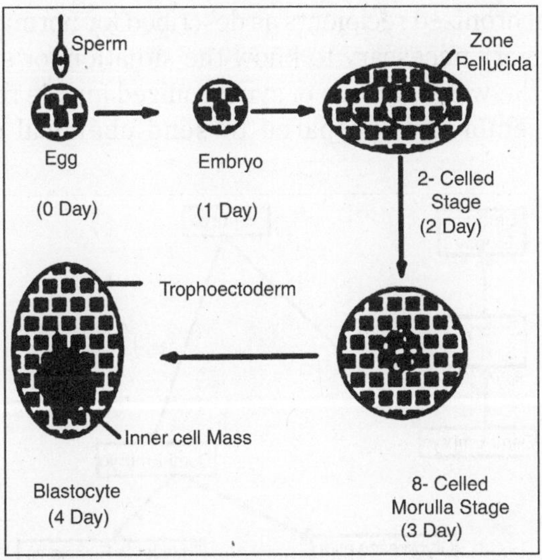

Fig. Developmental Stages of Embryo (Diagrammatic).

Embryo Splitting (Demi-Embryo)

Embryo of blastocyst stage is differentiated into two regions, trophoectoderm and inner cell mass (ICM). Trophoectoderm is a single layer of trophoblast cells lining the inner side of zone surrounding the blastoceal (a fluid filled cavity). Zona becomes the foetal membrane bf placenta. The ICM is the mass of cells which develops into foetus. The embryos of blastocyst stage can be split into equal two halves and transferred to females to produce identical twins. Thus, embryo splitting technology has increased the rate of pregnancy. To carry out embryo splitting, the blastocyst stage embryos are transferred for a few minutes into a cell culture medium consisting of hypertonic sucrose and bovine serum albumin.

The medium of high osmotic strength enters into cell membrane of the embryo (zona pellucida) and cells contract due to exosmosis. The bovine albumin serum attaches to zona pellucida and provides negative charge to the embryo. Then it is transferred into a plastic Petri dish containing standard cell culture medium. The outer membrane of embryo is negatively charged and Petri dish positively. Embryo sticks to the surface of Petri dish due to development of electrostatic interaction of two charges. The Petri dish is kept on stage of an inverted microscope. A micromanipulator equipped with fine surgical blade is used to cut the embryo roughly into two halves with minimal damage to the cells. The hemispherical mass of half embryo, demi-embryo or semi-embryo reforms spheres.

However, it is not necessary for successful implantation of demi-embryo that it should be enclosed in a zona pellucida. The demi-embryos are transferred into oviduct of synchronized recipients as described for normal embryo transfer. At this stage it is very necessary to know the situation for successful embryo implantation that the wall of uterus of synchronized female recipient is waiting for embryo, and embryo is prepared to send chemical signals from the trophoblast.

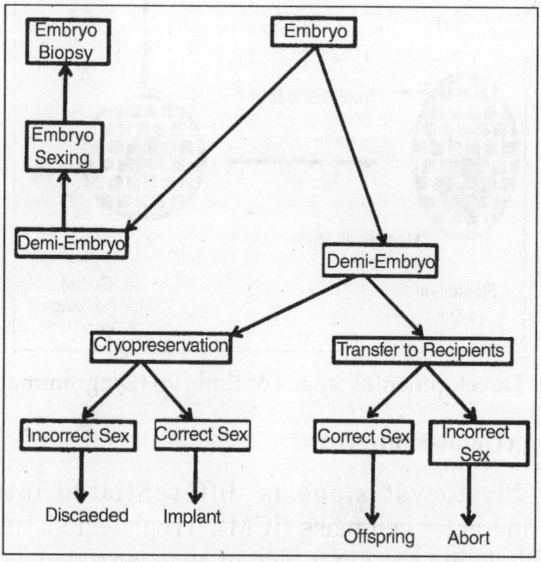

Fig. An Outline of Embryo Splitting.

To increase the rate of pregnancy, the embryo can be cut into equal three or four pieces, and transplanted in synchronized females. But too small trophoblast will not induce the uterus for pregnancy. So far it is unknown how cells of ICM are required for successful pregnancy, but it is clear that increasing in embryo splitting will have less probability of pregnancy.

Embryo biopsy (removal of small number of cells for genetic analysis) should be combined with splitting so that the twins which will be produced should be identical and of known genotype. Biopsy is very necessary in breeding so that sex and genetic diseases could be detected. In case the embryo has any genetic disease, it can be prevented from implantation in recipient females.

Embryo Sexing

Before the implantation of embryo its sex should be detected from the biopsy sample. The principle for sexing is very common. The presence of Y chromosome makes the offsprings male and that of X makes female. The second method is the use of polymerase chain reaction (PCR) machine in sex detection. PCR amplifies DNA sequence of Y chromosome and reaction products can be seen directly. Handy side *et al.* isolated single blastomere from early embryo from a womb, amplified DNA sequences of Y chromosome and carried out embryo sexing before implantation into uterus. It is true that the PCR was first commercially implemented for embryo sexing of livestock. Now the sexed embryos are available commercially but these are very costly.

INFERTILITIES IN HUMANS

There is a large number of reasons for developing infertilities in humans. By using IVF many more infertilities have been treated.

Male Sterility

In a fertile human the total count of sperms should be 15-20 million per ml. When the number of sperms decreases, the person is called oligospermic and the condition as oligospermia. Similarly, when there is very low number of motile sperms, the patient is called azoospermic and the phenomenon as azoospermia. Generally in azoospermic humans the number of non-motile sperms is high. In addition, for IVF, the presence of motile sperms should be around 10-20,000 per ml.

Female Sterility

There are several reasons and types of female infertility; few of them are:

- Non-functional (inaccessible or absent) ovaries (some times ovaries of female are non-functional or inaccessible, therefore, healthy oocytes will not be produced. In such cases oocytes can be procured from donor for IVF),
- Non-functional (or absent) uterus (in this condition the oocytes are obtained from female and subjected to IVF, and embryo transferred to surrogate mother for pregnancy and further development),
- Idiopathic infertility (no definite cause is known for idiopathic infertility. Possibly it may be due to abnormal fertilization or failure in fertilization. Therefore, IVF followed by embryo transfer will cure this infertility),
- Tubal infertility (fallopian tubes receive oocytes from follicles of ovaries for fertilization. Any deformity in fallopian tubes results in tubal infertility. IVF followed by embryo transfer in uterus has replaced the function of fallopian tubes. Therefore, the patients will not be subjected to surgical treatment).

WHO BENEFITS IVF

It is not likely that any patient of any age or any group of the society can go for treatment for IVF. There are several prerequisite conditions applicable for male and female partners.

The male should not be azoospermic; however, if he is azoospermic and desires for a baby, he can procure semen either from sperm bank or elsewhere.

Besides, the female partner should also be healthy and fit for IVF because it is she who is to bear the pregnancy.

Therefore, the female must be such that:

- Any surgical procedure should not create any trouble to her,
- Her ovaries should be accessible for oocytes,
- Her uterus should accept and bear pregnancy for the required period of about nine months,
- Her cervical canal should be easily negotiable for embryo transfer.

HOW THE PATIENTS FOR IVF TREATED

First of all the female aspirant for IVF is suggested to keep the records of her menstrual cycle upto six months. This information supports the doctors to take fruitful decision for the possibility of IVF. The female patient is hospitalized to ensure the date for collection of urine for the presence of luteinizing hormone (LH) surge, and to fix date for administrating human choripnic gonadotrophin (hCG). The hCG controls the final stage of formation of follicle and oocytes. The oocytes may be recovered during any menstrual cycles *i.e.* natural cycle, stimulated cycle or controlled cycle. According to the nature of three cycles, the treatment for IVF is recommended.

During natural cycle, spontaneous LH surge is found out at 3-6 h intervals by taking urine or plasma sample. Under the influence of LH surge fully developed follicles of both the ovaries burst and ova collected in fallopian tubes. In natural cycle, only one ova at a time is formed. Secondly, the stimulated cycle is created by administrating clomiphene and/or human menopausal gonadotrophin (hMG). Clomiphene (150 mg) per day is given for the 5th to 9th day of the cycle. However, the duration may be prolonged for 10 minutes but there will be more risk of uterus receptivity and abortion. The hMG alone or along with Clomiphene is also given that does not allow the high rate of abortion.

But several abnormalities are also caused by hMG. Consequently multiple follicles and oocytes, are stimulated instead of one. At this time hCG is also given to female patients so that inhibition of LH surge can be checked. The disadvantages associated with these are the ultrasonographic assessment of the hormonal effect and development of some of abnormal oocytes that result in abnormal pregnancy.

Thirdly, hCG is administered at the optimum stage of maturation that arrests the development of follicle. However, it is difficult to predict the optimum time for hCG administration that varies for different preovulatory follicles. Patients need not stay in hospital for a long duration because the doctor can (to his own and patient's convenience) recover oocytes through laparoscopy. His success rate in oocyte recovery, embryo transfer and pregnancies are achieved by the stimulated and controlled cycles.

Indicators for Ovary Stimulation

It is very necessary to monitor the stimulation of ovary so that the oocytes can be collected at right time of development.

There are several parameters which are used as indicators for stimulation of ovary, these include:

- Prediction of LH surge on the basis of menstrual cycle,
- Rise in temperature of body during preovulatory and postovulatory days,
- Changes in cervical mucus as it is secreted five days before LH surge, goes maximum a day before LH surge and falls sharply thereafter,
- Estimation of progesterone,
- Levels of oestrogen in urine, etc.

Oocyte Recovery and Culture

Oocytes are recovered through recently developed equipment called laparoscopy. Using microscopic equipment the follicles can be observed and aspirated with the aspirating apparatus which consists of 23 cm long needle made up of stainless steel (with internal diameter of 1.6 mm and external diameter 2.1 mm) and 52 cm long tubing. The total space between needle and tubing is 1.2 ml. The ovulating follicles are selected for aspiration. It consists of ova of less than 1.5 cm. From such follicles fluid (4-12 ml) is aspirated by using needle and tubing. After aspiration wall of follicle gets collapsed. Fluid is of straw coloured some times it contains blood also. When there appears blood, heparinised culture medium should be added in tube so that blood clotting may be prevented. However, if fluid contains blood, ova could not be easily identified and isolated. The aspirate is observed under the microscope for the presence of oocytes. If ova are not present in aspirate, the heparinised culture medium should be reflushed into follicles to get oocytes. The oocytes are identified under the microscope and incubated for 5-10 h depending on maturity of oocytes and follicles. The media used for oocyte culture and IVF are modified Ham's F10 medium, modified Whitten's medium, Earl's solution and Whittingham's T6 medium. Any one of these media may be used.

Semen Preparation

The semen is procured either (from husband of female partner or from semen bank or elsewhere. Husband's semen is collected at site when required through masturbation 60-90 minutes prior to insemination. Semen is subjected to centrifugation; sperm pellets are recovered and transferred to culture medium for suspending it again. It is then centrifuged and semen pellets are retransferred to culture medium. The sperms are incubated at 37°C for 30-60 minutes. The most active spermatozoa float at surface. Therefore, samples for IVF should be taken from surface of culture medium containing spermatozoa.

IN VITRO FERTILIZATION AND EMBRYO TRANSFER

Insemination is done by mixing about 1 ml culture medium containing oocytes with 1 ml of semen preparation (that contains 10,000-50,000 spermatozoa/ml). Oocytes are incubated for 12-13 h. Thereafter, they are observed keeping them in a Petri dish for the presence of two pronuclei and two polar bodies. Besides, the abnormal fertilized oocyte contains more than two pronuclei and granulation in cytoplasm. First cleavage occurs after 24-30 h of insemination. Delay in cleavage after insemination confirms that the embryo is abnormal. Thus, the abnormal embryo is discarded; only normal embryo is used. After fertilization, embryo should not be detained for el long time in vitro. It should be transferred to treated recipient female when it is in 2-4 cell stage. However, success has been achieved with transfer of 1-16 celled embryos. However, early embryo fails to survive in uterus. This stage is very critical, therefore, much care should be taken.

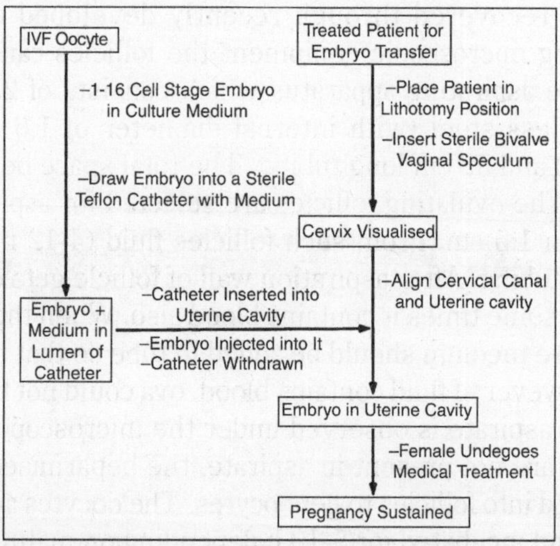

Fig. Outline of Embryo Transfer in Treated Female Human.

Some times, multiple embryo transfer is also done. It may increase the chance of embryo survival, but results in multiple pregnancy to which the female does not like. The transferred embryo stays in fallopian tube up to 8-16 cell stage, thereafter moves to uterus. Embryo of prolonged stage in vitro should not be transferred because of less chance of survival and pregnancy.

Embryo transfer should be done in an operation theatre. No anaesthesia should be given to female. She is given only Diazepam (Valium) of 10 mg orally before embryo transfer. The embryo is to be transferred through cervical canal; therefore, the patient is placed in lithotomy (*i.e.* in the position of knee and chest with tilting the head down) so that the fundus should be in lower position than cervical canal. The cervix is visualized by inserting a sterile bivalve vaginal speculum. Then uterine, cavity is aligned to cervical canal by manipulating the uterus. Distance between fundus and external cervical is earlier measured through ultrasonography. The embryo is drawn into a teflon catheter along with 10 ml of tissue culture medium. The catheter is gently inserted into culture medium. It should be made sure that the embryo has been transferred or not. The catheter is observed under microscope. Absence of embryo in catheter ensures for its transfer in uterus. Wrong placement of catheter or high amount of medium will result in tubal pregnancy or expulsion of embryo from uterine cavity. Therefore, care should be taken for proper handling. After embryo transfer, the female is allowed to take rest for about 2-7 h and abstain from intercourse for about 10 days. The patient follows the advice of doctor and undergoes for regular check up.

IN VITRO FERTILIZATION (IVF) TECHNOLOGY

The term in vitro means in glass or in artificial conditions, and IVF refers to the fact that fertilization of egg by sperm had occurred not in uterus but out side the uterus at artificially maintained optimum condition. In recent years the IVF technology has revolutionized the field of animal biotechnology because of production of more and more animals as compared to animal production through normal course.

For example, an animal produces about 4-5 offsprings in her life through normal reproduction, whereas through IVF technology the same can produce 50-80 offsprings in her life. Therefore, the IVF technology holds a great promise because a large number of animals may be produced and gene pool of animal population can also be improved. In India M.L. Madan, an animal embryo-biotechnologist at National Dairy Research Institute, Karnal (Haryana) has got success in producing more calves in cows. The IVF technology is very useful.

It involves the procedure:

- Taking out the eggs from ovaries of female donor,
- In vitro maturation of egg cultures kept in an incubator,
- Fertilization of the eggs in test tubes by semen obtained from superior male,
- Implantation of seven days old embryos in reproductive tract of other recipient female which acts as foster mother or surrogate mothers.

These are used only to serve as animal incubator and to deliver offsprings after normal gestation period. The surrogate mothers do not contribute any thing in terms of genetic make up since the same comes from the egg of donor mother and semen from artificial insemination.

In Vitro Maturation (IVM) of Oocytes

The immature oocytes are incubated in vitro so that they can be mature. However, immature oocytes should be taken out from follicles because they cannot mature in it but degenerate. Therefore, full potential of superovulation and all the oocytes can be utilized by IVF technology. Moreover, metabolic and hormonal requirement for oocytes during IVM should be found out so that the present rate of maturation (20%) could be improved. In majority of cases ovarian follicles never reach maturity and degenerate due to unknown causes. Possibly there may be genetic defects associated with them.

Culture of in Vitro Fertilized Embryos

IVF ofeggs is carried out in small droplets (microdroplets) of culture medium. Each microdroplet comprises of about 10 oocytes. The medium should be supplemented with penicillamine, hypotaurin, and epinephrine because they facilitate penetration of sperms into oocytes. Moreover, one dose of sperm is given that consists of about one million sperms per ml of medium. Thus, IVF embryo must be maintained at in vitro conditions for a few days so that it may develop into blastocyte. It takes about seven days for sheep and goats and eight days for cattle. There are many laboratories where about 60 per cent IVF embryos of catties are cultured to blastocyte stage.

The term delivery from cultured embryo is very low due to occurrence of high loss during first two months of pregnancy. This may be due to abortion of foetuses arising from the presence of genetic defects. It should be noted that before birth about 80 per cent genes play a key role in differentiation and development of foetuses. The oocytes which are forced to mature in vitro occasionally bears some defects. Some times environmental mutagenesis occurs in eggs, sperms or embryos. Artificial culture media should be improved as oxygen may have toxic effect. Therefore, gas atmosphere should be carefully controlled. The other most successful method of IVM is to place the fertilized zygotes into agar (so that it may wrap around it) and implant them in oviduct of synchronized sheep or rabbit where the environment for early development of embryo is perfect. For early bovine embryo the oviduct of rabbit and sheep has been used as in vitro culture system. Hundreds of cattle eggs can be put into oviduct of a sheep and many of these are recovered after a week. A good quality of embryo at the late morula/blastocyte stage of development with a yield of about 40 per cent or more has been recorded by Lu *et al*.. Brackett *et al*. reported the birth of first IVF calf after getting success in fertilizing the eggs recovered from ovulated cow. Thereafter, hundreds of IVF calves have been born in Japan, India, U.K., etc.

EMBRYO CLONING

A clone is a population of cells or organisms derived asexually from a single ancestor. They are genetically identical to each other to their common ancestor. Cloning means the production of exact genetic replica copies of an individual. They can not be considered as an offspring but simply the copy of a given individual. Much work has been done on cloning in plants and microorganisms. However, the techniques used in plants can not be applied for animals.

Moreover, many animals from a single genetically superior embryo can be produced. Still there is no method of finding out which embryos are capable of cloning. It is useless to clone an embryo if it is not superior.

Quadriparental Hybrids

During 1960s, Beatrice Mintz at Cancer Research Institute, Philadelphia (USA) demonstrated the interesting experimentation. She carried out fusion of embryos of two different species of mouse. This resulted in the formation of a single embryo which finally developed into a normal healthy animal having four parents. Embryo A was derived from the cross of male x female of one species, and embryo B derived after cross of male x female of the other species. The embryo A and B were united together and produced a single mass. In this experiment Mintz removed zona pellucida membrane of two early embryos and placed them in a suitable culture medium. The embryonic cells of the two embryos of blastula stage united randomly into a single mass of double sized embryo (blastocyst). A fresh membrane developed around the embryo. Then the embryo was transferred into the uterus of a foster mother. The foster mother was mated with a sterile male to bring her into proper stage for implantation. The first offspring with four parents was born in 1965. Similarly, exciting experiments have been done on another mammal. Following the same technique a hybrid of goat and sheep named geep was produced.

At present two types of techniques for embryo cloning *viz.*, nuclear transplantation (transfer) and embryonic stem cells, are being developed.

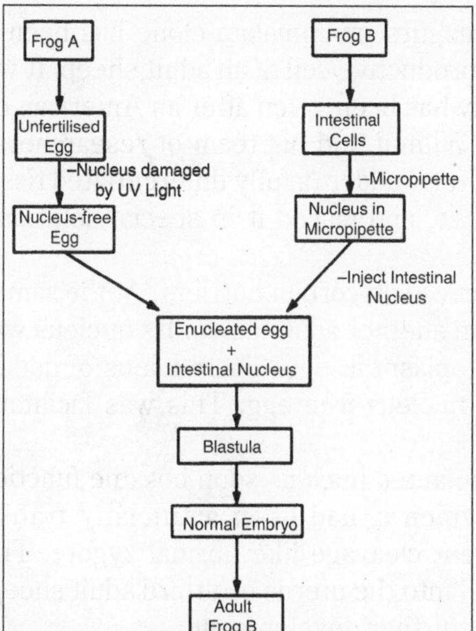

Fig. Nuclear Transplantation in Frog (Diagrammatic).

Nuclear Transplantation

Nuclear transplantation (also nuclear transfer) involves removal of a single blastomere from a cleavage stage embryo with a fine micropipette of glass, and placing it under the outer membrane of an unfertilized mature enucleated oocyte (whose haploid nucleus has been removed by using micropipette or destroyed by UV light). For the first time in 1955, Robert Briggs and Tom King at Cancer Research Institute, Philadelphia (USA) carried out nuclear transplantation experiment on embryonic cells of frog. They transferred nucleus of undifferentiated blastula (a stage soon after fertilization of egg) into an enucleated egg cell. They noticed the normal development of the embryo.

When they performed serial transplantation of differentiated nucleus from late gastrula (a stage after blastula) into a nucleus-free unfertilized egg, abnormal embryos were formed. This shows that cell nucleus is differentiated with embryo development.

In 1960s, J.B. Gurdon at Oxford University, U.K. transferred differentiated intestinal nucleus of a frog into nucleus-free unfertilized egg of different amphibian species (Xenopus laevis). The embryo developed into tadpole and matured into frog. This new enucleated cell developed into normal embryo. Any damage to the donor nucleus during transplantation leads to abnormal development.

DOLLY-The First Mammalian Clone

'Dolly', the worlds' first mammalian clone has been created from a fully differentiated non-reproductive cell of an adult sheep. It was born in February, 1996. The name Dolly has been given after an American country singer, Dolly Parton. In 1995, Ian Wilmut and his team of researchers at Roslin Institute, Edinburgh, Scotland, took udder (a fully differentiated tissue) from six year old sheep, Fin Dorset Ewe, and placed it in special solution that controlled cell cycle of cell division.

The cell was deprived off certain nutrient. At the same time an unfertilized egg was obtained from another adult sheep. Its nucleus was carefully removed leaving the intact cytoplasm in egg. The nucleus of udder cell was taken out and transferred into nucleus-free egg. This was facilitated by applying mild electric sock.

The newly transplanted nucleus soon became functional according to the new cytoplasm in which it had been artificially transferred. This viable combination underwent cleavage like normal zygote. This so called embryo was then transplanted into the uterus of a third adult sheep (surrogate mother/ foster mother) for its further development.

Finally, a normal healthy little lamb, Dolly was born in February, 1996 which was genetically similar to the clone mother from which nuclear DNA was taken out. It does not have any similarity with that sheep from which egg was taken out or surrogate mother because they did not contribute any genetic character.

Thus, Dolly has only a single parent because she has born asexually, a characteristic feature found in lower forms of animal life, not in mammals. Although behind this great success the rate of success is very slow, yet it has given some hope to embryo-biotechnologists to bring about refinement. Out of 277 nuclei transferred singly to enucleated egg, only 29 eggs grew into embryos. Out of these, only 13 embryo could be successfully transplanted into surrogate mothers. Of these only one ewe was successful in giving birth to an offspring, Dolly.

The significant considerations that can be derived from this experiment are that:

- The genes of differentiated cells have inherent totipotency,
- The interplay between the regulatory system of the genome in nucleus and the cytoplasmic factors of egg may make a cell totipotent,
- Possibly the cytoplasm of enucleated egg makes the transplanted nucleus totipotent like that of normal fertilized egg nucleus,
- The maternally derived information in egg cytoplasm has an important role in cleavage which usually occurs after fertilization.

Hence, it is the egg cytoplasm but not the nucleus that regulates cleavage. Because the udder cell nucleus has limited potential for mitosis; it is the egg cytoplasm that interacted intracellularly with udder cell nucleus and stimulated to undergo repeated mitosis,

- Carbon copy of the adult sheep could be produced without involving sperms from male partner,
- The cloned animal produced via nuclear transplan-tation technique will be capable of restoring fertility as in 1998 Dolly gave birth to a little lamb named Bonny.

Embryonic Stem (ES) Cells

Cloning of mice could not be done as in sheep via nuclear transplantation. This was due to acceleration of developmental programmes of embryo. However, it is evident that before first embryonic division the cell has started its process of differentiation.

Therefore, for cloning of mice an alternative approach has been made *i.e.* the use of ES cells. A blastocyst of mouse is place in culture condition. The inner cells that form future foetus continue to divide and remain in undifferentiated totipotent state as ES cells. There is a peptide growth factor known as leukaemia inhibitory factor (LIF) which establishes and maintains ES cell lines.

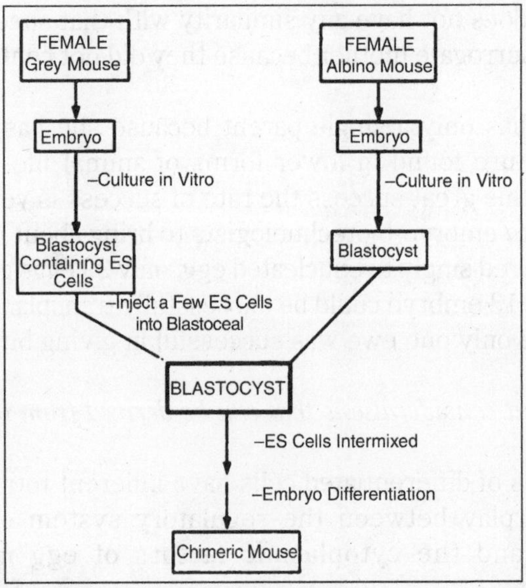

Fig. Production of Chimeric Mouse by Embryonic Stem Cell Transplantation.

The ES cell lines will be very useful in the area of production of transgenic animals. However, the ES cells are used in two different ways: a small number of ES cells can be injected into blastocoel space of a blastocyst. The ES cells get mixed with inner mass of cells of blastocyst to produce a chimera mouse which is a mixture of two cell genotype having the patches of different coloured fur. Crossing of male and female chimera will allow selection of homozygous mice derived from ES cells.

In Vitro Fertilization and Embryo Transfer in Humans

Due to several complicacies associated with humans, there has been the birth of in vitro fertilization technology. The IVF technology is a boon to childless couples. Initially this technique was pioneered by Prof. Robert Winston during 1970s which was later applied by P. Steptoe and R. Edwards for the production of first test tube baby, Louse Joy Brown, who was born on July 25, 1978. Since then more than 25,000 test tube babies have been produced so far through out the world.

4

Hormones in Animal Production

Hormone-dependent sex differences in growth rate have been known for a long time. It has also been known that growth rate and FCE (feed conversion efficiency) are higher in intact males than in castrates. It was natural, then, that the availability of hormones and other natural or synthetic substances displaying hormonal activity led to experiments aiming at their use to increase production. Beginning in the mid-1950s, DES (diethylstilboestrol) and hexoestrol were administered to cattle increasingly in the US and the UK respectively, either as feed additives or as implants, and other types of substances also gradually became available.

In general, such treatment has resulted in 10–15% increases in daily gains, similar improvements in FCE and improvement of carcass quality (increased lean/fat ratio). Thus there has been a substantial reduction in the amount of energy required per unit weight of protein produced, and the economic implications of this have been great. While the use of hormonally active substances in animal production rose, opposition to their use also increased, because of the theoretical possibility that residues in edible tissues might endanger consumers. The factors leading to the ban on DES in the US, first imposed in 1973, have been described.

Several reports confirm that DES endangers the health of animals and man, when repeatedly used in large doses . However, as regards risks due to the presence of residues in meat produced according to regulations, no documented deleterious effects have ever been reported in man, either from DES or any other substance with hormonal activity. A distinction should be made between the hormones as such, for which the metabolism in the body is relatively well known, and synthetic or other substances for whose metabolic inactivation the body may not possess the enzymes necessary. When natural hormones are used in animal production, claims of zero-tolerance residue levels are not meaningful, since these compounds occur in detectable and highly variable concentrations in body fluids as well as in the tissues of all animals, treated or not. For other substances with hormonal activity the situation is different. However, when residue levels are extremely low, it seems reasonable to weigh the potential risks against the undisputed positive effects some of these compounds have in animal protein production.

This paper will discuss types of substances with hormonal activity currently in use or under investigation, their effects, mechanism of action, metabolism/elimination, tissue levels, risks to the consumer and their economic importance. Finally, other avenues to increased animal production as alternatives to use of hormones will be briefly envisaged. For the sake of simplicity the term hormone will be used, even if incorrectly, to cover all substances with hormonal activity, whether natural or synthetic. Since much information on the question collected before 1975 has been reviewed previously, the main emphasis will be placed here on research since that time.

METHODS OF MEASURING HORMONE

It should be clearly understood that this discussion concerns only measurement methods intended to detect residues remaining in flesh and viscera. These methods are, however, not very reliable, and other biological material must be examined to determine whether the animals have been treated. Specialists agree that there are still difficulties in testing for residues in the tissues of animals intended for use as food. Many different methods have been proposed and discussed at the EEC level. While the trend is increasingly towards the use of radio-immunoassay, it is undeniable that the techniques are highly complex and still far from being perfected. DES and similar substances should preferably be looked for in urine and faeces. Instructions to this effect were issued in the United States in 1979 and confirmed in 1980. Account should also be taken of the fact that no difference may be found between animals properly implanted with anabolic agents and untreated animals.

This problem has received insufficient attention. The proceedings of the FAO/WHO Joint Meeting of March contain reports emphasizing this lack of difference and the relatively high, but physiological, levels of testosterone and progesterone in the flesh of bulls and gestating cows, respectively. This being the case, how can the natural presence of these substances be distinguished from their presence due to the use of natural products? Further, it should not be forgotten that many animal feeds contain phytohormones which can cause modifications in the morphology and histology of certain organs. After summarizing and stating the principles of the four main groups of useful techniques enumerated by Kroes *et al*. and Richou-Bac and Pantaléon—histological, biological, physico-chemical and radio immunoassay methods—this paper will study in detail the techniques that can be applied in practice for routine control.

- Histological methods, used specially in young bovines, consist in the observation of modifications in the male and female genitalia. The prostate and the bulbo-urethral glands are particularly examined in male calves and lambs, while in females attention is focused on the glands of Bartholin, the teats, the vagina and the cervix. Modifications are obvious when DES and zeranol, as well as hexoestrol, have been used. In pigs the effects are less discernable. It is generally agreed that the histological method makes it possible to conclude that an oestrogen has been used at some earlier time. The use of natural steroid oestrogens, in combination with testosterone or progesterone, leads only to slight histological modifications. The phyto-oestrogens may also have some effect and lead to modifications; this is the case of Fusariuminfected maize.

- The biological methods used are those of Astwood, which measures the increase in uterus weight of pre-puberal mice and rats, and of Allen and Doisy, based on modifications of the cytology of the vagina of adult mice after ovarectomy. The latter test, applied by Stob *et al.* in 1954, is the most sensitive, but it is responsive only to compounds with oestrogenic activity. Martin reports that the use of concentrated extracts placed in the vagina of mice also gives good results.
- Physico-chemical methods, which make it possible to identify with precision the anabolic agent sought for, require a fairly long extraction, followed by thin-layer or gaseous-phase chromatography. Hans and Abraham report that the latter techniques, combined with mass spectometry, can now be contemplated. Sensitivity is high. Further off is the possible use of liquid/liquid chromatography. Thin-layer chromato-graphy is currently used, and Verbeke gives details in an EEC document, while Waldschmidt and Schuller and Stephany have described the method.
- Radio-immunoassay, first described for the oestrogenic steroids by Jiang and Ryan, is based on competition between tritium-labelled and unlabelled antigens for the corresponding specific antibodies. This technique, increasingly used, is highly sensitive, making it possible to detect 0.05 to 0.005 ppb of DES and 0.1 ppb of TBA. It is, however, complex and long, although it is being improved. It has the same advantages as all uses of radioactive compounds. The technique has been well studied by Hoffmann and Hoffmann and Karg.

ANIMAL PRODUCTS

Dairy products are generally defined as foodstuffs produced from milk. They are usually high-energy-yielding food products. A production plant for such processing is called a dairy or a dairy factory. Raw milk for processing generally comes from cows, but occasionally from other mammals such as goats, sheep, water buffalo, yaks, or horses. Dairy products are commonly found in European, Middle Eastern and Indian cuisine, whereas they are almost unknown in East

Asian cuisine. A dairy is a facility for the extraction and processing of animal milk—mostly from goats or cows, but also from buffalo, sheep, camels —for human consumption. As an adjective, the word *dairy* describes milk-based products and processes, dairy cattle, dairy goat. A dairy farm produces milk and a dairy factory processes it into a variety of dairy products.

In New Zealand English a *dairy* means a corner shop, milk bar or Superette and *dairy factory* is the term for what is elsewhere a *dairy*. In the UK a *dairy* is a processing facility that turns milk into a range of products. In the U.S. a *dairy* can be the storage and processing facility for milk, the facility that processes and distributes the milk or the store that sells milk, butter, cheese and suchlike.

Milk-producing animals have been domesticated for thousands of years. Initially they were part of the subsistence farming that nomads engaged in. As the community moved about the country so did their animals accompany them. Protecting and feeding the animals were a big part of the symbiotic relationship between the animal and the herder. In the more recent past, people in agricultural societies owned dairy animals that they milked for domestic or local (village) consumption, a typical example of a cottage industry.

The animals might serve multiple purposes. In this case the animals were normally milked by hand and the herd size was quite small so that all of the animals could be milked in less than an hour—about 10 per milker.

Fig. Dairy Products

With industrialisation and urbanisation the supply of milk became a commercial industry with specialised breeds of cow being developed for dairy, as distinct from beef or draught animals. Initially more people were employed as milkers but it soon turned to mechanisation with machines designed to do the milking. Historically, the milking and the processing took place close together in space and time: on a dairy farm. People milked the animals by hand; on farms where only small numbers are kept hand-milking may still be practiced.

Hand-milking is accomplished by grasping the teats (often pronounced *tit* or *tits*) in the hand and expressing milk either by squeezing the fingers progressively, from the udder end to the tip, or by squeezing the teat between thumb and index finger then moving the hand downward from udder towards the end of the teat. The action of the hand or fingers is designed to close off the milk duct at the udder (upper) end and, by the movement of the fingers, close the duct progressively to the tip to express the trapped milk. Each half or quarter of the udder is emptied one milk-duct capacity at a time.

The *stripping* action is repeated, using both hands for speed. Both methods result in the milk that was trapped in the milk duct being squirted out the end into a bucket that is supported between the knees (or rests on the ground) of the milker, who usually sits on a low stool. Traditionally the cow, or cows, would stand in the field or paddock while being milked.

Young stock, heifers, would have to be trained to remain still to be milked. In many countries the cows were tethered to a post and milked. The problem with this method is that it relies on quiet, tractable beasts, because the hind end of the cow is not restrained.

Processing

When it became necessary to milk larger numbers of cows, the cows would be brought to a shed or barn that was set up with bails (stalls) where the cows could be confined while they were milked. One person could milk more cows this way, as many as 20 for a skilled worker. But having cows standing about in the yard and shed waiting to be milked is not good for the cow, as she needs as much time in the paddock grazing as is possible.

It is usual to restrict the twice-daily milking to a maximum of an hour and a half each time. It makes no difference whether one milks 10 or 1000 cows, the milking time should not exceed a total of about three hours each day for any cow.

As herd sizes increased there was more need to have efficient milking machines, sheds, milk-storage facilities (vats), bulk-milk transport and shed cleaning capabilities and the means of getting cows from paddock to shed and back. Farmers found that cows would abandon their grazing area and walk towards the milking area when the time came for milking. This is not surprising as, in the flush of the milking season, cows presumably get very uncomfortable with udders engorged with milk, and the place of relief for them is the milking shed.

As herd numbers increased so did the problems of animal health. In New Zealand two approaches to this problem have been used. The first was improved veterinary medicines (and the government regulation of the medicines) that the farmer could use. The other was the creation of *veterinary clubs* where groups of farmers would employ a veterinarian (vet) full-time and share those services throughout the year.

It was in the vet's interest to keep the animals healthy and reduce the number of calls from farmers, rather than to ensure that the farmer needed to call for service and pay regularly. Most dairy farmers milk their cows with absolute regularity at a minimum of twice a day, with some high-producing herds milking up to four times a day to lessen the weight of large volumes of milk in the udder of the cow. This daily milking routine goes on for about 300 to 320 days per year that the cow stays in milk. Some small herds are milked once a day for about the last 20 days of the production cycle but this is not usual for large herds.

If a cow is left unmilked just once she is likely to reduce milk-production almost immediately and the rest of the season may see her *dried off* (giving no milk) and still consuming feed for no production. However, once-a-day milking is now being practised more widely in New Zealand for profit and lifestyle reasons. This is effective because the fall in milk yield is at least partially offset by labour and cost savings from milking once per day. This compares to some intensive farm systems in the United States that milk three or more times per day due to higher milk yields per cow and lower marginal labour costs.

Farmers who are contracted to supply liquid milk for human consumption often have to manage their herd so that the contracted number of cows is in milk the year round or the required minimum milk output is maintained. This is done by mating cows outside their natural mating time so that the period when each cow in the herd is giving maximum production is in rotation throughout the year.

Northern hemisphere farmers who keep cows in barns almost all the year usually manage their herds to give continuous production of milk so that they get paid all year round. In the southern hemisphere the cooperative dairying systems allow for two months on no productivity because their systems are designed to take advantage of maximum grass and milk production in the spring and because the milk processing plants pay bonuses in the dry (winter) season to carry the farmers through the mid-winter break from milking.

It also means that cows have a rest from milk production when they are most heavily pregnant. Some year-round milk farms are penalised financially for over-production at any time in the year by being unable to sell their overproduction at current prices.

INDIA'S TRADITIONAL DAIRY PRODUCTS

From burfi to kulfi, from *kalakand* to *shrikhand*, from *gulabjamun* to *chumchum* extends the delectable world of Indian milk delicacies. This article brings together information culled from diverse sources on age-old, milk-based specialties from different regions. Also included is information culled from diverse sources on the Indian milk-based sweets and other specialties from different regions of the country.

Since time immemorial, a significant proportion of milk has been used in India for preparing a wide variety of dairy delicacies — an unending array of sweets and other specialties from different regions of the country. In the process, the basic limitation of milk — its perishable nature — has been tastefully overcome. Its processing aims to extend the shelf-life of milk, while converting it into mouth-watering tit-bits.

Thus, diverse methods to prepare as well as preserve milk products have been developed. An estimated 50 to 55 per cent of the milk produced in India is converted into a variety of traditional milk products, using processes such as coagulation (heat and/or acid), desiccation and fermentation. Over the millennia, these processes have largely remained unchanged, being in the hands of halwais, the traditional sweetmeat makers, who form the core of this cottage industry.

Although 46 per cent of the milk produced in the country is consumed as liquid milk, increase in consumption can be stimulated. Milk plays an important role in the national diet. In Indian households, the life of milk is extended from 12 to 24 hours by repeated boiling. It is preserved by souring with the aid of lactic cultures, which imparts an acid taste, particularly refreshing in hot climate.

Curd

The first of these products developed was *dahi* (curds or yogurt), obtained by fermenting milk. In the process, the digestibility of milk constituents improves.The product is widely consumed along with meals. The surplus dahi is used as the intermediate product and churned into *makkhan* (butter), while the liquid whey — *chhach* or *mattha*— is consumed as a refreshing beverage or converted into *kadhi*, a spicy dish served hot with rice.

Nutritional benefits: When food is supplemented with 250 gms of dahi a day, the status of thiamine improves. *Dahi* also increases the pyruvic and lactic acids among children on a typical poor rice diet. Due to fermentation of milk, a greater amount of phosphorus and calcium is made available to the digestive system by their precipitation in the lower intestines due to the acid condition induced by Lactobacillus sp. The consumption of sour milk also results in increased efficiency of the human body to cope with a sudden influx of lactic acid in the system.

Dahi can also be consumed by people who suffer from lactose intolerance. Thus, dahi in its different forms as lassi, kadhi, shrikhand, etc. also contributes significantly to improve the nutritive contents of an average Indian diet.

Ghee and Makkhan

A week's collection of makkhan(butter) is converted into *ghee*. *Dahi* has a shelf-life of a day or two; *makkhan*, a week; and *ghee*, about a year. Ghee, thus, became the main dairy product for extended preservation. A system of collection of ghee from villages for trade gradually developed. Ghee mandis (trading centers) have been in existence for centuries. India produces some 900,000 tonnes of *ghee*, valued at Rs 85,000 million.The value of the resultant *lassi* is Rs 25,000 million.

For most uses, the wholesome flavour of *ghee* is its chief attraction. It is served in melted form and mixed with rice or lightly spread on *chappatis*. It is widely used for shallow frying and deep frying of food. Innumerable Indian sweetmeats based on cereals, milk solids, fruits and vegetables are cooked, by preference, in ghee.

Butter-milk is a by-product in the preparation of *makkhan*. It is estimated that about 55 kg of butter-milk is produced for every kg of ghee. While most of it is consumed by villagers and their families, some quantity is either given away or fed to cattle. The reason for this is lack of any market for it in rural areas. Butter-milk is rich in milk protein and calcium, and forms a nutritive and refreshing beverage.

Nutritional benefits: Makkhan and ghee contribute as much as one-third of the fat in the Indian diet. Ghee is produced mainly for consumption directly as food and as an ingredient of food preparations including sweets. Over the centuries, people have cultivated a liking for the aroma and flavour of ghee and prefer it over vegetable oils, the other traditional cooking medium. The vegetarian habits of many Indians preclude from their diet animal fats such as tallow or lard, used in the West.

Thus, ghee forms an important source of fat in the vegetarian diet. *Ghee* and *makkhan* are important sources of vitamins A, D, E and K. They also contain small amounts of essential fatty acids such as arachidonic and linoleic. Considerable losses of Vitamin A and carotene can occur during cooking, the latter being more rapid. Below 125°C, Vitamin A is fairly stable, but above this temperature it is rapidly destroyed. Some 10-20 per cent of carotene is lost during the normal cooking operations.

Khoa and Chhana

One major milk product in common use is *khoa*, obtained by rapidly evaporating milk in shallow pans to a total solids of about 70 per cent and capable of being preserved as such for several days. It is used as an ingredient in making different kinds of traditional mithais (sweets) such as *peda, burfi* and *gulabjamun*. Some 900,000 tonnes of khoa valued at Rs 45,000 million is produced in the country.

Yet another milk product of significance is *chhana*, a product of acid coagulation of hot milk and draining out of whey. It is used in preparing different kinds of sweets such as rasagollas. As they are especially popular in the Eastern region, they are called Bengali sweets. Approximately 1,200,000 tonnes of chhana, valued at Rs 6,000 million is produced in India.

The value of khoa and chhana produced is probably twice the value of all milk handled by the organized sector in the country. The traditional dairy products sector in India, like its agricultural counterpart, is grossly undermanaged. It, however, provides economic opportunities that even the Western dairy world would be envious of. The value of khoa and chhana-based sweets could possibly exceed Rs 130,000 million. India's high-value, high-volume market for traditional dairy products and delicacies is all set to boom further under the technology of mass production. This market is the largest in value after liquid milk and is estimated at US $3 billion in India and US $1 billion overseas. More and more dairy plants in the public, cooperative and private sectors in India are going in for the manufacture of traditional milk products.

This trend will undoubtedly give a further stimulus to the milk consumption in the country and ensure a better price to primary milk producers. Simultaneously, it will also help to productively utilize India's growing milk surplus.

Eggs as Dairy

Eggs are sometimes categorized as dairy, defining dairy as "food that is produced by animals (other than meat)" rather than as milk specifically. The Open Directory Project at one point listed cooking eggs as a subcategory of cooking dairy products. Defining dairy as limited to milk products, however, is more common.

Industrial Product

Cream and Butter

Today, milk is separated by large machines in bulk into cream and skim milk. The cream is processed to produce various consumer products, depending on its thickness, its suitability for culinary uses and consumer demand, which differs from place to place and country to country. Some cream is dried and powdered, some is condensed (by evaporation) mixed with varying amounts of sugar and canned. Most cream from New Zealand and Australian factories is made into butter. This is done by churning the cream until the fat globules coagulate and form a monolithic mass. This butter mass is washed and, sometimes, salted to improve keeping qualities. The residual buttermilk goes on to further processing. The butter is packaged (25 to 50 kg boxes) and chilled for storage and sale. At a later stage these packages are broken down into home-consumption sized packs. Butter sells for about US$3200 a tonne on the international market in 2007 (an unusual high).

Skim Milk

The product left after the cream is removed is called skim, or skimmed, milk. Reacting skim milk with rennet or with an acid makes casein curds from the milk solids in skim milk, with whey as a residual. To make a consumable liquid a portion of cream is returned to the skim milk to make *low fat milk* (semi-skimmed) for human consumption. By varying the amount of cream returned, producers can make a variety of low-fat milks to suit their local market. Other products, such as calcium, vitamin D, and flavouring, are also added to appeal to consumers.

Casein

Casein is the predominant phosphoprotein found in fresh milk. It has a very wide range of uses from being a filler for human foods, such as in ice cream, to the manufacture of products such as fabric, adhesives, and plastics.

Cheese

Cheese is another product made from milk. Whole milk is reacted to form curds that can be compressed, processed and stored to form cheese. In countries where milk is legally allowed to be processed without pasteurisation a wide range of cheeses can be made using the bacteria naturally in the milk. In most other countries, the range of cheeses is smaller and the use of artificial cheese curing is greater. Whey is also the byproduct of this process.

Cheese has historically been an important way of "storing" milk over the year, and carrying over its nutritional value between prosperous years and fallow ones. It is a food product that, with bread and beer, dates back to prehistory in Middle Eastern and European cultures, and like them is subject to innumerable variety and local specificity. Although nowhere near as big as the market for cow's milk cheese, a considerable amount of cheese is made commercially from other milks, especially goat and sheep.

Whey

In earlier times whey was considered to be a waste product and it was, mostly, fed to pigs as a convenient means of disposal. Beginning about 1950, and mostly since about 1980, lactose and many other products, mainly food additives, are made from both casein and cheese whey.

Yogurt

Yoghurt making is a process similar to cheese making; only the process is arrested before the curd becomes very hard.

Milk Powders

Milk is also processed by various drying processes into powders. Whole milk and skim-milk powders for human and animal consumption and buttermilk (the residue from butter-making) powder is used for animal food. The main difference between production of powders for human or for animal consumption is in the protection of the process and the product from contamination. Some people drink milk reconstituted from powdered milk, because milk is about 88 per cent water and it is much cheaper to transport the dried product. Dried skim milk powder is worth about US$5300 a tonne (mid-2007 prices) on the international market.

Other Milk Products

Kumis is produced commercially in Central Asia. Although it is traditionally made from mare's milk, modern industrial variants may use cow's milk instead.

TRANSPORT OF MILK

Historically, the milking and the processing took place in the same place: on a dairy farm. Later, cream was separated from the milk by machine, on the farm, and the cream was transported to a factory for butter making. The skim milk was fed to pigs. This allowed for the high cost of transport (taking the smallest volume high-value product), primitive trucks and the poor quality of roads.

Only farms close to factories could afford to take whole milk, which was essential for cheese making in industrial quantities, to them. The development of refrigeration and better road transport, in the late 1950s, has meant that most farmers milk their cows and only temporarily store the milk in large refrigerated bulk tanks, whence it is later transported by truck to central processing facilities.

MILKING MACHINE

The milking machine is a nearly automatic machine installation for milking cows. It is not a single unit, but rather an assembly of components designed to handle as many as 200 cows an hour. The system consists of the cluster (the assembly that is manually attached to the cow), a milk tube, a pulse tube and pulsator, a vacuum pump or blower, and perhaps a recorder jar or milk meter that measures yield. Together, the system allows milk to flow into a pipeline in preparation for shipping to a processing plant.

The cluster consists of teatcups, a shell and liner device that actually performs the milking action, and a claw or manifold that spaces the teatcups and connects them to the milk and pulse tubes. The milk tube carries the milk and air mixture away from the cow's udder to receiving tanks. The pulse tube, or airline, carries the varying air pressure from the pulsator device to the tanks, drawing the milk and fluids out of the cows as well. In operation, milk is drawn from the cow's teats because a vacuum is created within the cup device, forcing the milk through the teat canal. The pulsator alternates the pressure, first creating a vacuum (milk phase), and then applying air, which causes the flexible liner in the cup to collapse and massage the teat (rest phase). The alternating process of milk-and-rest is continued in a rhythmic pattern for the cows' health and good milk productivity.

Early attempts at milking cows involved a variety of methods. Around 380 B.C., Egyptians, along with traditional milking-by-hand, inserted wheat straws into cows' teats. Suction was first used as a basis for the mechanized harvesting of milk in 1851, although the attempts were not altogether successful, drawing too much blood and body fluid congestion within the teat.

To encourage further innovations, the Royal Agricultural Society of England offered money for a safe, working milking machine. Around the 1890s Alexander Shiels of Glasgow, Scotland, developed a pulsator that alternated suction levels to successfully massage the blood and fluids out of the teat for proper blood circulation. That device, along with the development of a double-chambered teatcup in 1892, led to milking machines replacing hand milking. After the 1920s machine milking became firmly established in the dairy industry. Today, the majority of all milking is processed by machine.

THE MANUFACTURING PROCESS

The milking machine components are created and assembled in several major manufacturing plants throughout the world using traditional processes and procedures. Stainless steel and plastic are used for containers and liners and cast iron and steel for vacuum pumps, controls, and metering devices.

Receiving

Here stainless steel is received in large sheet or tube form. Stainless steel is used to fabricate components that will come in contact with milk. The sheets are protected from scratches by a vinyl lining, which will be removed later after forming and machining. Milking machine components are created using stainless steel and plastic for containers and liners, and cast iron and steel for vacuum pumps, controls, and metering devices. At this point, the stainless steel from the foundry has a dull finish.

Cutting, Machining and Forming

The steel sheets may be sent directly to a polishing station where large machines create a smooth finish. Once the initial finish is achieved, the sheets may move on rollers to numerically controlled punch machines, where they are cut into shapes for various parts. From there, some parts are formed or bent into shape on large brakes. Mounting panels for controls, structural components, and small detail parts for the claws and pumps are made at this stage.

Creating the Shell and Claws

The shell that contains the liner is fabricated as an extrusion. This process involves forcing steel tubing over a die under heat and pressure to form the elongated rigid piece. The claw is made as a series of smaller extruded pieces of tubing that form the manifold, which spaces the teatcups in a cluster formation. The extrusions are then manually welded together in fixtures according to the desired size. Both the completed claws and shell have a dull finish that will require polishing.

Making the Receivers

Receiver tanks are stainless steel tanks that receive milk from the milk lines, generally holding from 15-26 gallons (60-100 1). Although some are customised plastic, most are steel with the heads or ends spun on specialty machines. This production technique shapes a thin steel disc as it is being turned in a lathe. The disc is shaped as it is forced over a steel shape or mandrel.

Once the operator forms the ends of the receiver as cups, another technician will weld the body to the head, leaving orifices or openings for milk input and output. Individually manufactured, the receivers will also be polished by hand. Some receivers will have translucent plastic panel inserts so that dairy farmers can visually gauge the cow's milk production.

Polishing

All major stainless steel components are polished to the familiar finish associated with food-handling equipment. The manufacturers use a variety of mechanized belts, cloths, and spinning wheels in what becomes a very labour intensive process to meet government and industry sanitary standards.

This polishing is in addition to the polishing the large stainless sheets undergo following receiving. Workers wield an assortment of hand polishers and attachments to shine all contact and protective surfaces, from the claws to receivers to pipelines.

Making the Vacuum Pumps or Blowers

These "drivers" of air and fluids through the system are manufactured by select vendors. The unit is basically a sealed iron case with a set of timing gears inside and an impeller, resembling two blades or paddles on a shaft, that spins at over 3,000 revolutions per minute to create a vacuum that will draw fluids in the line. The making of the pumps requires raw castings to be poured for the various parts. The rough parts are hand-machined on a bench by technicians and then placed on a CNC rotating cutter table where the cutting stone makes two precise passes along the pre-determined edges to cut away any rough surfaces.

The assembly of the major components—along with seals, bearings, and shafts—is critical because a variation of just 0.002 inch can change the balance and affect the pump. The components are placed on a coordinate measuring machine where the profile is checked so that all parts fit properly. The technicians, using feeler gauges, will set proper clearances before locking the critical gear parts in place and bolting the unit together. Then they will run the pumps in a test booth before releasing them for distribution.

Assembly

The manufactured and purchased parts are placed in several different areas where the milking machine subassemblies come together. Technicians will select the parts according to the system design specifications, often customised to meet certain operations. It is at this stage that pressures and loads will gauge equipment performance. Final assembly of all machine components will not occur until after shipment to the farm and installation in bams, where often elaborate stalls and stations will be used to maximize the milk harvest.

Workers at assembly will also make initial line connections from valves to pumps to meters, checking for tolerances and poor fits. Vacuum pumps or blowers, the heart of the milking system, are tested so that both milk and accompanying air can be efficiently removed. The pumps will be tested under simulated loads.

Vacuum controllers, which admit the necessary air to maintain the proper vacuum level, are selected. The air lines and milk lines are selected to size and connection hardware grouped. Controls, also purchased, are prepared for installation.

STANDARDS

Milking machine manufacturers are subject to a variety of standards, some self-imposed. In addition to inspections throughout the manufacturing process, all installations are set up by trained dealers and electrical contractors. Equipment designers follow Association of Agricultural Engineer standards and sanitary guidelines established by a dairy industry council.

The Future

Advances in technology have introduced several new innovations to milking machines. Automatic detacher units that connect loosely to the milking claw allow cows to move and shift freely during milking. Based on the rate of milk flow, the detacher can also detect the end of milking, shutting the vacuum and actually removing the claw from the cow.

Automatic backflushing units are also gaining popularity. These units and systems send chemical and rinse solutions through pipelines and clusters to reduce the risk of infection and mastitis (udder inflammation).

The use of automatic identification systems, such as electronic transponder cow neck-tags, have enabled dairy farmers to keep track of milk production by individual cows.

Robotics are at the forefront of milking technology, especially in Europe. Automatic attachment devices have been created but not quite perfected. This new innovation will require little manual labour, and the machines will oversee much of the milking process from the time a cow enters a milking centre until it leaves to graze.

Milking Shed Layouts

Bail-style sheds: This type of milking facility was the first development, after open-paddock milking, for many farmers. The building was a long, narrow, *lean-to* shed that was open along one long side. The cows were held in a yard at the open side and when they were about to be milked they were positioned in one of the bails (stalls). Usually the cows were restrained in the bail with a breech chain and a rope to restrain the outer back leg.

The cow could not move about excessively and the milker could expect not to be kicked or trampled while sitting on a stool and milking into a bucket. When each cow was finished she backed out into the yard again. As herd sizes increased a door was set into the front of each bail so that when the milking was done for any cow the milker could, after undoing the leg-rope and with a remote link, open the door and allow her to exit to the pasture. The door was closed; the next cow walked into the bail and was secured. When milking machines were introduced bails were set in pairs so that a cow was being milked in one paired bail while the other could be prepared for milking.

When one was finished the machine's cups are swapped to the other cow. This is the same as for *Swingover Milking Parlours* as described below except that the cups are loaded on the udder from the side. As herd numbers increased it was easier to double-up the cup-sets and milk both cows simultaneously than to increase the number of bails. About 50 cows an hour can be milked in a shed with 8 bales by one person.

Herringbone Milking Parlours: In herringbone milking sheds, or parlours, cows enter, in single file, and line up almost perpendicular to the central aisle of the milking parlour on both sides of a central pit in which the milker works. After washing the udder and teats the cups of the milking machine are applied to the cows, from the rear of their hind legs, on both sides of the working area. Large herringbone sheds can milk up to 600 cows efficiently with two people.

Swingover Milking Parlours: Swingover parlours are the same as herringbone parlours except they have only one set of milking cups to be shared between the two rows of cows, as one side is being milked the cows on the other side are moved out and replaced with unmilked ones. The advantage of this system is that it is less costly to equip, however it operates at slightly better than half-speed and one would not normally try to milk more than about 100 cows with one person.

Rotary Milking Sheds: Rotary milking sheds consist of a turntable with about 12 to 100 individual stalls for cows around the outer edge. A "good" rotary will be operated with 24–32 stalls by one (two) milkers. The turntable is turned by an electric-motor drive at a rate that one turn is the time for a cow to be milked completely. As an empty stall passes the entrance a cow steps on, facing the centre, and rotates with the turntable. The next cow moves into the next vacant stall and so on.

The operator, or milker, cleans the teats, attaches the cups and does any other feeding or whatever husbanding operations that is necessary. Cows are milked as the platform rotates. The milker, or an automatic device, removes the milking machine cups and the cow backs out and leaves at an exit just before the entrance. The rotary system is capable of milking very large herds—over a thousand cows.

Automatic Milking sheds: Automatic milking or 'robotic milking' sheds can be seen in many European countries. Current automatic milking sheds use the voluntary milking (VM) method. These allow the cows to voluntarily present themselves for milking at any time of the day or night, although repeat visits may be limited by the farmer through computer software. A robot arm is used to clean teats and apply milking equipment, while automated gates direct cow traffic, eliminating the need for the farmer to be present during the process. The entire process is computer controlled. There is a description of an automatic system here—

Supplementary accessories in sheds: Farmers soon realised that a milking shed was a good place to feed cows supplementary foods that overcame local dietary deficiencies or added to the cows' wellbeing and production. Each bail might have a box into which such feed is delivered as the cow arrives so that she is eating while being milked. A computer can read the eartag of each beast to ration the correct individual supplement.

The holding yard at the entrance of the shed is important as a means of keeping cows moving into the shed. Most yards have a powered gate that ensures that the cows are kept close to the shed. Water is a vital commodity on a dairy farm: cows drink about 20 gallons (80 litres) a day, sheds need water to cool and clean them. Pumps and reservoirs are common at milking facilities.

USE OF HORMONES IN ANIMAL PRODUCTION

A reading of many popular, and even official, papers reveals that different types of anabolic agents, whatever their structure, are grouped together under the general name of hormones, among them those whose activity as oestrogens is emphasized and those which reinforce masculine traits. Published material and research results make it possible to distinguish three types of anabolic agents. The first type includes those not found in mammals and birds but obtained exclusively through organic synthesis: diethylstilboestrol (DES) or stilboestrol, dienoestrol, hexoestrol and their derivatives. These compounds, active when administered orally or as implants, are harmful for laboratory animals, whether given directly or as residues in the flesh and offal of treated animals. The use of these substances, a potential danger to human health, is prohibited in many countries, particularly since the methods of measuring DES levels in meat are unreliable, so that measurements must be made on the excreta. The second type of anabolic agents consists of the natural steroids, normally present in all animals: testosterone, oestradiol, oestrone, progesterone. Secreted by the endocrine glands, these substances and their derivatives, although absorbed by the intestines, are deactivated in the liver. They produce measurable effects only when implanted.

The third type, found in the higher plants and moulds, consists of substances with a more or less clear oestrogenic and, at times, anabolic activity. Among these phytoestrogens, only zearalenone has been isolated for use in animal production: as zeranol, it gives good results, but has an effect on genital formation and the level of thyroxine in plasma. It should be added that Sharaf and Gomaa attribute oestrogenic effects to Vitamins E, B_6, C and perhaps A, while Nelson *et al.* have studied the same effects in O-p'-DDT. This paper discusses only the consequences for the consumer of the use in animal production of the first two types: the synthetic agents such as DES and the natural steroids oestradiol, progesterone and testosterone. First mention should be made of cases of cancer of the vagina and hypogonadism observed in the offspring of women treated with DES.

Such observations are verified by research based on the so-called relay methodology, based on the fact that methods of analysis for a residue and its metabolites fail to reflect practical realities, since they may lead to findings of residues without bioavailability or *vice versa*. Further, the individual biological evaluation of each metabolite fails to reflect the overall consumption of all of them. The relay, a farm animal, changes this situation. In order to determine whether this was true of the first two types of anabolic agents, rats and mice of both sexes received balanced diets containing 20% of the flesh or 6% of the liver of male and female calves implanted with DES (24 mg in each of two implants), oestradiol and progesterone (20 + 200 mg in two doses) or oestradiol + testosterone (20 + 200 mg in two doses). Animals were implanted at the beginning of the growth period and 60 days later, 38 days prior to slaughter. Chemical measurement of residues showed DES levels of 30 g/kg in the flesh, but none in the liver. Levels of natural steroid hormones did not exceed levels in untreated animals. Feeding rats and mice on the flesh of DES-implanted calves led to growth in both sexes, sterilized the females and sharply atrophied the testicles, seminal ducts and penis of the males. Liver behaviour was less clear. Pokrovski *et al.* and Nesterin obtained similar results with flesh. It was also observed that DES persisted for nearly 40 days in the environment and was found in alfalfa.

Animals fed on the flesh and liver of calves implanted with oestradiol + progesterone or oestradiol + testosterone behaved in the same manner as controls fed on the flesh and liver of control calves. After 24 months, the percentage of tumors (primarily mammary adenomata, frequent in old Wistar rats) was the same in both groups. Trenbolone acetate (TBA), a testosterone derivative, leaves little or no residue. Measured by radio-immunoassay, Gropp *et al*. also studied their biological effects by the relay toxicity method, on Sprague-Dawley rats fed on the flesh of calves implanted with a mixture of 20 mg oestradiol-17β + 140 mg TBA, as well as with doses 10 and 25 times higher. The results of this study, which continued for 110 weeks and covered two generations, showed the absence of any unfavourable influence on growth, organ weight, reproduction, teratogenicity and cancerogenicity. Only females fed on the flesh of control calves to which TBA had been added in doses 25 times normal, had significantly lower growth($P < 0.05$). The biochemical constants in the blood were the same in both groups.

These experiments show that compounds of natural steroid male and female hormones, and their derivatives, appear to be without danger for the consumer. They are useful in animal production, obviating a number of often costly and polluting therapeutic operations. On the contrary, the use of DES-type non-steroid anabolic agents has serious disadvantages. It is difficult to detect their residues in meat. Relay toxicity results show that analyses and biological behaviour fail to agree. The indiscriminate prohibition of all types of anabolic agents may be leading to the use of DES and similar substances. Analyses of urine in which they can be identified most easily, have revealed them in 20% of cases. In the United States, McClung estimates that 500 000 bovines are being treated illegally. In Europe, illegal use appears to be fairly general, but is tending to decrease. The use of DES and similar substances should thus be prohibited, while that of the steroids and their derivatives should be authorized under veterinary control. This was the view of many writers as early as 1967, and their conclusions were confirmed by the FAO/WHO Joint Committee of 1975 and a recent international symposium. Any new anabolic agent should be severely tested before it is distributed widely. There is a need to develop practical, economic and rapid — but very exact — methods of detecting residues, particularly those originating in products chemically different from the natural steroid hormones and their derivatives.

EXCESSIVE USE OF HORMONES IN ANIMAL HUSBANDRY

Hormones which are intended for occasional therapeutic use are injected into young animals such as calves to fatten them faster. Like athletes taking steroids, this practice results in "fast flesh", but the side effects are unclear. Some single out specific hormone levels and combinations as dangerous to human consumers, while others insist that any hormone additive is potentially dangerous. Practically, it is an open question whether beef raised with no hormones at all can compete with the quality and market price of beef raised with some hormones, although consumers are now reacting against the flabby, tasteless quality of "hormone-plumped up" meat. The practice of using hormones to improve the growth rates of animals and reduce the fat content of the meat became fashionable during the 1970s, when farmers found they could cut costs as well as satisfy consumer demand for leaner meat. Bovine stromatotrophin (BST) is a synthetic, genetically engineered substance based on a naturally occurring bovine growth hormone and injected into dairy cows to increase production of milk. The hormone has been widely criticized for its detrimental effects on animal welfare and could pose a health hazard to those who drink the milk. Whilst widely used in the USA, Canadian and European resistance to the use of the hormone hinges in part on the failure to agree on an international standard for use. The sale and marketing of BST has been banned in the European Union since 1990. The moratorium expired at the end of 1999 but a renewal was requested from the EU Council of Ministers. Milk from cows treated with recombinant bovine growth hormone (rBGH in the US, rBST in Europe) has significantly elevated levels of IGF-I (insulin-like growth factor-I), a growth factor responsible for a wide variety of biological functions. In addition, the IGF-I in the milk of rBGH-treated cows is potentially more bioactive than the naturally occurring form, and this bioactivity may be increased further by pasteurization.

Children may be particularly vulnerable to adverse effects, because of their rapid growth rate and their more permeable intestinal tract. In addition, other segments of society may also be at an increased risk, such as those with celiac disease, Crohn's disease, autism, cirrhosis, cow's milk allergy and people taking certain medications The hormones 17β-oestradiol, progesterone, testosterone, zeranol, trenbolone and melengestrol acetate (MGA) promote growth in cattle. According to a 1999 European Commission report, these hormones have adverse developmental, neurobiological, genotoxic and carcinogenic effects on people who consume them, either via the parent compound, or via metabolites. 17β-oestradiol both initiates and promotes cancerous tumours. Prepubertal children are at the greatest risk. Growth hormones have been banned in the European Union since 1988.

However, many banned substances continue to enter from South America and eastern Europe. Organized crime controls the traffic throughout Europe with production sources through chemical companies and the pharmaceutical industry. Hormones banned in the EEC/EU are held by some farmers to be necessary to the production of high-quality beef at competitive prices. At a time of beef surplus, the EEC/EU banned all hormones, reputedly to discourage farmers from producing beef. There is not a fundamental conviction among farmers in the EEC/EU that the total ban is necessary or economically viable, and this makes the ban difficult to enforce, however needed it may be for public health reasons. In 1999, the USA was using threat of prohibitive tariff protection on European imports if Europe does not drop its restrictions on hormone treated beef exports from the USA. It called the arguments against use of hormones "phony science" Stilbene oestrogens used in animal production as anabolic agents are orally active, persist in food, pose environmental problems because of their low biodegradability. Diethyl stilbesterol (DES) is a known carcinogen. On reaching adolescence, the children of mothers who had taken DES in pregnancy were developing cancer, the girls in the vaginal tract and the boys in the prostate, bladder or testicles. Sometimes the cancer showed as early as seven, sometimes as late as the mid-twenties. Abnormalities showed in 1978 in Milan in children who were fed veal which had been treated with DES. Knowledge of the toxicology of many other used substances is still rudimentary

- Some synthetic hormones, such as diethyl stilbenes, are notorious cancer causing agents.
- Under European law, substances can be administered to livestock only for veterinary purposes or on grounds of hygiene, unless they are not harmful to their health or welfare. This does not seem to be the case as regards BST.
- There is no documented case where use of antibiotics in livestock has been proved to cause treatment failure in humans. Without growth promoters Europe would need more land to grow animal feed, be swamped by an extra 7 million cubic meters of waste manure each year, and face additional emissions of up a third of methane, nitrogen, phosphorus and ammonia, because growth promoters make animals' dietary systems more efficient.

ALTERNATIVES TO THE USE OF HORMONES

Growth rates are influenced by many factors, especially genetic constitution and feeding. Over time, selection as well as improvements in management systems, feed composition and feeding programmes have contributed much to increasing productivity in meat as well as milk.

Although it is difficult to evaluate the exact relative contributions of these factors, the overall improvements have been dramatic. An example is the increase in milk yield per head in US dairy cattle. In the period 1944–1975, the number of dairy cows decreased by 33%, while the average yield per cow increased by 60%.

These gains represented a saving of about 23 billion kg of total digestible nitrogen per year, the volume of milk produced remaining relatively constant. The saving is equivalent to about 1.1 billion bushels of maize. Data illustrating progress in beef production over the years are scarce, but increases in productivity similar to those for milk production are unlikely.

In addition to the use of hormones, many avenues are still open for increasing productivity in meat and milk production, including breeding programmes, regulation of rumen fermentation, optimalization of the balance between the indirect and direct feeding of the ruminant organism proper, and disease control.

BREEDING PROGRAMMES

Systematic selection of high-quality sires, combined with an increase in the number of offspring from high-yielding females through embryo transfer, may bring about further improvements in beef and milk production. In many countries, development along these lines has hardly begun.

However, the establishment of effective breeding associations and the strict organization of programme planning and execution are prerequisites for realizing the potentials in this sector.

REGULATION OF RUMEN FERMENTATION

The microbial systems in the rumen are extremely complex, and the balance between the various strains of bacteria is susceptible to changes brought about by many factors. The recent introduction of substances such as monensin offers great promise in altering the fermentation pattern to the benefit of productivity by increasing FCE. Since the very extensive breakdown of carbohydrates and protein represents loss of much energy, research is currently being conducted in many laboratories in order to find new methods of increasing FCE.

OPTIMALIZATION

To a large extent, feeding a ruminant means feeding the rumen microbes which then themselves serve as feed for the organism proper. This is indirect feeding, expensive in energy. On the other hand, the ruminant possesses, in the postruminal part of its digestive tract, all the enzymes necessary for utilizing all types of nutrients except cellulose. The rumen microbes are necessary for the utilization of cellulose, which globally represents an enormous source of energy.

However, it is possible to sustain an adequate microbial population in the rumen even when ruminal breakdown of part of the easily digestible nutrients is prevented.

Enabling nutrients to bypass the rumen will increase the utilization of feed for production, and also create a more adequate supply of amino acids. Increased rumen bypass of nutrients can currently be brought about by several means, including formaldehyde and heat treatment of protein-rich feeds. A third method, aiming more at specific substances that may be rate-limiting for production (*e.g.* certain amino acids), or of significance in treatment of diseases, is protection against rumen degradation by such means as incorporation into the ration of long-chain fatty acid mixtures in the form of small pellets. In the future, new methods of increasing rumen bypass will undoubtedly contribute significantly to increased productivity of ruminants.

DISEASE CONTROL

Whatever management system is adopted, effective disease control is essential for productivity. In many areas of the world, infectious and parasitic diseases inflict heavy losses on animal production. A recent study has disclosed nearly a one-to-one relationship between investment in agricultural research and annual productivity of edible protein in ruminants. An increase of about 45% in scientist/man years and a corresponding increase in funding for research and development is considered sufficient to raise productivity in this sector by 50%. Investment in disease control is an important aspect of this work. Annual world mortality losses from disease exceed 50 million cattle and buffalo, and 100 million sheep and goats. Non-lethal diseases are believed to lead to an equivalent reduction in production. Thus, investment in disease control holds great promise for future augmentation of animal protein production.

In these perspectives, the significance of hormones in animal production may seem marginal, leading to the question of what priority to give to the various efforts to increase productivity and production. In the global context it is, however, at least at present, impossible to adopt one approach to the exclusion of others. As long as preparations exist that combine positive effects on yield and feed utilization with low or non-existing risk to the consumer, there will be a market for them.

What is more, the use of hormonally-active substances in the future may not be limited to those currently available. Common to the present compounds, natural or synthetic, is that they are degraded in the body only to a limited extent. An entirely different situation exists for proteid hormones, which are broken down completely to amino acids, leaving no residues whatever. An example is the growth hormone which not only stimulates growth but also milk secretion, even in high-yielding cows. This anabolic hormone is currently available only in small quantities for research. However, a recent breakthrough in the use of recombinant DNA technique has made large-scale microbial production of species-specific peptide hormones a realistic possibility. Combined with the development of miniaturized automatic delivery systems for subcutaneous use, a new era may be visualized as regards the use of hormones in animal production.

HORMONE PREPARATIONS

HORMONES OF ENDOGENOUS ORIGIN

These comprise the "classical" steroid sex hormones, oestradiol-17β, testosterone and progesterone. The two former are used either in the free form or as esters, mainly those of propionic or benzoic acid. Esterification generally causes prolongation of the half-life of the compounds in the body by 40 to 50%. The natural hormones having low bioavailability when administered orally, owing to rapid conjugation and metabolic transformation in the liver, they are therefore administered by subcutaneous implantation.

HORMONES OF EXOGENOUS ORIGIN

Of the oestrogens, the stilbene derivatives diethylstilboestrol (DES) and hexoestrol possess high biological activity and have been used most widely. They are active orally as well as by implantation. Other orally active oestrogens include ethynyl-oestradiol, a more slowly metabolized derivative of the true hormone, with higher activity.

An oestrogen with an entirely different structure is zeranol, a derivative of a resorcylic acid lactone occurring in the fungus Giberella zeae. The synthetic androgens comprise a large number of substances, most of which are steroids. Of these, trenbolone acetate (TBA) possesses strong anabolic properties and has received much attention during recent years, used alone or in combination with an oestrogen. Another anabolic steroid is methyl-testosterone. Of synthetic gestagens, only one will be mentioned here: melengestrol acetate, which stimulates growth in heifers but not in steers, and which can also be used for the suppression of oestrus. Numerous other gestagens also exist, but at present few other than progesterone and melengestrol acetate are used to stimulate growth.

In addition to these substances, numerous others exist, and some of them are used more or less frequently in clinical veterinary medicine. However, clinical applications of hormones are not considered to be of consequence to the consumer, since such treatment is much less frequent than the use of hormones to promote growth.

Hormone preparations in current use as growth stimulants are listed in Table, which also shows modes of application, dosages, etc. It will be noted that almost all preparations currently in use are based on implantation, the site usually being the base of the ear, or less frequently, the dewlap.

RANGE OF APPLICATION

In cattle the use of hormones is limited to veal calves and beef cattle. Veal calves are produced mainly in continental Europe, to an extent of about 8 million per year.

Research has demonstrated that hormone treatment improves growth rate, nitrogen retention and FCE during the five- to six-week period before slaughter. Beef cattle, including steers as well as heifers, were treated in large numbers, especially in the USA and the UK, with DES or hexoestrol, administered orally, until the use of these compounds was restricted. During the last several years, practice has changed dramatically in the direction of increased use of implants of natural steroids, synthetic anabolic steroids and the phyto-oestrogen zeranol.

In sheep, especially in wether lambs, some increase in gain has been reported (11), but results are somewhat ambiguous. In swine, hormone treatment may increase growth rate, FCE and lean/fat ratio of the carcass in male castrates. Poultry generally do not appear to respond to oestrogens by increased gain but by changes in lipid deposition.

In male and female turkeys, androgens have recently been reported to increase growth rate as well as FCE . When DES was used as a feed additive, a usual procedure was to start treatment of steers at a body weight of 360 kg and continue administration for 120 to 170 days. Since restrictions on its use were imposed, most preparations have been administered as implants, whose effect is usually limited to 80 to 100 days. Practice varies with management systems. Animals may be implanted at live weights from 270 to 450 kg. Depending upon the age and weight at the time of implantation, the animals are either slaughtered at the end of this first period, or fed for an additional period, either without further treatment or after a second implant to act for another 80 to 100 days.

Most types of implants in use are not removable, but removable types have recently been tested and their effects described. When tested in steers, no reduction in performance was recorded when the implants were withdrawn 32 and 39 days before slaughter. Implantation is subcutaneous, usually at the base of the ear, thus eliminating the risk that residues of the implantation site will be present in edible tissue.

EFFECTS OF HORMONES

VEAL CALVES

In veal calves, hormone treatment may begin at a body weight of about 65 kg, the animals being slaughtered at about 170 kg. Implants of 20 mg oestradiol-17β + 200 mg progesterone in males and 20 mg oestradiol-17β + 200 mg testosterone in females resulted in a 20% increase in daily gain and 21% higher nitrogen retention in the period studied . In other studies, improvements of 10 to 12% in gain and 10% in FCE have been reported.

Nitrogen retention is about 70% in the very young veal calf, but decreases gradually to below 40% at the age of about 15 weeks. For ages of 10 to 15 weeks, the average conversion of feed protein to body protein is about 40%; this rate can be increased to about 60% by hormone treatment.

The effective preparations were DES, oestradiol-17β, and the combination of TBA + oestradiol-17β. More recently, positive effects have been reported for zeranol alone (36 mg) and for zeranol (36 mg) + TBA (140 mg), with increases in nitrogen retention of the same order as for DES and E_2 + TBA. When zeranol + TBA was implanted at the age of 56 days, the growth rate up to day 106 increased by 18% .

STEERS

The most extensive studies of the effects of hormones on growth and FCE have been carried out on steers, under strictly controlled conditions as well as in the field. Since 1975, most studies have involved implants of oestrogens alone, androgens alone, or combined oestrogen/androgen preparations, although many trials have also been based on oestrogen/progesterone combinations during recent years. Oestrogen implants have included DES, hexoestrol, oestradiol-17β and zeranol. DES implants have, as in previous studies, resulted in an increase of about 12% in gain and in improvement in FCE of the order of 10%. Hexoestrol implants, usually in doses of 30 to 60 mg, have been shown in numerous experiments to lead to considerable improvement in growth rate and FCE. In 19 trials carried out over the years on experimental husbandry farms in the UK, the overall average increase in gain produced by 45 or 60 mg hexoestrol implants was 0.16 kg a day, and in only 2 of the trials was it less than half that figure. Oestradiol - 17β implants alone (30 mg) have resulted in a 24% increase in gain and a 13% improvement in FCE.

Zeranol implants, usually at 36 mg, have consistently improved gain as well as FCE. In a series of 21 UK trials over several years, the average response to zeranol implants alone was an increase in daily gain of 0.15 kg. Only in one trial was there no response . Similar results have been obtained in Ireland. Positive effects on gain in steers have been observed under a variety of experimental conditions, under controlled feeding, on ad lib feeding of standardized ratios, and on pasture.

TBA implants administered alone at a dose of 300 mg have also had positive effects on growth, even if combination with an oestrogen has yielded better responses (vide infra). In a series of 8 trials in the UK, the average additional daily gain amounted to 0.09 kg, with considerable variation among trials . Similar results have been reported from Ireland. Combined preparations. A number of trials have been carried out with implants containing two hormones.

The combination of an oestrogen with an anabolic steroid, or with progesterone, has met with the greatest responses. Synovex-S has consistently increased gain as well as FCE, with responses averaging about 20% and 17% respectively. Hexoestrol + TBA (usually 30 or 45 mg hexoestrol + 300 mg TBA) has resulted in marked increases in gain, of the order of 30% and in FCE of the order of 20%. Oestradiol-17β + TBA (20/140 mg) has given similar results, as has Zeranol + TBA (36/300 mg), also recently tested.

Hormone preparations have also been tested in combination with substances such as monensin, which increase FCE by promoting propionic acid formation in the rumen. Results have varied from no effect to marked additional gain. Reimplantation, tested under various forms of management with varying results, has not gained general acceptance. Lamming has stated that "repeat implantation of hormone is not likely to produce the benefits obtained from its initial use, since a second dramatic change in the endocrine balance of the animal is not likely to occur. In addition, double implantation increases the possibility of exceeding the optimum dose rate and the chance of deleterious side effects occurring." The evidence for highly significant positive effects on the growth rate and FCE of steers is thus beyond dispute, the most marked effects being provoked by implants combining an oestrogen with an androgen of high anabolic activity.

BULLS

Since the entire male animal produces its own anabolic androgen, testosterone, an effect of additional hormones similar to that for steers is not to be expected. The number of trials with bulls is also limited. Positive effects on gain have been reported using DES alone and combined oestrogen/TBA implants; in other studies, no effect on gain has been recorded, while a certain increase in the deposition of fat in the carcass has been observed.

HEIFERS

Recent trials with beef-producing heifers have mostly been based on the use of an androgen, although oestrogens have been tested, alone or in combination. Thus, zeranol has been reported to increase gain, while in other trials no response has been observed. TBA administered alone (300 mg) has led to increases in weight gain and FCE of the order of 36% and 25% respectively.

In other trials, combinations of an oestrogen with TBA or testosterone have yielded significant growth responses. In general, it appears that the effect of TBA alone in heifers corresponds closely to the effect of combined oestrogen/TBA implants in steers.

SHEEP

Trials have mainly concerned wether lambs, and positive effects of hormonal treatment have been reported using DES, hexoestrol and zeranol, although other reports have indicated that zeranol yields no significant effects (74, 75). Wether lambs implanted with TBA + oestradiol-17β have shown increases in gain, carcass weight and FCE. In general, however, the results obtained in sheep thus far do not warrant the same clear-cut conclusions as for steers and heifers.

SWINE AND POULTRY

There is little evidence that existing hormonal preparations influence the growth rate and FCE to an extent that would be interesting from a practical point of view. The lean/fat ratio in male castrate and female pig carcasses may be increased by the use of oestrogen/androgen combinations. In poultry, redistribution of fat in the body is a known effect of oestrogens. Recent research indicates improved growth rate and FCE using androgens in young male and female turkeys.

UNDESIRABLE SIDE EFFECTS IN TREATED ANIMALS

Reported side effects of hormone treatment for growth stimulation are few and generally concern the use of oestrogens in steers. Changes in body conformation such as feminization and raised tail-heads were described as early as 1958. Similarly, bulling has occurred with increased frequency, although in most animals it is limited to the first few days after implantation.

However, it has been reported from Kansas that 2.2% of all steers fed in pens have to be removed, at an estimated loss of $23 per head. In a study of the effect of reimplantation of oestrogens in steers, all animals were given a 30 mg DES implant at a live weight of 260 kg, and then reimplanted 91 days later, with either 30 mg DES or Synovex S. Following the second implant, the frequency of the steer-buller syndrome was 1.65% for the DES-DES group, and 3.36% for the DES-Synovex S group. The economic advantage of using DES + DES was estimated at $1.15 per head. The steer-buller syndrome is a special problem in feedlots.

5

Genetically Engineered Animals and the Protection of Human Health

Numerous transgenic domestic animals have already been created for the expression of proteins in their derived products, such as milk, blood or eggs. However, none of these animals has yet been used commercially, although they represent a cheaper means of expressing proteins than other industrial methods. The resulting substances are subject to specific regulatory procedures and their properties are tested in the same way as substances resulting from more "conventional" technology.

WHAT IS A TRANSGENIC ANIMAL?

There are various definitions for the term transgenic animal. The Federation of European Laboratory Animal Associations defines the term as an animal in which there has been a deliberate modification of its genome, the genetic makeup of an organism responsible for inherited characteristics.

CONTAMINATION OR SPREAD OF A NOVEL PATHOGEN

One of the main problems associated with transgenic animals created for the production of biopharmaceuticals is the theoretical risk of micro-organisms acquiring the transgene. This has however not been observed to date. Another problem is the possible generation of pathogens virulent to humans or animals through the recombination of the transgene vector and nonpathogenic endogenous proviruses in the transgenetic animal.

This can be foreseen for retroviruses that efficient vectors for inserting genes into germline cells and for adenoviruses. It would be possible for many species and has in fact been observed among pigs. There is therefore significant risk of generation of a pathogenic recombinant virus. A similar risk arises with the use of lentiviruses, notably BIV and FIV, as gene vectors. Recombination with a virus such as HIV is however considered highly unlikely because of the large genetic distance that exists between them. However, the use of a lentivirus could also lead to the generation of a pathogenic recombinant.

Disposal of Animals and Prevented Entry of Unwanted Animals into the Food Chain

Animals used for the expression of proteins should not enter the human or animal food chain, at least not without authorization.

Animals that do not contain the product of the transgene would not pose any risk to human health, but should nevertheless be totally segregated in the absence of specific, appropriate regulation.

Transgenic animals without appropriate expression of the desired trait, unwanted male animals and animals at the end of their service life should not be used, advertently or inadvertently, as food or feed.

XENOTRANSPLANTATION

This biotechnology would help solve the sometimes serious problem of insufficient supply of human organs or tissue for transplantation. Trials with the transplantation of pig cells and tissue to humans have been carried out in several countries, including the USA, Russia and those of the European Union.

This technique is thus of particular interest to human health, and extensive research has been–and is being–carried out to overcome the inherent immunological and physiological problems. The only animal currently taken seriously for xenotransplantation is the pig. Primates present serious problems of health safety and ethical concerns.

Introduction of Exogenous Infectious Agents into the Receiving Organism

Xenotransplantation provides a new pathway for the transmission of existing or new human pathogens. Of main concern are viruses, including exogenous viruses and porcine endogenous retroviruses, as bacteria and parasites are generally easy to eliminate.

Some viruses can be eliminated through strict management (vaccination, isolation, close epidemiological surveillance, caesarean section), but comprehensive elimination is rendered difficult by the large number of possible viruses and inadequate methods of virus detection. It is however suggested that such problems equate with those of allotransplantation, where human pathogens can also be transferred during the transplant of human organs or tissue. Given the potential benefits of these emerging techniques, the public would however seem to be willing to accept the attendant risks.

Introduction of Porcine Endogenous Retroviruses

Pigs possess a number of retroviruses in their genome which, depending on their subtype, can affect human cells to varying degrees. No transfer of disease or dissemination of the virus in the tissues have been observed in humans receiving transplants. Nor has any tested porcine endogenous retrovirus caused disease in species involved laboratory trials. It is stressed that the virulence of porcine endogenous retroviruses among mice could be limited by genetic defects. However, these viruses represent a high risk because of their adaptability, variability and resemblance to known human pathogenic viruses. The risk of a nonpathogenic retrovirus becoming virulent is mooted but disputed. Some scientists argue that a series of consecutive events would be needed for this to happen, lessening the chances of such a pathogenic virus emerging. Others call for further information.

Animal biotechnology is a field that is under constant evolution. Many transgenic animals have been successfully created for a variety of purposes, and the prospects are enormous. Many techniques exist for the insertion of genes into host cells.

They all have advantages and drawbacks, and most have limitations linked to the difficulty of controlling the expression of transgenes inserted at random into the genome of cells; the variability of resulting genomes and unexpected side effects; the population of "surplus" animals; the possible mobilization of genetic sequences; and the possible emergence of new pathogens. Maintaining genetic diversity is another problem. With regard to human health, further research is required into the risks associated with introducing animals created for biopharmaceuticals into the food chain and the risks of generating new human diseases from xenotransplantation.

TRANSGENIC ANIMALS

Much has been described about the methods of transfer of genes/DN A fragments/chromosomes into animal cells in different sections of biotechnology either through microinjection, drug delivery system, electroporation or mediated by viruses. Cell manipulation has been aimed at production of novel chemicals, pharmaceutical drugs or improvement in animal breed. The other purpose has been to study the structure and function of genes through molecular markers and genome mapping.

Improvement in livestock has already given good results for increased milk production in catties, increased growth rate of livestock and fish, production of valuable proteins in large quantities in milk, urine and blood of livestock, wool production in transgenic sheep.

Thus the transgenic animals can be used as bioreactors for large scale production of valuable recombinant chemicals such as hormones, interferons, proteins, etc. Thus, manufacturing of recombinant drugs through transgenic animals is called 'molecular farming' or 'molecular pharming'.

STRATEGIES FOR GENE TRANSFER

Desired foreign genes are transferred into animal cells/embryos via virus, microinjection, targeted gene transfer methods, etc. as discussed below:

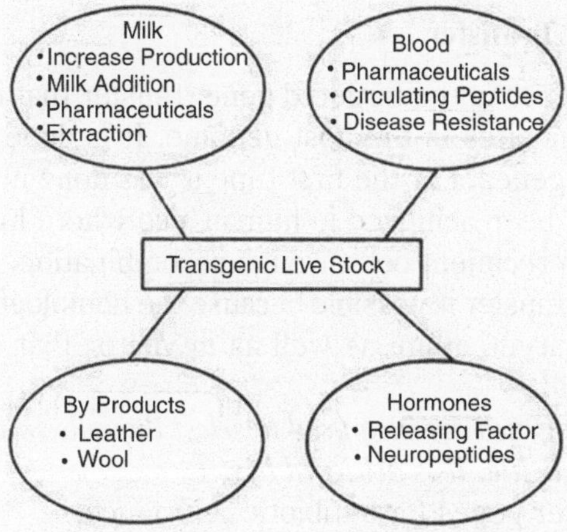

Fig. Areas of Investigation of Domestic Livestock via Vransgenic Technology.

Transfection of Animal Cells/Embryos

During 1970s, attempts were made to produce transgenic animals. The first success was achieved in 1976 when mouse embryos were infected by retrovirus. The retroviruses are supposed to be an efficient vector that transfect the animal cells and deliver its genes leading to production of transgenic cells/ animals. The other viruses used for this purpose are vaccinia virus, adeno-associated virus, herpes virus and bovine papiloma virus.

The transfected cultured mammalian cells have been used for diagnostics of oncogene (*i.e.* cancerous gene) as well as for gene therapy.

The steps for detection of cancer gene follow:

- Isolation of DNA from tumour cell line,
- DNA fragementation into 30-50 kb long pieces through mechanical shearing,
- Dissolution in phosphate buffer followed by precipitation by adding $CaCl_2$,
- Pouring of this solution onto a layer of mouse 3T3 cells; foci of cells developed,
- Use of transfected cells in detection of cancer causing genes.

The other application of transfected cells is in gene therapy. Genes of desired function are inserted in cultured cells. The later is placed in patient's body to rectify the malfunctioning gene.

Transfer Through Microinjection

Microinjection method has also been developed and variously used in production of transgenic animals. So far gene transfer has been successfully carried out in several classes of animals *viz*., fish, birds, insects, mammals, etc.

The Targeted Gene Transfer

The other approach is the targeted gene transfer that involves transfer of genes at homologous sites in the host genome. It is done just to replace the wild type of mutant genes. For the first time it was done in bacteria and yeast. In 1985 success has been achieved in human also where human ?-globin gene was transferred into recipient cell through recombination.

Targeted gene transfer is possible because the homologous DNA sequences arc present at the targeted site as well as in vector that carries the desired gene of foreign origin.

Besides, marker genes are also used to select the cells in which gene has been transferred at targeted site. It is achieved by:

- Using marker genes for antibiotic resistance,
- Hypoxanthine phosphoribosyl transferase (HPRT),
- Polymerase chain reaction.

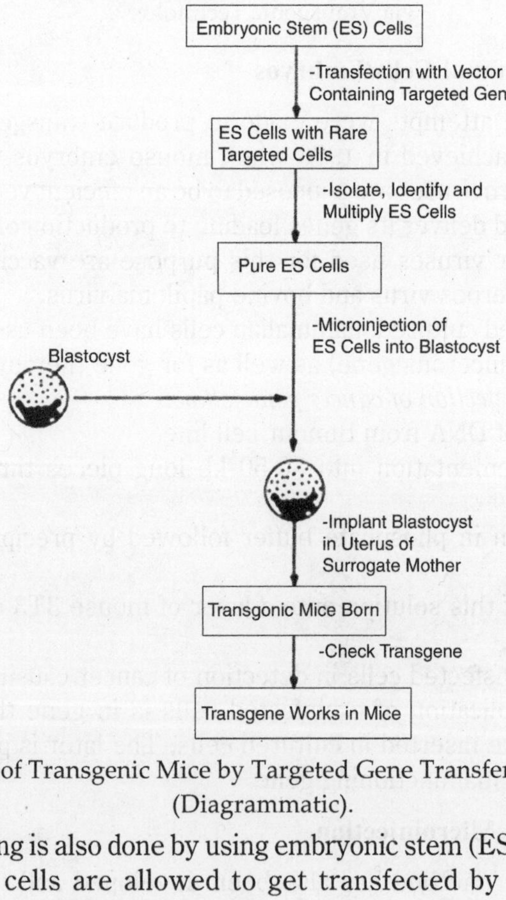

Fig. Production of Transgenic Mice by Targeted Gene Transfer through ES Cells (Diagrammatic).

Gene targeting is also done by using embryonic stem (ES) cells as described earlier. The ES cells are allowed to get transfected by vector containing desirable genes. In transfected cells targeting of gene to specific site by homologous recombination occur. Thereafter, the transfected cells are identified and isolated from bulk. They are multiplied and introduced into blastocyst through microinjection. The blastocyst is transferred into uterus of a surrogate mother for further developmental stages. The animals, which are born, are checked for the presence of transgene. The transgenic animal is crossed with a normal one to study the inheritance of introduced foreign gene. An outline of production of transgenic mice is shown in Figure.

TRANSGENIC MAMMALS

For the first time in 1982, there appeared a report on the transfer of human growth hormone gene of rat fused to the promoter of mouse metallothioneine I gene. It was done by microinjection method. As a result of presence of a novel gene, there has been a drastic increase in body weight of mice. Since then a large number of transgenic mammals and other animals have been produced such as cow, pig, rabbit, goats, sheep, fish, etc. The purpose of production of transgenic animals has been to produce more protein in milk and meat, disease resistance, leaner meat, good quality wool and more specifically improvement in genetic traits. They are also used as bioreactor for molecular farming.

TRANSGENIC SHEEP

So far it is not clear why the rate of transgenesis in sheep is very low *i.e.* 0.1 to 0.2 per cent. It needs improvement by regular check through the biotechnological methods (using PCR, etc). Method of production of transgenic sheep is the same as described for transgenic mice). Due to commercial appeal, the Pharmaceutical Proteins Ltd, Cambridge (UK) provided fund to J.P. Simons for the production of transgenic sheep.

In 1988, Simons reported first production of transgenic sheep. He first produced two transgenic ewe that consisted of about 10 copies of human antihaemophilic factor IX gene in the form of cDNA. It was fused with about 10.5 kb long β-lactoglobulin (BLG) gene. Moreover, the BLG gene is important for the expression of gene in mammary gland. The ewes secreted human alpha-1 antitrypsin (hα-1AT) *i.e.* human factor IX in milk because the gene had tissue specific expression. Inspite of low expression of transgene, the hα-lAT is active. The transgenic ewes were born in summer 1986. Again they were mated at the end of the year 1986. Single lamb was born from each ewe in 1987 that inherited BLG-factor IX transgene. Due to the presence of both the genes factor IX was secreted in milk.

In 1991, Alan Colman and coworkers at Edinburgh produced five transgenic sheep, four female and one male. The transgene was ovine-α-lactoglobulin promoter fused to hα-1AT gene. The concentration of h α-1AT in milk was recorded to about 35 grams per liter. The biological activity of protein derived from milk was the same as that of plasma derived antitrypsin.

TRANSGENIC FISH

Fish are the important source of fat and proteins for humans and the delicious diets are prepared in certain societies. Therefore, demand of good quality fish is increasing gradually. Gene transfer in embryos of several species of fish such as medaka fish, salmon, carp, zebrafish, goldfish, trout and cattlefish has been successfully achieved. The novel desired gene is introduced into early embryo through microinjection because pronuclei are not easily visible. In some fish transgene is microinjected into the nucleus of oocyte. The introduced gene replicates at the time of development of embryo. In most of the fish fertilization is accomplished out side in water. Therefore, embryo of different developmental stages can be collected easily. In those cases where microinjection technique is not successful, transgenes are incorporated into embryos through electroporation technique. Inheritance of transgene occurs in the Mendalian way. Stuart *et al*. found that spermatozoa of transgenic trout and zebrafish containing foreign DNA transmitted the transgene to the next generation. Normal oocyte fertilized with sperms of transgene fish gave rise to offsprings of which 10-50 per cent of Fl offsprings contained foreign genes. The number of transgenic fish increased in F2 generation.

Animal Bioreactors and Molecular Farming

Transgenic animals are used as bioreactor for mass production of drugs and proteins called molecular farming (pharming). It is a new industry. The Transgenic Scien (an American Company) has produced transgenic mice which secreted in milk about 0.5 grams/liter of hGH. Due to expression of transgene, this mice hormone has no adverse effect.

Adverse effect of cattle and pig derived human hormones have been observed. In addition, the pigs that secrete extra hormone, hGH are sterile. This company has started scaling up of production of hGH through transgenic rabbit because of high concentration of protein in milk and short gestation period of rabbit. A transgenic lamb has been produced by the Institute of Animal Physiology and Genetic Research, Edinburgh (UK). This lamb contains alpha-antitrypsin (αAT) gene. In human the deficiency of αAT causes fluid accumulation in lungs resulting in death of patients. The ?AT inhibits the enzyme elastase that digests foreign particles and clears the lungs. In the absence of αAT elastase is not inhibited and, therefore, it digests lung tissue. In lactating milk of transgenic lamb, presence of αAT can be tested. Transgenic sheep can also be used for production of αAT commercially.

APPLICATION OF MOLECULAR GENETICS

Methods have been developed to isolate animal genes and to characterize them. The first approach is to identify DNA sequence associated with economic trait loci.

The second step is to incorporate the identified gene into three maps:

1. A physical map of chromosome (where DNA sequences are assigned to specific sites on specific chromosomes),
2. A linkage map (where linkage of different genes and markers is assigned in the same chromosome),
3. A genetic map (where inheritance of economic traits is corrected with the inheritance of genes and markers).

The unknown and novel genes associated with genetic diseases can also be identified through their expression. Then the protein sequences are identified through amino acid sequencing or injecting the protein into suitable animal (*e.g.* mouse, rabbit, chicken, etc.) to raise the antibodies.

The animal production industry is gradually reaching towards revolution due to the development of detailed linkage and genetic map, more knowledge of expression and regulation of genes and techniques for large propagation. The application of molecular genetics in animals currently being used are: breeding selected traits into livestock, animal cell culture, and production of transgenic animals.

Selected Traits and their Breeding into Livestocks

There are several genetic traits known which can be bred into livestock if required, for example disease resistant traits. Some traits are determined from the coordinated expression of many genes but not single gene. The discipline of quantitative genetics unravels the total number of genes contributing to a trait, contribution of individual trait and its location on the chromosomes. Altogether they are termed as quantitative trait loci (QTLs), therefore, herds are surveyed for reference animals that have been bred for a broad range of phenotypic variations.

Diagnosis, Elimination and Breeding Strategies of Genetic Diseases

An increase in homozygosity of recessive genes and the frequency of specific alleles at QTLs has been resulted due to isolation of domestic animals in small breeding group. Consequently most breeds of livestocks comprises of variant genes which are harmful or lethal at homozygous stage (when it contain two variant genes). Such allele may be found in 15 per cent cases with 5 per cent homozygotes. Several variant genes responsible for genetic diseases have been identified. Diagnostic tests are available which permit animals as normal, carrier (*i.e.* heterozygous with one normal allele or one variant allele) or affected (*i.e.* homozygous with two variant alleles). Now PCR is used to amplify the region of genes which has been affected. Thus amplified DNA fragement is used to identify the variants. It is well known fact that genetic disease occurs when a gene malfunctions and enzyme is not expressed. When both alleles are defective in homozygote these results in phenotypic defects. The carriers (heterozygote) are phenotypically normal. Therefore, the genetic defects can be diagnosed by genetic tests and rectified biotechnologically during pregnancies that affect foetus. The second approach for elimination of defect from a herd is the use of normal (homozygous) animals for mating. Use of MOET with embryo biopsy and splitting on the best animals is the other method of spread of genetic disease and curing them. Pre-identified carrier or affected embryos should not be transferred (this aspect has been discussed earlier).

Application of Molecular Markers in Improvement of Livestock. During the last two decades the progress in recombinant DNA technology and gene cloning has brought in revolutionary changes in the field of basic as well as applied genetics. Several new approaches for genome analysis have been made. Now it is possible to uncover a large number of genetic polymorphism at the DNA sequence level and to use them as markers for evolution of genetic basis for the observed phenotypic variability. Usually a marker is considered as a constituent that determines the function of construction. Variations occurring at different levels *i.e.* at the morphological, chromosomal, biochemical or DNA level can serve as genetic markers. Thus genetic markers can be defined as any stable and inheritable variation that can be measured or detected by a suitable method, and that can be used to detect the presence of a specific genotype or phenotype other than itself. The markers revealing variations at DNA level are referred to as 'molecular markers'. So far an unlimited number of molecular markers is known since the first demonstration of DNA level polymorphism in 1974, which is called restriction fragment length polymorphism (RFLP). The molecular markers are classified into two broad categories, the hybridization-based markers and the PCR-based markers.

The Hybridization-Based Markers

This includes the traditional RFLP analysis. During RFLP analysis well labeled probes for important genes (*e.g.* cDNA or genomic sequence) are hybridized onto filter membrane containing restriction enzyme digested DNA. Then these are separated by gel electrophoresis and subsequently transferred onto these filters by Southern blotting. Thereafter, the polymorphisms are observed as hybridization bands. The individuals that carry different allelic variants for a locus will show different banding pattern. Hybridization can also be carried out with the probes (*e.g.* genomic or synthetic oligonucleotide) for the different families of hypervariable repetitive DNA sequences such as minisatellite, simple repeats, variable number of tandem repeats (VNTR) and microsatellite to reveal highly polymorphic DNA fingerprinting pattern.

The PCR-Based Markers

There is no need of probe-hybridization step. The PCR-based markers have lead to discovery of several useful methods which are easy to screen. On the basis of types of primers (*i.e.* primers of specific sequences targeted to particular region of genome or primers of arbitrary sequences) used for PCR.

These markers further can be subdivided into two groups, the sequence-targeted PCR assay, and the arbitrary PCR assay:

1. The sequence-targeted PCR assay. In this assay system, a particular fragment of interest is amplified using a pair of sequence-specific primers. In this category, PCR-RFLP or cleaved amplified polymorphic sequence (CAPS) analysis is a useful technique for screening of sequence variations that give rise to the polymorphic restriction enzyme (RE) sites. A specific region of DNA encompassing the polymorphic RE sites is amplified. The amplified DNA fragment is digested with respective RE. The variations in sequence are screened by several approaches.
2. The arbitrary PCR assay. In this assay randomly designed primer is used to amplify a set of anonymous polymorphic DNA fragments. When primer is short, there is high probability of priming taking place at several sites in genome that are located within amplifiable distance and are in inverted orientation. Polymorphism detected by using this method 'is called randomly amplified polymorphic DNA (RAPD). Based on this principle several techniques have been developed which differ in number and length of primer used, stringency of PCR conditions and the method of fragment separation and detection.

Properties of Molecular Markers

In genetic analysis, many types of markers *viz.*, morphological, chromosomal, biochemical and molecular markers are used. Morphological (*e.g.* pigmentation and other features) and chromosomal (*e.g.* structural and numerical variations) markers usually show low degree of polymorphism, therefore, they are not very useful. Biochemical markers have been tried out extensively but have not been found encouraging as they are sex linked, age dependent and influenced by the environment. The molecular markers capable of detecting the genetic variations at the DNA sequence level, have removed the limitations. They possess unique genetic properties that make them more successful than the genetic markers. They are numerous, distributed on genome, follow typical Mendalian inheritance and are multiallelic giving heterozygosity of more than 70 per cent and unaffected by environmental factors.

For genetic analysis, molecular markers offer several advantages.

Mitra et al. (1999) have discussed the several advantages of molecular markers:

- The DNA samples can easily be isolated from blood, tissues *e.g.* sperms, hair follicle as well as archival preparations,
- The DNA samples can be stored for a longer time and readily be exchanged between the laboratories,

- The analysis of DNA can be carried out at an early age or even at the embryonic stage, irrespective of sex,
- Once the DNA is transferred onto a solid support *e.g.* filter membrane, it can be repeatedly hybridized with the different probes and heterologous probes and in vitro synthesized oligonucleotide probes can also be used, and
- The PCR-based methods can be subjected to automation.

Application of Molecular Markers

Polymorphism observed at the DNA sequence level has been playing a major role in human genetics for gene mapping, pre-and post-natal diagnosis of genetic diseases, and anthropological and molecular evolution studies. Similar approach for exploitation of DNA polymorphism as genetic markers in the field of animal genetics and breeding has opened vistas in livestock improvement programmes. Much interest has been generated in determining variability at the DNA sequence level of different livestock species, and in their assessment whether these variations can be exploited efficiently in conventional as well as transgenic breeding programmes. Molecular markers can play important role in livestock improvement through conventional breeding strategies. The various possible applications of molecular markers are: short-range (immediate) applications and the long-range applications.

Table. Molecular Markers Useful in Conventional Livestock Breeding.

Application	Marker system
Short-range/immediate Application	
Parentage determination	DFP, microsatellite
Genetic distance estimation	DFP, RAPD, microsatellite
Determination of zygosity / Freemartinism	RFLP, PCR-RFLP, microsatellite RFLP,
Sex determination	PCR-RFLP, DFP, microsatellite
Identification of disease carrier	RFLP, CAPS, microsatellite
Long-range Application	
Gene mapping	Type II markers e.g. VNTR, minisatellite, microsatellite, RAPD
Marker-assisted selection	Any marker having direct/indirect association with the performance traits/QTL under question

Transgenic Breeding Strategies

The current breeding strategies of livestock largely rely on the principle of selective breeding. In this method genetic improvement is brought about by increasing the frequency of advantageous alleles of many loci. The actual loci are rarely identified. In these methods genes cannot be moved from distant sources like different species or genera due to reproductive barrier.

The recent development in molecular biology has given rise to new technology called transgenesis, which has removed the breeding barriers between different species or genera. Transgenesis has opened up many vistas in understanding behaviour and expression of a gene. It has made possible to alter the gene structure and modify its function. There are many applications of transgenesis, but the most convincing one is the development of transgenic dairy animals for the production of pharmaceutical proteins in milk, and animals with altered milk composition.

The transgenesis first starts with identification of the genes of interest. In this context, molecular markers can serve as reference point for mapping relevant genes. After successful production of transgenic animals, appropriate breeding methods could be followed for multiplication of transgenic herd. Molecular markers can also be used for identification of the animals carrying the transgenes. Most of the QTL are polygenic in nature and transgenesis presently single gene traits are being manipulated. The technology holds future promises in moving polygenic QTL across the breeding barriers of animals. However, it is expected that molecular markers wili serve as a potential tool to geneticists and breeders to evaluate the existing germplasm, and to manipulate it to create animals of desired traits as demanded by the society.

TRANSGENIC ANIMALS

A transgenic animal is an animal whose hereditary DNA has been augmented by addition of DNA from a source other than parental germplasm through recombinant DNA techniques. Transfer of genes or gene constructs allows for the manipulation of individual genes rather than entire genomes. There has been dramatic advances in gene transfer technology in the last two decades since the first successful transfer was carried out in mice in 1980. The technique has now become routine in the mouse and resulting transgenic mice are able to transmit their transgenes to their offspring thereby allowing a large number of transgenic animals to be produced. Successful production of transgenic livestock has been reported for pigs, sheep, rabbits and cattle. The majority of gene transfer studies in livestock have, however, been carried out in the pig. Although transgenic cattle and sheep have been successfully produced, the procedure is still inefficient in these species.

Transgenesis offers considerable opportunity for advances in medicine and agriculture. In livestock, the ability to insert new genes for such economically important characteristics as fecundity, resistance to or tolerance of other environmental stresses would represent a major breakthrough in the breeding of commercially superior stock. Another opportunity that transgenic technology could provide is in the production of medically important proteins such as insulin and clotting factors in the milk of domestic livestock. The genes coding for these proteins have been identified and the human factor IX construct has been successfully introduced into sheep and expression achieved in sheep milk. Moreover, the founder animal has been shown to be able to transmit the trait to its offspring. To date, the majority of genes transferred into sheep have been growth hormone encoding gene constructs. Unfortunately, in most cases the elevated growth hormone levels have resulted into a clinical diabetes situation leading to an early death of the transgenic sheep. Transgenic sheep have recently been generated which express the visna virus envelope gene.

The first reports of the production of transgenic animals created a lot of excitement among biological scientists. In the field of animal breeding, there were diverse opinions on how the technology might affect livestock genetic improvement programmes. Some believed that it would result in total reorganisation of conventional animal breeding theory while others considered the technology as an extension of current animal breeding procedures which, by broadening the gene pool, would make new and novel genotypes available for selection. Application of the technology in animal improvement is still far from being achieved. However, consideration needs to be given to its potential role in this field. Smith *et al*. presented a comprehensive evaluation of strategies for developing, testing, breeding and disseminating transgenic livestock in the context of quantitative improvement of economic traits.

An important contribution of transgenic technology is in the area of basic research to study the role of genes in the control of physiological processes. The understanding of the molecular control of life processes has important implications for both medicine and agriculture. For example, the generation (through mutation of an endogenous gene) of an organism which lacks a specific gene is a powerful tool to investigate the function of the gene product. This type of genetic analysis has been facilitated by the availability of in vitro cultures of embryonic stem cells from mice. Recent advances in *in vitro* technology (*in vitro* fertilisation and maturation) will increase the number of zygotes available for gene transfer purposes. This, plus the utilisation of embryonic stem cell and primodial germ cell technologies should enhance the efficiency of gene transfer in cattle and sheep considerably.

Genetic Characterisation of Animal Genetic Resources

Developing countries are endowed with the majority of the global domestic animal diversity-landraces, strains or breeds. Some livestock breeds in these countries are in immediate danger of loss through indiscriminate crossbreeding with exotic breeds. The importance of indigenous livestock breeds lies in their adaptation to local biotic and abiotic stresses and to traditional husbandry systems.

However, most of these animal genetic resources are still not characterised and boundaries between distinct populations are unclear. In such cases breeds are defined on the basis of subjective data and information obtained from local communities. Reliance on these criteria as the basis for classification for utilisation and/or conservation may be misleading. Additionally, historical evidence is not always accurate, relying as it often does on subjective judgements. Archival research can reveal much about the original type of a breed or strain but it is molecular genetic evidence which is factual and precise. It is in this sphere that biotechnology has an important role.

Genetic uniqueness of populations is measured by the relative genetic distances of such populations from each other. Polymorphism in gene products such as enzymes, blood group systems and leukocyte antigens which have traditionally been used for measuring genetic distance are being rapidly replaced by polymorphism at the level of DNA, both nuclear and mitochondrial as a source of information for the estimation of genetic distances. The first DNA polymorphism to be used widely for genome characterisation and analysis were the restriction fragment length polymorphism (RFLP) which detect variations ranging from gross rearrangements to single base changes.

Minisatellites sequences of 60 or so bases repeated many hundreds or thousands of times at one unique locus within the genome have been used to generate DNA fingerprints typical of individuals within species. Microsatellites repeats of simple sequences, the commonest being dinucleotide repeats are abundant in genomes of all higher organisms, including livestock. Polymorphism of microsatellites takes the form of variation in the number of repeats at any given locus and is generally revealed as fragment length variation in the products of polymerase chain reaction (PCR) amplification of genomic DNA using primers flanking the chosen repeat sequence and specific for a given locus. Ease of identification and of sequence determination and need for only small amounts of DNA, are some of the advantages of microsatellites. Additionally, because microsatellite polymorphism can be described numerically, they lend themselves to computerised data handling and analyses. Microsatellites can be used in non-PCR systems in a way similar to minisatellite probes.

Randomly amplified polymorphic DNA (RAPD) has been extensively used for genetic characterisation of a wide range of organisms. The technique uses short (up to 10 bases) primers to amplify nuclear DNA in the PCR. The procedure does not require knowledge of the sequence of DNA under study; primers are designed randomly. The basis of the polymorphism detected by this method is that products are either generated in PCR or not.

Complete sequencing of the genome is the ultimate form of genetic characterisation. Sequencing has traditionally been expensive and laborious, but with the advent of automated sequencing this is changing rapidly. However, sequencing is unlikely to be used as a technique of choice for genetic characterisation. Nei and Takezaki reviewed statistical methods for estimating genetic distances and for constructing phylogenetic trees from DNA sequence data and concluded that different analytical methods may produce different results. Teale *et al.* commented on what considerations to be made in using DNA polymorphism data for genetic distance estimation and cautioned that great care has to be taken in selecting characterisation methods and in interpreting the resulting data. While recognising the importance of the uniparental mode of inheritance of mitochondrial DNA in detecting underlying population structure not discernible from analyses of nuclear DNA, Loftus *et al.* concluded that mitochondrial DNA analysis may not be sufficient to resolve breed differences within Africa. MacHugh *et al.* suggested that microsatellite polymorphism may be more suitable when trying to discriminate between closely related populations. Regardless of which method is used, the ultimate goal in genetic characterisation for conservation is to obtain a measure of available diversity.

Conservation of Animal Genetic Resources

The terms conservation, preservation, ex situ and in situ are used here according to the definition given by FAO. There are several ways, differing in efficiency, technical feasibility and costs, to conserve animal genetic resources. Developing and utilising a genetic resource is considered the most rational conservation strategy. However, there are cases where ex-situ approaches are the only alternatives. *Ex-situ* approaches include: maintenance of small populations in domestic animal zoos; cryopreservation of semen (and ova); cryopreservation of embryos; and some combinations of these. Brem *et al.* reviewed biotechnologies for *ex-situ* conservation.

Cryopreservation of gametes, embryos or DNA segments can be quite an effective and safe approach for breeds or strains whose populations are too small to be conserved by any other means. The safety of these methods has been demonstrated by background irradiation studies. For example, studies based on irradiation of mouse embryos exposed to the equivalent of hundreds of years of background mutation showed no detectable damage. Regeneration of offspring following transfer of frozen-thawed embryos has been successful for all major domestic species, except the buffalo. In cattle, the transfer of frozen-thawed embryos is now a commercial practice and embryo survival rate after thawing can be as high as 80% with a pregnancy rate of about 50%. Cryopreservation of oocytes followed by successful fertilisation and live births have been achieved in the mouse. Cryopreserved bovine oocytes have been successfully matured and fertilised in vitro and zygotes developed to blastocyst stage. These trends strongly suggest that long-term cryopreservation of mammalian oocytes is possible.

Respective pregnancy rates of 58 and 50% for fresh and frozen-thawed *in vitro* produced embryos have been reported. Also, calves have been produced from transfer of both split and frozen-thawed *in vitro* produced embryos. Economic aspects of genetic conservation in farm animals has been assessed by Brem *et al.*. The study concluded that costs of *ex situ* live animal conservation was moderate to high while costs of long-term cryopreservation of gametes were low.

Development in genetic engineering, cryobiology, cell biology and embryology will provide techniques that may enhance our ability to preserve germplasm *in vitro*. Techniques such as transfer of DNA within and between species and the production of viable transgenic animals are far from practical application. However, biotechnology will certainly contribute newer and cheaper methods for preservation such as storage of catalogued DNA. At present, other than live animal and embryo preservation, the other techniques do not allow preservation of genomes in a form which can be reactivated *in toto* at a later stage, but they permit the preservation of individual genes or gene combinations for possible future regeneration.

Conservation of indigenous animal genetic resources should be one of the priority livestock development activities for developing countries. The critical importance of these resources to their owners in developing countries need not be emphasised.

Their importance to developed countries is also becoming evident as indicated by the increasing importation of tropical germplasm by these countries. It is highly likely that these resources will become of increasing importance to the industrialised countries either as sources of unique genes or when environmental concerns necessitate change in production systems. Developed countries should, thus, assist in the conservation and development of these resources. Technology for cryopreservation of semen and embryo is sufficiently developed to be applied in developing countries. What is missing is financial support to implement conservation programmes. Such support has been provided for world-wide conservation activities for plant germplasm. There is also a strong case for support of animal genetic resources conservation.

ANIMAL GENETIC ENGINEERING

There are some serious issues related to genetic modification of animals using animal genetic engineering techniques. One is not sure of the consequences of these genetic modifications and the further interaction with the environment. Proper clinical trials are also necessary before one can use it for commercial purposes. In the recent past people have raised objections on some of the methods used *e.g.* the transfer of a human genes into food animals, use of organisms containing human genes as animal feed. Some religious groups have expressed their concern about the transfer of genes from animals whose flesh is forbidden for use as food into the animals that they normally eat. Transfer of animal genes into food plants that may be objectionable to the vegetarians.

Besides this, there are several other aspects of this issue have to be sorted out:

- What will be the consequences, if a modified animal will breed with other domestic or wild animals thereby transferring the introduced genes to these populations?
- What are the health risks to human on consumption of genetically modified animals and their products?
- With the production of disease resistant animals, what will be the effect on ecology?
- There is also wide spread concern about the risks of human recipients getting infected with animal viral diseases after a xenotransplantation., which might infect the population at large.
- There are also concerns about the risk that drug resistance gene markers used in genetic engineering procedures might inadvertently be transferred and expressed.

The need of the hour is to formulate clear guidelines which should be followed while using genetic engineering techniques in bio-medical research. *e.g.* products from transgenic organisms should be clearly marked to give choice to people who follow dietary restrictions due to religious beliefs. In fact all the ethical and moral issues raised by some aspects of biotechnology should be addressed by open discussion and dialogue.

QUALITY CONTROL CONSIDERATIONS

Quality is important in all aspects of tissue culture since the quality of materials used *i.e.* media and other reagents) will affect the quality of the cultures and products derived from them.

The main areas of quality control that are of concern for tissue culture are:

- The quality of the reagents and materials
- The provenance and integrity of the cell lines
- The avoidance of microbial contamination

REAGENTS AND MATERIALS

A potential source of contamination is reagents and materials, in particular bovine serum which has been identified as a source of bovine viral diarrhoea virus (BVDV). Porcine trypsin is also a potential source of Mycoplasma hyorhinis. Good quality reagents and materials are available from numerous manufacturers of tissue culture media and supplements. In addition manufacturers including Sigma will carry out a range of quality control tests including screening for mycoplasma and BVDV and supply a Certificate of Analysis with their products. These state the product and lot numbers and forms a vital part of record keeping and tracking of reagents used in the production of cell stocks. It is advisable to further test key reagents such as FBS to ensure that they are 'fit for purpose' due to batch-to-batch variation. Manufacturers of sterile plastic ware (flasks, centrifuge tubes, pipettes) designed for tissue culture use are also supplied with Certificates of Analysis for each batch produced, which should be kept for future reference.

PROVENANCE AND INTEGRITY OF CELL LINES

The sourcing of cell lines can have an important effect on quality since freshly imported cell lines are a major source of contamination.

The advantages of obtaining cell lines from a recognized source such as a culture collection are:

- Contaminant free
- Fully characterized and authenticated in terms of DNA profile and species of origin
- Supplied with a detailed data sheet.

Once cell lines have been obtained from a reputable source it is important to implement master and working cell banking procedures and the associated quality control steps such as routine testing for microbial contaminants and confirming the identity of cultures.

AVOIDANCE OF MICROBIAL CONTAMINATION

Potential sources of contamination include other cell lines, laboratory conditions and staff poorly trained in core areas such as aseptic techniques and good laboratory practice.

Thus the use of cells and reagents of known origin and quality alone is not sufficient to guarantee quality of product (cell stock or culture products); it is necessary to demonstrate quality throughout the production process and also in the final product. Routine screening aids the early detection of contamination since all manipulations are a potential source of contamination.

The 3 main types of microbial contaminants in tissue culture are:

1. Bacteria and Fungi
2. Mycoplasma
3. Viruses

Bacterial and Fungal Contamination

Bacterial contamination is generally visible to the naked eye and detected by a sudden increase in turbidity and colour change of the culture medium as the result of a change in pH.

The cell culture may survive for a short time but the cells will eventually die. Daily microscopic observation of cultures will ensure early detection of contamination and enable appropriate action to be taken as soon as the first signs of contamination become apparent. In addition specific tests for the detection of bacteria and fungi should be used as part of a routine and regular quality control screening procedure.

Mycoplasma Contamination

Mycoplasmas are the smallest free-living self-replicating prokaryotes. They lack a cell wall and lack the ability to synthesize one. They are 0.35m in diameter and can be observed as filamentous or coccal forms. There are 5 major species that are tissue culture contaminants, namely *M. hyorhinis*, *M. arginini*, *M. orale*, *M. fermentans* and *Acholeplasma laidlawii*.

The effects of mycoplasma infection are more insidious than those of bacteria and fungi inducing several long term effects.

These include:

- Reduced growth rate
- Morphological changes
- Chromosome aberrations
- Alterations in amino acid and nucleic acid metabolism.

However, despite these well-documented effects the presence of mycoplasma is often not tested for with the consequence that in such laboratories the majority of cell lines are positive for mycoplasma. Mycoplasma contamination is difficult to detect requiring the use of specialist techniques. In the past only specialist laboratories, such as culture collections, have performed these tests. However a variety of commercial kits are now available although the performance characteristics of these kits can be extremely variable. A combination of these should be used as part of a routine and regular quality control screening procedure. Culture collections such as ECACC are able to test cultures if required.

Viral Contamination

Some cell lines contain endogenous viruses and secrete virus particles or express viral antigens on their surface (*e.g.* EBV transformed lines). These cell lines are not considered contaminated. However, bovine serum is a potential source of bovine viral diarrhoea virus (BVDV) contamination. Use of infected serum will lead to contamination of cell lines with the virus. Contamination of cell lines with BVDV may cause slight changes in growth rate but since this virus is non-cytopathic macroscopic and microscopic changes in the culture will not be detected. Suppliers of bovine serum are aware of this and screen sera accordingly and generally serum is sold as BVDV tested.

ENVIRONMENTAL MONITORING

It is good practice to monitor the laboratory environment where cell cultures and their products are prepared. Class II microbiology safety cabinets should be checked every 6 months to ensure that they are working efficiently.

However it is also advisable to monitor the number of contaminants within the cabinet by periodically placing open settle plates (blood agar bacteriological culture plates) on the cabinet work surfaces. In addition settle plates should be used to assess airborne microbial burden at selected points around the laboratory. Plates should be left open for a period of 4 hours.

After this time they should be covered, placed in sealed boxes and incubated at 32°C and 22°C for up to 7 days. At the end of this period the plates should be examined for the presence of microbial growth. The position of each plate in the cabinet should be recorded and results stored for trend analysis. Acceptable limits should be defined in terms of "alert" levels and "action" levels, the actual values being dependent on the criticality of the work and the levels of cleanness that can be achieved under normal operating conditions.

WHAT TO DO IN THE EVENT OF CONTAMINATION

One hugely under-estimated problem in tissue culture is the routine use of antibiotics. Continuous use of antibiotics is unnecessary and can lead to the development of resistant strains that are difficult to eradicate and may require the use of more exotic antibiotics that may be toxic to the cell cultures. In addition the use of antibiotics may mask a low level of contamination.

Once a contamination has been detected, whether is it due to bacteria, fungi or mycoplasma, the recommended course of action is to discard the culture and continue the work with earlier stocks that are known to be free of contaminants or obtain fresh stocks from a recognized source. However if this is not possible eradication of the contaminant may be attempted with the use of antibiotics. In addition culture collections such as ECACC will attempt to eradicate any contaminants if required. Please contact ECACC for further details. Viral infections are virtually impossible to remove from cultures since they do not respond to antibiotic treatment. Also, since they are intra cellular parasites it is not possible to remove them by centrifugation or other separation techniques. If virus free stocks or a virus free alternative is not available then a thorough risk assessment should be undertaken prior to continuing work with the infected cell line.

TECHNOLOGY OF GENETIC ENGINEERING

Genetic engineering is a radical new technology, one that breaks down fundamental genetic barriers — not only between species, but between humans, animals, and plants. By combining the genes of dissimilar and unrelated species, permanently altering their genetic codes, novel organisms are created that will pass the genetic changes onto their offspring through heredity.

Scientists are now snipping, inserting, recombining, rearranging, editing, and programming genetic material. Animal genes and even human genes are being inserted into plants or animals creating unimagined transgenic life forms. For the first time in history, human beings are becoming the architects of life. Bio-engineers will be creating tens of thousands of novel organisms over the next few years. The prospect is frightening.

Genetic engineering poses unprecedented ethical and social concerns, as well as serious challenges to the environment, human health, animal welfare, and the future of agriculture. The following is just a sampling of concerns:

Genetically engineered organisms that escape or are released from the laboratory could wreak environmental havoc. Genetically engineered "biological pollutants" have the potential to be even more destructive than chemical pollutants. Because they are alive, genetically engineered products are inherently more unpredictable than chemical products — they can reproduce, migrate, and mutate. Once released, it will be virtually impossible to recall genetically engineered organisms back to the laboratory. A report published by 100 top American scientists warned that the release of gene-spliced organisms "...could lead to irreversible, devastating damage to the ecology."

Gene-splicing will likely result in unanticipated outcomes and dangerous surprises. Biotechnology is an imprecise science and scientists will never be able to ensure a 100 percent success rate. Serious accidents are bound to occur. Researchers conducting experiments at Michigan State University recently found that genetically altering plants to resist viruses can cause the viruses to mutate into new, more virulent forms, or forms that can attack other plant species. Some other scary scenarios: foreign genes from genetically engineered plants could be carried by pollen, insects, wind, or rain, and flow into other crops, as well as wild and weedy relatives. Disaster would follow if genetically engineered crop traits, such as insect and virus resistance, found their way into weeds, for instance. Genetically altered plants could produce toxins and other substances that might harm birds and other animals. Genetic engineering of plants and animals will almost certainly endanger species and reduce biological diversity. By virtue of their "superior" genes, some genetically engineered plants and animals will inevitably run amok, overpowering wild species in the same way that introduced exotic species, such as kudzu vine and Dutch elm disease which have created problems in North America. What will happen to wild species, for example, when scientists release into the environment carp, salmon, and trout that are twice as large, and eat twice as much food, as their wild counterparts? Another danger lies in the creation of new kinds of crops and domesticated animals. Once researchers develop what is considered to be the "perfect tomato" or "perfect chicken" these will be the ones reproduced in large numbers; "less desirable" species would fall by the wayside. The "perfect" animals and plants could then be cloned (reproduced as exact genetic copies), reducing even further the pool of available genes on the planet.

Genetically engineering plants to be herbicide-tolerant will lead to increased use of chemicals in agriculture and further contamination of the environment. Biotech companies love to say that genetic engineering will end the use of dangerous chemicals in agriculture. But the leaders in biotechnology are the giant chemical companies like Monsanto, Du Pont, and Rhone-Ponlenc; they aren't interested in losing profits from the sale of chemicals. These companies are genetically engineering plants to be resistant to herbicides that they manufacture so they can sell more herbicides to farmers who, in turn, can apply more poisonous herbicides to crops to kill weeds. In fact, crops genetically engineered to be herbicide-tolerant account for nearly half of the applications for field testing submitted to the U.S. Department of Agriculture (USDA) since 1988. Even genetically engineering crops to produce their own pesticides presents dangerous problems. Pests will eventually evolve that are resistant, then stronger chemicals will be needed to get rid of the pests. And what will happen when the pesticide gene spreads to weeds and other unwanted plants?

The genetic engineering of crops and food-producing animals can produce toxic and allergic reactions in humans. Someone allergic to peanuts or shellfish, for example, would have no way of knowing if a tomato or other food had been altered with proteins from these substances, and could have a fatal reaction by eating such genetically altered foods. In addition, genetic engineers can take proteins from bacteria they find in the soil, the ocean — anywhere — and incorporate them into human food. Such substances have never been in the food supply before, so their toxic or allergenic characteristics are unknown.

Genetically engineered products do not have a good track record for human safety. In 1989 and 1990, a genetically engineered brand of L-tryptophan, a common dietary supplement, killed more than 30 Americans and permanently disabled or afflicted more than 5,000 others with a potentially fatal and painful blood disorder, eosinophilia myalgia syndrome, before it was recalled by the FDA. The manufacturer, Showa Denko K.K., Japan's third largest chemical company, had used genetically engineered bacteria to produce the over-the-counter supplement. It is believed that the bacteria somehow became contaminated during the recombinant DNA process. There were no labels on the product to identify the product as having been genetically engineered. The patenting of genetically engineered foods, and widespread biotech food production, will eliminate farming as it has been practiced since the beginning of humankind's appearance on the planet. If the trend is not stopped, the patenting of transgenic plants and food-producing animals will soon lead to tenant farming in which farmers will lease their plants and animals from biotech conglomerates and pay royalties on seeds and offspring. Eventually, within the next few decades, agriculture will move off the soil and into biosynthetic industrial factories controlled by chemical and biotech companies. Never again will people know the joy of eating naturally produced, fresh foods. Hundreds of millions of farmers and other workers worldwide will lose their livelihoods. The hope of creating a human, sustainable agricultural system will be destroyed.

The genetic engineering and patenting of animals reduces living beings to the status of manufactured products and will result in much suffering. In January

1994, then-USDA Secretary Mike Espy announced that USDA scientists had completed genome "road maps" for cattle and pigs, a precursor to ever more experimentation on live animals. In addition to the cruelty inherent in such experimentation (the mistakes are born with painful deformities, crippled, blind, and so on), these "manufactured" creatures have no greater value to their "creators" than mechanical inventions. Animals genetically engineered for use in laboratories, such as the infamous "Harvard mouse" which contains a human cancer-causing gene that will be passed down to all succeeding generations, were created to suffer. A purely reductionist science, biotechnology reduces all life to bits of information (genetic code) that can be arranged and rearranged at whim. Stripped of their integrity and sacred qualities, animals who are merely objects to their "inventors" will be treated as such. Currently, more than 200 genetically engineered "freak" animals are awaiting patent approval from the federal government.

No one is regulating genetically engineered organisms adequately or properly testing them for safety. In 1986, Reagan-era policymakers stitched together a patchwork of pre-existing and only marginally appropriate statutes to ease the way for new biotechnology products. But these laws were created years ago to deal with chemicals — not the unpredictable living products of genetic engineering. To date, no suitable government apparatus has been set up to deal with this radical new class of potentially overwhelming environmental and health threats. The FDA's policy on genetically altered foods illustrates the problem. In May 1992, then Vice President Dan Quayle, and head of the Competitiveness Council, announced the U.S. Food and Drug Administration's newly developed policy on biotech foods: genetically engineered foods will not be treated differently from naturally produced foods; they will not be safety tested; they will not carry labels stating that they have been genetically engineered, nor will the government keep track of foods that have been genetically engineered. As a result, neither the government nor consumers will know which whole or processed foods have been genetically engineered. Vegetarians and followers of religious dietary restrictions face the prospect of unwittingly eating vegetables and fruits that contain genetic material from animals — including humans. And health risks will be discovered only by trial and error — by consumers. USDA oversight is no better. This agency has the conflicting task of both promoting and regulating agriculture, including genetically engineered plants and animals used for food. Indeed, the USDA is a primary sponsor of biotech research on plants and animals.

By patenting the genes they discover and the living organisms they create, a small corporate elite will soon own and control the genetic heritage of the plant. Scientists who "discover" genes and ways of manipulating them can patent — and thus own — not only genetic engineering techniques, but the very genes themselves. Chemical, pharmaceutical, and biotech companies such as DuPont,

Upjohn, Bayer, Dow, Monsanto, Cib-Geigy, and Rhone-Poulenc, are urgently trying to identify and patent plant, animal, and human genes in order to complete their take-over of agriculture, animal husbandry, and food processing. These are some of the same companies that once promised a carefree life through pesticides and plastics. Would you trust them with the blueprints of life?

Genetic screening will likely lead to a loss of privacy and new levels of discrimination. Already, people are being denied health insurance on the basis of "faulty" genes. Will employers require genetic screening of their employees and deny them work on the basis of the results? Will the government have access to our personal genetic profiles? One can easily imagine new levels of discrimination being directed against those whose genetic profiles reveal them to be, for example, less intelligent or predisposed to developing certain illnesses.

Genetic engineering is already being used to "improve" the human race, a practice called eugenics. Genetic screening already allows us to identify and abort fetuses who carry genes for certain hereditary disorders. But within the next decade, scientists will likely have a complete map of the human genome to work with. Will we abort fetuses on the basis of non-life-threatening impairments such as myopia, because someone is predisposed towards homosexuality, or for purely cosmetic reasons? Researchers at the University of Pennsylvania have applied for a patent to genetically alter sperm cells in animals so traits passed down from one generation to the next can be changed; the application suggests that this can be done in humans too. Moving from animal eugenics to human eugenics is one small step. Everyone wants the best for their children; but where do we stop? Inadvertently, we could soon make the efforts of the Nazis to create a "superior" race seem bumbling and inefficient.

The U.S. military is building an arsenal of genetically engineered biological weapons. Although the creation of biological weapons for offensive purposes has been outlawed by international treaty, the U.S. continues to develop such weapons for defensive purposes. However, genetically engineered biological agents are identical whether they are used for offensive or defensive purposes. Areas of investigation for such weapons include: bacteria that can resist all antibiotics; extra-hardy, more virulent bacteria and viruses that live longer and kill faster; and new organisms that can defeat vaccines or natural human or plant resistances. Also being studied are the development of pathogens that can disrupt human hormonal balance enough to cause death, and the transformation of innocuous bacteria (such as are found in human intestines) into killers. Some experts believe that genetically engineered pathogens that can target specific racial groups are being developed as well.

GENETICS OF TUBERCULOSIS IN IRISH DAIRY COWS

Tuberculosis has been an on going problem for the Irish cattle industry for many years. Prior to the introduction of the 'test and slaughter' policy in

1954 bovine tuberculosis was wide spread, with an animal infection level of 3.0% (140,000 reactors per annum). Progress was rapid during the initial stages of the programme, when animal infection level was reduced to 0.4% (23, 000 reactors) by 1965 when Ireland received 'official tuberculosis free status'. However progress has stalled at this point; in 2006 animal infection level was 0.4% (26,000 reactors). Current control measures, which include annual testing of every bovine each year, have effectively minimised cattle-to-cattle transmission of infection. However, there are three main factors constraining the eradication of tuberculosis in Ireland including: the continuing expansion of the national herd, the frequency of movement during the production cycle, and the presence of a wildlife reservoir, the Eurasian badger *Meles meles*.

Furthermore, the current programme represents a very substantial cost to the industry and government (approximately □80 million per annum). Selection for increased resistance to tuberculosis susceptibility provides an alternative strategy that could enhance existing control programmes. However, this strategy requires additive genetic variation for resistance to tuberculosis. There is some evidence at the genus level that *Bos indicus* is more reistant to tuberculosis than *B. taurus*. Furthermore, there is evidence that certain family lines are more resistant than others in; a heritability ranging from 0.06-0.08 has been estimated in cattle in Latvia. Similarly, a heritability estimate of 0.06 has been published for an analogous disease, paratuberculosis (Johne's) disease in Dutch dairy cattle. Nevertheless, there is a lack of information on genetic parameters for tuberculosis resistance in cattle, which is largely a result of the lack of suitable data. Ireland is uniquely placed to address this issue, given the availability of three national databases (including the: AHCS [Animal Health Computer System], CMMS [Cattle Movement Monitoring System] and ICBF [Irish Cattle Breeding Federation database]) with linkable animal and herd level records of animal breeding and disease control. Furthermore, the sound epidemiological and immunological knowledge generated in Ireland and abroad were an invaluable resource in the development of a trait definition, data editing criteria and for the selection of herd and animal level risk factors to be included in analyses.

The objective of this study was to estimate the genetic variation in tuberculosis resistance in Irish dairy cows. A total of 15,321 animals (including 1,268 tuberculosis positive animals) from 724 herd test dates during the period of 1st November 2002 and 31st October 2005 were available for inclusion in the analysis. All animals were Holstein-Friesian and had an identified sire. Tuberculosis positive animals were defined as animals with a skin change where the *Mycobacterium bovis* purified protein derivative response was 4 mm greater than that of the *M. avium* purified protein derivative response. Animal linear, animal threshold and maternal effects linear models were used to estimate the additive genetic and residual variance for tuberculosis susceptibility. Fixed effects in the models included, herd episode, month of test, age at test, year of calving, month of calving, proportion of Holstein genes, heterosis and recombination loss.

Heritability of tuberculosis susceptibility was 0.0317 (SE=0.0106). The animal threshold model estimated an almost four fold larger heritability (h^2=0.1189 (0.0281). Threshold models have been shown to be theoretically better, and are more accepted for all or nothing (reactor/non reactors) trait heritability estimation, indicating the higher heritability estimate should contribute to greater genetic improvement. The heritability estimate from the maternal effects model was 0.0285 (SE=0.0110), which was less than the estimate of the animal linear model, indicating that maternal inheritance dose not influence tuberculosis resistance in cows. The low heritability estimates observed are characteristic for disease traits, and are similar the earlier estimates of 0.06-0.08 and 0.06 for tuberculosis and paratuberculosis resistance respectively. This study has demonstrated that exploitable genetic variation for tuberculosis susceptibility exists among Irish dairy cows, and that selecting for increased resistance should be possible.

The results from this study will be used to generate of breeding values for the industry, allowing breeders to identify Irish and foreign animals that are genetically resistant to infection with tuberculosis. Furthermore, while selecting for increased resistance to tuberculosis, breeders maybe indirectly selecting for general health, i.e. heightening resistance against a variety of pathogens that infect cattle in a similar manner. The breeding values estimated from this study will also be used in a collaborative project in Trinity collage Dublin, to identify the major genes associated with tuberculosis resistance. This additional knowledge will provide for more (accurate) information on the genetic merit of animals, which will be used to enhance selection programmes in the future.

Increasing the resistance of our national herd to tuberculosis would prove synergistically beneficial with existing eradication efforts, through reduced animal/herd incidence, resulting in reduced animal/environmental bacterial loads, ultimately reducing cattle-cattle, cattle-environment, cattle-wildlife, and environment-wildlife transmission of the disease, a major step toward the eradication of tuberculosis, and attainment of the much coveted official bovine tuberculosis free status in Ireland.

ANIMAL HEALTH AND DISEASE DIAGNOSIS

Successful control of a disease requires accurate diagnosis. This has been greatly improved in recent years through developments in biotechnology. The most recent major development, the finding that it is possible to immortalise individual antibody-producing cells by hybridisation to produce antibodies of a given class, specificity and affinity (*i.e.* monoclonal antibodies) has provided a tool that permits the analysis of virtually any antigenic molecule. The use of monoclonal antibodies has revealed that the failure of vaccines (*e.g.* of rabies) to provide protection in all parts of the world was due to the diversity in the antigenic composition of the causative virus. The (monoclonal antibody) technology is relatively simple and can readily be applied in developing countries. Monoclonal antibodies are currently supplied to developing countries directly or in the form of kits and simple reagents for completion of the tests. For example, kits for rinderpest virus diagnosis used in African countries come in this form.

The ability to generate highly specific antigens by recombinant DNA techniques has made it possible for an increasing number of enzyme-linked immunosorbent assays (ELISA) to have the capacity to differentiate between immune responses generated by vaccination from those due to infection.

This has made it possible to overcome one of the major drawbacks of antibody detection tests: the fact that, because antibodies can persist in animals for long periods, their presence may not indicate current infection. ILRAD has developed a technique to overcome this problem in diagnosis of trypanosomiasis. The parasite antigen detection test uses monoclonal antibodies raised in laboratory mice to capture the parasite antigens which are then revealed by their reaction with a second layer antibody to which is conjugated an easily detectable enzyme. This test reveals current infections and facilitates differentiation between the major trypanosome species. This has important implications for disease control, especially because of the association of different parasite species with different epidemiological and disease circumstances.

The advent of PCR has enhanced the sensitivity of DNA detection tests considerably. For example, PCR used in combination with hybridisation analysis, has been shown to provide a sensitive diagnostic assay to detect bovine leukosis virus.

Other diagnostic techniques include nucleic acid hybridisation (NAD) and restriction endonuclease mapping (REM). As has been indicated above, one of the most valuable features of these molecular techniques is their specificity and sensitivity. A good example of the specificity of NAD is its application in distinguishing infections caused by *peste des petite ruminants* (PPR) virus from rinderpest, diseases whose symptoms are clinically identical and which cannot be distinguished antigenically with available serological reagents. This technique also allows comparison of virus isolates from different geographical locations. Detailed overview of biotechnological tools for diagnosis of livestock diseases has been provided by Bourne and Bostock and Robinson and McEvoy.

Other available diagnostic techniques which may have application in small ruminants and/or cattle include: nucleic acid probes (NAP) for heartwater *(Cowdria ruminantium), chlamydia psitacci,* Paratuberculosis and Bluetongue; restriction endonuclease reaction (RENR) for diagnosis of *Corynebacterium pseudotuberculosis*; PCR for characterising subtypes of Bluetongue from different geographical regions; monoclonal antibodies (MAB) for differentiating false positive anti-Brucella titres caused by *Yersinia enterocolitica* and true positive anti-Brucella titres in latent infected animals; and MAB for diagnosis of Toxoplamosis, Pasteurellosis, *Mycoplasma* spp, PPR and Boder Disease. Examples of tropical diseases for which diagnostic tests are available are presented in Table. Teale reviewed the diagnostic techniques that are currently available or which could be developed in the near future.

Table. Tropical Diseases for which Probes and Monoclonal Antibodies (MAB) are Available.

Viral diseases	Bacterial diseases
Rinderpest	Cowdriosis
Peste des petite ruminants	Contagious bovine pleuropneumonia
Bluetongue	Contagious agalactia
African horse sickness	Contagious caprine pleuropneumonia
Foot-and-mouth disease	Anaplasmosis (A. centrale and A. marginale)
Vesicular stomatitis	Haemorrhagic septicemia (only MAB available)

VACCINES

Conventional means of controlling major livestock diseases include chemotherapy, vector control, vaccination, slaughter of infected stock, and other management practices (including grazing management and controlled stock movements).

Vector control requires continuous application of pesticides. These are often unaffordable to farmers in developing country. Moreover, where these drugs or pesticides are used, resistance by parasites is often encountered and reinfection following administration of drugs against parasitic diseases usually occurs. Additionally, in many cases drugs are not readily available locally. In some cases where they are available, they are ineffective, either because they have been partly preserved or they are not genuine.

Immunisation remains one of the most economical means of preventing specific diseases. An effective vaccine can produce long-lasting immunity. In some cases, vaccination can provide lifetime immunity. Moreover a small number of doses is usually required for protection. Level of infrastructure and logistical support required for a large-scale vaccination programme is such that a successful vaccination campaign can be implemented in remote rural areas. In general, vaccines offer a substantial benefit for comparatively low cost, a primary consideration for developing countries.

Vaccines have conventionally been produced by several methods some of which have become rather static with regard to efficacy, safety, stability and cost. Very effective vaccines against animal diseases such as rinderpest and pig cholera have been in use for more than 20 years and have helped to significantly reduce the incidence of these diseases world-wide.

However, vaccines of questionable efficacy also exist. Impotency, instability, adverse side effects, and reversion of attenuated organisms to wild (disease-causing) forms represent some of the problems. However, research strategies for the development of better, cheaper and safer vaccines are constantly being sought. Through the use of monoclonal antibodies and recombinant DNA technologies, it is now possible to define and produce immunogenic components much more rapidly. These technologies are increasingly being used to clarify the pathogenetic mechanisms and immune response to microbial diseases. This should lead to the production of more effective vaccines in the future. To date many candidate vaccines have been produced by these techniques. However, only few of these are being produced commercially. Table summarises some vaccines developed by recombinant DNA technology. There are other vaccines under various stages of development.

The following are of particular relevance to small ruminants: recombinant vaccines against Bluetongue and Rift Valley Fever may soon be available for field use; a hybrid virus vaccine against Orf, a thermostable recombinant vaccine against PPR and recombinant vaccines against *Taenia ovis* and Echinococcus; tick gut antigens and *Haemonchus* gut antigens produced by molecular techniques offer ways for prophylaxis against these parasites.

Progress is also being made towards the development of vaccines against Babesiosis and Theileriosis in cattle which may indicate prospects for similar immune-prophylaxis becoming available for sheep and goats as well.

Table. Example of Some Novel Animal Vaccines

Species	Disease	Producer
Avian	Coccidiosis,	Genex and A.H. Robins
	Newcastle virus	Codon and Salsbury labs
Bovine	Papilloma virus	Molecular genetics
	Viral diarrhoea	California Biotechnology
	Brucellosis	Ribi ImmunoChem
	Rinderpest	USDA and University of California, Davis
Swine	Parvovirus	Applied Biotechnology
	Dysentery	Codon
Equine	Influenza, Herpes	California Biotechnology
		Applied Biotechnology
Companion	Canine parvovirus	Applied Biotechnology

Physiology of Lactation and Growth

Recombinant bovine somatotropin (BST) is a genetically engineered synthetic analog of the natural growth hormone. Since the 1970s there have been a number of studies on the effects of BST on milk yield, reproductive performance and health as well as its likely effect on humans who consume such milk. Under good management and feeding, regular BST administration to lactating dairy cows increases milk yield by 15-30% and also increases efficiency of milk production.

BST is now readily available and is already in commercial use in the United States. However, the appropriateness of BST use to increase milk production in the USA is doubtful as the country already has a milk surplus and the public is also concerned about its effects on health. Indeed, the benefit of BST in the USA is perceived to be towards reducing the production costs of large dairy farms, but this could push smaller farmers out of the market. The appropriateness of BST use for developing countries is still a matter of debate. Those supporting its introduction argue that its use in commercial dairy farms could increase the national milk output. Those opposed point out the fact that BST does not improve milk yield in indigenous non-dairy breeds and that its use on crossbred and exotic dairy cattle will require more feeding, and that provision of adequate nutrition is already a problem for most dairy operations in developing countries. Whether or not BST affects reproductive performance has not yet been conclusively established. Although most reports have indicated nonsignificant effects of BST on reproductive performance, most such studies have been on single lactations; the effect of BST on lifetime productivity needs to be investigated. Some studies have shown that feed conversion efficiency declines in subsequent lactations.

The study by Phipps *et al.* also indicated a decline in the incremental amount of milk yield from BST application in subsequent lactations. Additionally, the effect of BST on reproductive performance is likely to be more adverse in the presence of higher biotic and abiotic stresses, including nutritional stresses. There is also need to examine the economics of BST application in view of the known association of its use with mastitis and other infections. BST use is thus bound to be associated with increased use of antibiotics and other veterinary drugs. Thus, in evaluating the potential role of BST in developing countries, one needs to consider not only the possible response levels of the cattle in these countries to BST treatment, but also the cost of BST, the amount and cost of other incremental inputs required for effective use of BST, and the milk prices. Ultimately, the main technical constraint to BST use in developing countries will not only be its cost, but the absence of an efficient delivery system; current use of the technology requires regular injections.

Porcine somatotropin (PST) and recombinant growth hormone stimulatory peptides (*e.g.* growth hormone releasing factor, GRF) along with BST have been shown to increase growth rates by 8-38% in cattle, sheep and pigs. In almost all cases, administration of exogenous growth hormones have been associated with increased carcass protein and reduced carcass fat. Other growth-promoting agents (*e.g.* anabolic steroids and beta agonists) have been shown to have even larger effects but public concerns over the possible residual effects in meats have led to their being banned in most developing countries. The use of anabolic implants is, however, permitted in some countries such as the USA.

TRANSGENIC ANIMALS AND DERIVED PRODUCTS

Concern about the quality and safety of food products has been growing since the early 1980s. Many hazards have been identified, studied and handled by instituting national regulations and international standards and by setting up regulatory and inspection systems in processing plants, livestock farms and slaughterhouses.

These hazards can be biological (bacteria, parasites, viruses, prions), physical (metals, etc.) and chemical (residues of pesticides or veterinary drugs, toxic substances, etc.). New hazards are now perceived in relation to the use of transgenic animals. Transgenic animals include those obtained through nuclear transfer, those with improved production traits and those developed for biomedical purposes. Animals for xenotransplantation are barred from the food chain by national regulations.

Products Derived from Cloned Animals

In 2002, some 1400 Holstein cattle produced by blastomere nuclear transfer (not somatic cells) were raised in the US. They and their subproducts (animal and milk) are marketed and enter the food chain. No result of analysis of products derived from these transgenic animals is so far available, nor is there any information on abnormalities of gene expression in these adult cloned animals or analytical information on the composition of milk or meat derived from them or their offspring. The animals are considered perfectly safe and no apparent ill effects have been noted among consumers. However, caution and risk analysis are recommended for such derived products, especially in the light of the problem of reprogramming the inserted nucleus. The safety of food from offspring of cloned animals is a controversial subject. Some see no risk as they originate from natural mating, while others point out that the offspring of a cloned animal has one half of the latter's genome. The cloning of animals from somatic cells is a recent technique, so it is difficult to make a firm decision on the safety of related milk or meat. Genetic variations appearing with nucleus differentiation and possibly causing modified genome expression are not known. The findings of studies on products of cloned animals will no doubt be available soon.

ALLERGENICITY AND HYPERSENSIBILITY

Most food allergens are proteic, well known and identified. As regards the safety of new proteins in foods derived from transgenic animals, the possibility that a new protein could cause an allergic reaction among certain individuals needs to be assessed. FAO has published an approach for assessing the potential allergenicity of new proteins extracted from foods derived from transgenic animals, using the decision tree reproduced in Figure.

If the new protein is produced from a transgene that is known to have allergenic potential, or if the animoacid sequence is similar to that of a known allergenic protein, then the allergenic potential of the new protein can be tested with the serum of individuals with known food allergies. The problem arises for proteins with unknown allergenic potential which cannot be easily assessed, as there is no existing human immunoserum. The problem is further compounded when the new protein derives from an existing protein that has no known allergenic potential. Thus, assessing the potential allergenicity of a protein from a transgene remains one of the major difficulties in the overall safety assessment of transgenic foods. Assessment of allergenicity requires an understanding of several factors, including the source of the transgene protein, its level of expression, its chemical and physical properties, and its similarities to any known allergen.

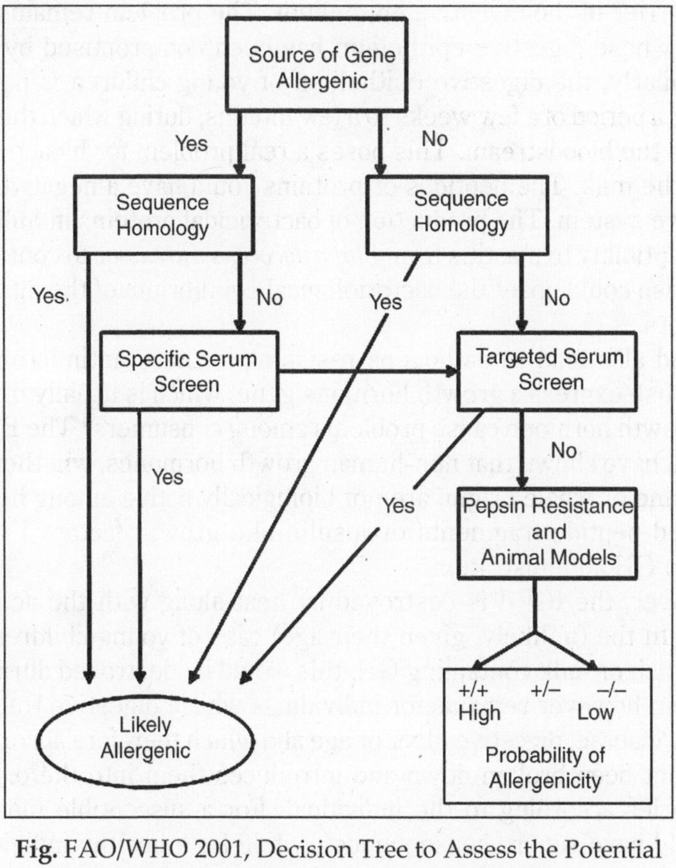

Fig. FAO/WHO 2001, Decision Tree to Assess the Potential Allergenicity of Foods Derived from Biotechnology.

All these factors have to be considered together to have an idea of the potential allergenicity of the new protein. The absence of appropriate and effective methods of analysis is a major problem, as is the feasibility of performing adequate assessments for an increasing number of transgenic products. In this connection, WHO and FAO are working on the development of a standardized protocol for the assessment of allergenicity. The animal model for allergenicity tests of products derived from transgenic animals is recognized as effective. In vivo allergenicity tests among humans would be the best option but raises ethical problems. Post-market surveillance of a product is also advocated, but modalities for doing this are difficult to determine.

Other Bioactive Compounds

Some transgenic animals are designed to express biopharmaceuticals. That these products could retain their bioactivity in the organism after consumption of the food is a safety concern. All proteins and polypeptides are normally broken down in the digestive tract, thus destroying their bioactivity. Only small peptides pass the barrier of the digestive epithelium. The problem remains, however, for people whose digestive epithelium has been compromised by disease or injury. Similarly, the digestive epithelium of young children is permeable to proteins for a period of a few weeks to a few months, during which these proteins could enter the bloodstream. This poses a real problem for bioactive peptides contained the milk. The peptides or proteins could have a negative impact on the digestive system. The production of bactericidal proteins in milk to reduce cattle susceptibility to mastitis from *Staphylococcus aureus* or to control bacterial disease in fish could upset the bacteriological equilibrium of the intestinal flora of consumers.

It could also lead to pathogens resistant to these antimicrobials. Many transgenic fish express a growth hormone gene, which is usually of fish origin. Can this growth hormone cause problems among consumers? The FDA reports that studies have shown that non-human growth hormones, whether of bovine, ovine, porcine or whale origin, are not biologically active among humans. Nor are digested peptide fragments or insulin-like growth factors I secreted in response to GH administration.

Moreover, the IGF-I is destroyed by heat along with the active growth hormones. In the (unlikely, given their age) case of young children being fed transgenic fish or milk containing GH, this would be destroyed during heating. The problem however remains for individuals whose digestive tract has been impaired by disease, digestive ulcer or age and which therefore absorbs proteins that have not been broken down and introduces them into the organism. So the risk varies according to the individual. For a susceptible individual, the hazard could have severe consequences, while for a healthy individual of any age, the risk is thought to be low to moderate.

Toxicity

Because proteins are partly broken down in the digestive tract, their direct toxicity is only a mild food safety concern. Moreover, any direct toxicity of proteins would normally be detected during pre-market assessment. The risk is therefore minimal.

The problem arises in connection with unintended and unanticipated effects of products derived from GM animals, linked not only to the proteins that are directly expressed by a transgene but also to any unintended by-products. For example, the insertion of a transgene could disturb a specific metabolic process and cause the expression of metabolites with some toxicity. What we need to know is whether there are effective methods of analysis and screening that can detect a toxic substance or a change in metabolism that introduces toxicity. Whole foods can have a tiny concentration of proteins expressed by a transgene or arising as a by-product of the transgene. Such toxicity can be very difficult to detect given the sensitivity limits imposed by laboratory animal numbers, the small quantities of substances tested in whole foods, and the possibility that any negative effect will seldom occur.

Methods include:

- Analysis of the structural/functional relationship of the substance and of its physical and chemical properties and comparison with existing databases of toxic agents to see if the molecule has similarities with a known toxic agent.
- *In vitro method:* Testing the stability of molecules under different conditions or the ability of the molecule to act on other molecules with specific functions. Few *in vitro* tests have been validated and extrapolated *in vivo* as this is difficult, for the method is still fraught with uncertainties.
- *Animal model*: The best indicator of possible toxicity. However, extrapolation remains uncertain, hence the use of safety factors. This model serves to determine acute, subchronic or chronic toxicity. The problem is that the dietary regime needs to be balanced if parallel side effects are not to be caused (high possibility of bias). A precise diet, specific doses and clear protocols are required.

Institutions, Scientific Expertise and International Standards

International agencies have been charged with establishing international standards for GMOs and derived products. Codex is developing principles for risk analysis applied to foods from GMOs. The first principle is to have a pre-market assessment involving case-by-case study of the direct and indirect consequences of inserting a new gene (toxicity, allergenicity, properties of compounds, stability of genome, nutritional effects and other impacts). Codex principles are however for reference only and not binding.

Evaluation of Safety of Foods Derived from Genetically Modified Animals

As concerns examination of the safety of GM foods, the risks associated with genetic engineering are the involuntary modification of the host genetic substance, which can induce changes in food components, including the toxic substances.

Evaluation of the safety of GM foods should therefore include the following procedures:

- Identification of the organism that has been modified and of the source organism(s) of the inserted gene(s). The origin and nature of all the genetic elements that have been introduced into the modified organism need to be identified, including the structural and regulatory sequences and all other parts of the vector sequences. The possibility of horizontal transfer of genes and related consequences also needs to be studied. If the transfer of genes resistant to antibiotics is intended, the possible health implications of concentrations of micro-organisms in the human digestive tract or other parts of the body need to be considered.
- Identification of primary and secondary genetic products, including a description of the characteristics of the inserted gene. Identification of hazards requires an understanding of the expressed inserted genes, the characteristics, concentration and location of expressed food products and the consequences of expression. For each protein resulting from genetic modification, the factors taken into account include possible toxicity and allergenicity from exposure to the expressed protein.
- Evaluation of the safety of new substances anticipated in the foods (*i.e.* proteins, glucides and lipids).
- Assessment of unanticipated effects on the composition of foods, including assessment of changes in the concentration of nutritive or toxic substances occurring naturally, identification of anti-nutritional factor compounds significantly altered in GM foods and assessment of the safety of compounds showing significant modified concentration.
- Evaluation of any toxin produced directly by modification.
- Evaluation of type of treated food, assuming normal manufacturing procedures.
- Evaluation of food consumption patterns, including identification of potential human population consuming GM foods and probable consumption levels, and evaluation of possible consequences if the consumption of modified foods differs from that of equivalent traditional foods.
- Assessment of allergenicity of new foods.

Current approaches to assessing the toxicity of specific proteins are based on internationally approved, acute or chronic testing on laboratory rodents or on fast-growth domesticated species, such as chickens. However, no repeated dose testing with GM foods or food products has yet been published. As regards allergenicity, most scientists now base their evaluations on a decision tree that has been designed to evaluate the potential allergenicity of foods from GM plants, in the light of current knowledge of what is defined as an allergen and how to better identify an allergen.

Concept of Substantial Equivalence

The OECD's principal contribution to the evaluation of safety of foods from genetically modified animals has been its publication *Safety evaluation of foods derived by modern biotechnology: concepts and principles*. According to this report, the most practical way of determining the relative safety of a new food product is to determine whether it is substantially equivalent to its conventional counterpart, where this exists. This concept was initially applied to organisms of terrestrial origin but, with certain qualifications, was then also applied to products of aquatic biotechnology. Substantial equivalence recognizes that the purpose of evaluation is not to guarantee absolute safety, but to determine whether the transgenic food presents more risks than its conventional counterpart, should this exist. It is generally agreed that such evaluation requires an integrated, specific and gradual approach. The factors taken into consideration when a transgenic food is compared to its conventional counterpart are (as detailed in the previous chapter on the evaluation of safety of foods derived from transgenic animals):

- Its identity, source and composition; its characteristics in relation to the "natural" product;
- The effects of processing and cooking;
- The processing stage, the dna itself and the products of protein expression of the introduced dna; its nutritional qualities;
- The effects on its function;
- Its potential toxicity and allergenicity, as well as possible secondary effects;
- Consumption of the transgenic food and the nutritional impact of its introduction.

Interactions between environmental factors and animal genotype also need to be taken into account. These variations can influence the phenotype and composition of the modified animal and need to be studied during the evaluation of equivalence. If the food derived from a GMO is judged to be substantially equivalent to its conventional counterpart, it is then considered to present the same level of risk. Otherwise, further tests are needed. Confirmation of substantial equivalence does not mean that the product of a genetically modified animal is identical to the conventional product. Comparison does not take all elements into account and substantial equivalence only provides assurance that the appropriate components of safety of the product will be present in equivalent quantities. As the comparative approach links the composition of new foods to existing products with a history of safe use, the consequences of introducing new foods into a diet can be foreseen. The concept means that all similarities between foods from genetically modified animals and conventional foods can be judged to be safe. Differences noted during comparison are subject to further meticulous examination which includes the testing of traditional nutritive elements and toxicological or immunological testing or long-term studies, depending on the differences noted. Foods from transgenetic animals might not have any conventional equivalent or the differences between these foods and their conventional equivalents may not be sufficiently characteristic. The history of safe use of the conventional product could not therefore be used to assess the safety of the new food under study.

Substantial equivalence cannot be used to determine the safety of foods for which there is no comparator. However, that a novel food does not bear much resemblance to its counterpart does not necessarily make it dangerous. Its safety needs to be determined by using traditional food safety approaches, with supplementary testing on the consequences of modification.

POST-APPROVAL MONITORING OF FOODS DERIVED FROM TRANSGENIC ANIMALS

It is widely accepted that foods and food products need to be thoroughly evaluated before being placed on the market, if consumers, animal health and the environment are to be protected. If the assessment raises safety concerns, the product will not be approved for commercial release. Some countries therefore fail to see any scientific justification for the post-approval monitoring of these products. Other countries, on the other hand, see the monitoring of possible negative or positive impacts of novel foods as a logical follow-up to the initial science-based risk assessment.

They argue that post-approval monitoring should be mandatory to consolidate the initial pre-market risk assessment, which was based on the then state of scientific knowledge. This could apply to any pre-market assessment. It is recognized that it is sometimes difficult to draw clear conclusions from epidemiological studies, especially when applied to food compounds. Post-approval monitoring is advocated as the only feasible way of demonstrating the absence of possible long-term or unexpected consequences of consuming a novel food. Implementing this strategy is not easy, however.

TRANSGENIC AGRICULTURE AND CROPS

Genetic engineering is an ever-expanding field in the world of science. In the last ten years, the process of genetically altering organisms has evolved from experimentation to commercial application. It is taking the place of selective breeding for isolating desired traits in different organisms. Transgenics has also led to a better understanding of gene function, allowing for better understanding of human diseases such as Alzheimer's, diabetes, and cancer. For some, such as the starving poor and those afflicted with genetically-caused diseases, transgenic organisms are a blessing. For others, such as poor farmers and the animals themselves, genetic engineering spells the end of the world.

The Federation of European Laboratory Animal Associations defines the term transgenic animal as an animal whose genome has been deliberately modified (Margawati). This definition also holds true for any other transgenic organisms, including plants and bacteria. In fact, the principle of transgenics is the introduction of an alien gene into an organism's DNA. Generally, the gene would not occur naturally in its host organism. Transgenic science was born in 1970 when experiments were conducted to produce different strains of mice; cells originating from one strain were introduced to the embryos of another strain by aggregation or injection during the blastocyst stage (C. River). The resulting offspring showed characteristics from each strain used. These first transgenic animals were created without the use of recombinant DNA, but the process was a building block for the mechanisms of genetic engineering used today (C. River). This process is actually quite similar to that of cloning, when one organism is a genetically identical to its parent.

TRANSGENIC ORGANISM

A transgenic organism is an organism which has been modified with genetic material from another species. The genetic modification is accomplished by inserting DNA into an embryo with the assistance of a virus, a plasmid, or a gene gun. The embryo is allowed to develop, and the mature organism will express the DNA which has been inserted into its genome. Transgenic organisms can also pass the modification on to future generations by breeding with other members of the same species.

Organisms may be genetically modified for any number of reasons. Genetic modification may be performed to make organisms more vigourous, to add resistance to specific threats, or for the goal of expressing a particular trait. It can be used for everything from adding nutrition to staple crops for the benefit of people in the developing world to making fish glow so that they can be sold as novelty pets.

While species cannot interbreed, as a general rule, DNA from one species can express in another. This is because DNA codes for the same thing no matter where it is, so if a researcher can determine which part of an organism's genome codes for a particular trait and the trait is compatible with the organism being modified, the DNA can be successfully inserted to cause the trait to be expressed in a different species.

Rice with Vitamin A

Rice does not contain very much vitamin A. In the poorer parts of Asia, where rice is almost the only food of the rural population, a vitamin A deficiency is common, leading to early blindness. Now Drs. Ingo Potrykus and Peter Beyer, two genetic engineers, have transferred the genes for vitamin A from other species into rice, creating a strain of rice which is rich in vitamin A — the amount of rice in a typical third world diet could provide about fifteen percent of the recommended daily allowance of vitamin A, sufficient to prevent blindness. Now that a few plants with this trait have been created, they are being cross bred with other varieties of rice using conventional breeding techniques, as has been done for centuries.

Such cross breeding could further increase the vitamin A content. The development of rice with vitamin A was carried out at the Swiss Federal Institute of Technology, making free use of patented technology and of the earlier research which had established the basic facts about how plants synthesize vitamins. The corporations holding the various patents all agreed to cost-free use of their patents as long as the rice was to be provided free to poor third-world farmers. The new rice strain was then turned over to the International Rice Research Institute, a non-profit organization based in the Philippines, where it will be evaluated for its adaptability to various growing conditions, food safety, and environmental impacts, etc.

The IRRI preserves thousands of varieties of rice with different genetic characteristics, so the new strain can be cross bred to produce varieties suitable for almost any locality. The result is that rural Asians can soon expect to retain normal eyesight. Genetic engineers also intend to produce a rice variety rich in iron, because iron-deficiency anemia is a common problem in the same rural populations. But this is a more difficult problem than increasing rice's vitamin A content. Rice contains a substance called phytate. Phytate prevents the body from absorbing iron, so it does little good to breed for increased iron content, and the rice plant cannot reproduce without adequate phytate in the grains. Dr. Potrykus hopes to be able to find a gene coding for a protein that will break down phytate when the rice is cooked.

No-Till Agriculture

The world's biggest environmental problem is loss of topsoil to wind and drainage. The US experienced its dust bowl during the early part of the 20th century as a result of ploughing up the prairie. The problem is much worse in tropical soils, which may have a thin, inches thin, layer of topsoil above a type of soil which, when ploughed, turns into a non-porous concrete-like substance. One field, one crop, once. The problem, both in the prairie and in the tropics, is deep ploughing, which kills weeds which would otherwise crowd out the desired crop. The solution is called low-till agriculture. The soil is broken up but not deeply ploughed. Weeds are killed instead by herbicides. A herbicide of choice should be cheap, quickly biodegradable and non-toxic. An excellent choice is a chemical called glyphosate, except that glyphosate kills the crops as well as the weeds. So genetic engineers found a gene which lets plants tolerate glyphosate, and transferred it into soybeans. Today, 63 per cent of the soybeans grown in the US are glyphosate tolerant, allowing soil saving no-till agriculture on half the US soybean acreage. There is another advantage to no-till agriculture. There are lots of plant residues beneath the ground, both root systems and humus transported by earthworms. Ploughing brings this material to the surface, where it can oxidize. Carbon dioxide is created, a greenhouse gas. So transgenic soybeans are a positive factor in postponing global warming. Approximately four tons of carbon dioxide are retained in the soil per acre per year. This saving is applicable to the accounting of CO_2 reductions in the Kyoto Treaty on Climate. In fact, any technology which reduces the need to plough, spray, or till crops will reduce carbon dioxide emission. Consider a tractor pulling a ten foot wide harrow over a square mile of agricultural land. Simple arithmetic shows that the tractor will travel 528 miles, all the while burning gasoline. Perhaps you are thinking that even if ploughing has disadvantages, herbicides don't sound very good either. The very word means "plant killer". But that is not the choice. Traditional soybeans are also grown using herbicides, most of which are far more toxic than glyphosate. On the average, the transgenic soybeans actually use 30 per cent less total herbicide than conventional soybeans. So environmentally, this is a no-brainer.

Witchweed Control

Farmers in east Africa are plagued by a devastating parasitic weed called Striga, or witchweed. Farmers are used to dealing with weeds that grow in the soil alongside the crop and compete for nutrients. From time immemorial, they have dealt with those weeds by pulling them up by hand. Less neighbour intensive methods like spraying and ploughing are now common. But none of these methods work for the witchweed. Striga attacks plants directly, underground, even before the weed has emerged above the soil surface. It sucks nutrients from the seeds and the roots of the crop.

In some parts of Africa, the striga parasite destroys as much as 80 per cent of the crop yield. But now that a herbicide resistance trait can be transferred to a crop, scientists in Israel and Kenya, working together, have demonstrated a new strategy for striga control. Before planting the crop, they soak its seeds in a herbicide. The seeds are unharmed, but they become poisonous to the striga parasite. The seed germinates and sprouts without interference. By the time the crop is harvested, the herbicide has decomposed and disappeared. Their demonstration used herbicide resistant transgenic corn. The same strategy would probably work with Africa's other important grains, sorghum and millet. Soaking seeds would use far less herbicide than spraying it on the ground, and the complex spraying apparatus would not be needed. This is a significant consideration in Africa, where so many farmers are too poor to own expensive equipment.

Saving the Banana

Wild bananas have seeds. They reproduce sexually, like beans and oak trees. But they aren't easy to eat. Bananas grown on plantations have no seeds. They are reproduced by taking cuttings from older banana plants. Cultivated bananas are seedless because they have three of each type of chromosome instead of the normal two of each type. Such plants are called "triploid". They are always sterile. Genetic triploid freaks arise from time to time in nature, but modern breeders can also use chemicals or electric shocks to create triploid mutant cells. Bananas have been cultivated for many thousands of years and there are about three hundred different banana varieties. Each variety was developed by crossbreeding wild bananas. Whenever a promising variety was been produced, the breeder caused it to be triploid, hence seedless.

That plant was cloned by propagating cuttings, and it became the parent of its variety. All bananas plants of the same variety are genetically identical, like identical twins. Of the three hundred varieties, only one single variety completely dominates international trade. It is called Cavendish. It is possible that you have never seen a banana other than a Cavendish banana. Certain kinds of fungus can infect and kill banana plants. In many parts of the world, Cavendish banana plants are being attacked by a fungus called black Sigatoka. Since wild bananas can reproduce sexually, they are not all identical and some wild bananas can resist black Sigatoka.

If the black Sigatoka fungus is present in a region, the resistant banana types become prevalent. But the Cavendish banana plants have no resistance. They are all identical so they all die. To grow bananas commercially, growers must spray their plants with fungicides. Year after year, the black Sigatoka fungus has been evolving resistance to these fungicides, so growers have to spray more and more fungicide each year. Approximately one third of the cost of raising a banana is the cost of spraying it with fungicides, and it gets more and more costly each year.

A form of black Sigatoka banana disease, now spreading around the world, can tolerate all known fungicides. Soon it will attack bananas in Central America and the Carribean islands, the heartland of banana culture. The Cavendish banana will become virtually extinct. Agronomists estimate that this will happen within ten years. This is not a fairy tale. It has happened before. Forty years ago, the most popular variety of banana was one called Gros Michel. But Gros Michel was susceptible to a fungus called "race 1 Panama disease". Now it is gone. Cavendish bananas, which are not susceptible to race 1 Panama disease, replaced them. (A related fungus, race 4 Panama disease, can infect Cavendish banana plants. At present it is only found in Malaysia and some nearby countries.) In turn, some other variety of banana could replace the Cavendish. It would look different and taste different but it would still be a banana. There is only one practical way to save the Cavendish banana. It must be given some combination of genes from wild bananas which are not susceptible to the fungus. But this can't be done by any natural technique. Cross-breeding can create other banana varieties because wild bananas reproduce sexually. But Cavendish bananas reproduce only by cloning. Conventional breeding would have to rely on rare mutated Cavendish banana plants which can produce seeds and which can therefore be crossbred, in theory. Even the mutated plants produce only a tiny number of viable seeds, as few as two or three seeds in a hundred pounds of bananas. No banana breeding experts think that they can breed fungus resistance into a Cavendish banana variety in only ten years. Biotechnology can rescue the Cavendish banana. Scientists in Belgium have used genetic engineering to transfer some fungal resistance genes from wild bananas. The transformed Cavendish plants are not susceptible to the fungus and they can then be reproduced into nursery stock by the usual method of taking cuttings.

The Eggplant in Winter

The edible part of an eggplant is formed from the ovary of its flower. In this way, it is like the edible flesh of an apple, a pepper or grape. When we eat these fruits, we discard the seeds. But the plants only transform their ovaries into fruits when they start to produce seeds, although in the case of an eggplant, its seeds are so tiny that we ignore them. Eggplants will only set seeds in warm weather, so to grow them in the winter in an unheated greenhouse, the grower must use a chemical to trick the plant into beginning fruit development without setting seed. Such fruits do not grow very large or very fast under these conditions. So eggplants are expensive in the winter.

But now scientists in Italy have transferred two genes into a variety of eggplant, which not only allows the plant to set fruit in cool greenhouse conditions without chemicals, but also increases productivity of the same plant in either hot or cold weather. The eggplant variety that the Italian scientists created is seedless. One of the two transferred genes is a switching gene which is turned on only in the ovary part of a flower. That gene turns on the other transferred gene, which makes a protein involved in synthesizing a growth hormone.

The growth hormone makes the ovary grow into the fruit, just as it would have done in a traditional eggplant making seeds. Neither gene requires either seed setting or warm weather. Where does one get seeds to produce large numbers of seedless eggplants? The transformed plants produce pollen, so they can be crossed with traditional eggplant varieties and the hybrid produced by that crossing has the seedless and self-starting property. The scientists report productivity increases of 37 per cent for the new variety, and they believe that the seedless type would be more marketable.

FRUSTRATE TRANSGENIC AGRICULTURE SYSTEM

Although there are legitimate reasons to oppose genetic engineered agriculture, or at least to demand the most careful controls, there are a community of opponents who have taken their opposition beyond what is ethical. I am not talking about a reasonable opposition expressing the concerns summarized earlier, but rather about an opposition for which the ends justify the means, including lies, vandalism, etc. I need to stress this point. There are many people, sincerely opposed to genetic engineered crops, whose ethics I do not question. I believe that most of the opponents of biotechnology would fall into that category. Yet, in too many cases, those sincere concerns are based entirely on misinformation which originated in deliberate lies and fear mongering.

We are here exposing the ethics of the people who have created the lies. These opponents must have some motive, and it seems that an alliance has emerged between at least four groups, each with its own agenda. First there are people who sincerely believe that genetic engineering is ethically wrong, and that anything they do to stop it from happening is therefore right. Second there are foreign governments and their constituencies who are worried about American domination of agriculture. Closely related to this are advocates of organic agriculture who seem to be engendering public fear to make their own products more salable. Third, there are environmental groups who have been misled by a radical fringe and have become willing to do anything to stop genetic engineering agriculture. The most conspicuous among these groups is Greenpeace. Finally, there are people opposed to capitalism or to large businesses dominating agriculture. Opposition from environmental groups is particularly frustrating to me. Most measures which benefit the environment require people to give up something, and they don't like to do that. Recycling is nearly painless and saves money, but many people won't make the effort. A few degrees adjustment of a thermostat could save vast amounts of energy, but most people would rather be comfortable. We end up settling for half measures. But here is a technology that benefits the environment without asking people to give up anything, and its biggest opposition comes from environmental groups.

MISINFORMATION ABOUT FOOD SAFETY

There is a deliberate campaign to frighten people about the safety of the food supply. This campaign has worked successfully in England and in much of Europe. Dr. Arpad Pusztai, who worked at the Rowett Institute in Aberdeen,

Scotland, performed an experiment. It began when a gene was transferred from a poisonous plant, the snowdrop, into a potato. The transferred gene specifies the production of a poisonous compound called lectin. Dr. Pusztai proceeded to experiment with rats. Some rats were fed with the raw potatoes which were genetically engineered to contain the poison. The rats in the control group were fed ordinary raw potatoes and were also given the amount of lectin poison which the first group of rats would have gotten from eating the transgenic potatoes. Both groups of rats developed malformed organs, and there was no statistically significant difference between the rats who consumed the poisonous potatoes and those who consumed the poison.

However, Dr. Pusztai claimed that his data showed that the rats who ate the genetically modified potatoes had more deformed organs. No scientific journal would publish Dr. Pusztai's interpretation, and his institution would not support him. He hired an independent statistician to review his data, who also considered the data to show no difference between the two groups. But Dr. Puzstai held to his opinion. Eventually the disagreement became serious enough that his connection with the Rowett Institute was ended. The opponents of genetic engineering, mostly in England, have blown this result into a cause celebre. Dr. Pusztai is portrayed as muzzled by the scientific establishment, although the British medical journal Lancet eventually published Pusztai's paper over the recommendations of its reviewers because of the widespread public interest.

The British tabloid press covers this story continuously, with lurid photographs of deformed rat organs. The potatoes genetically engineered to be poisonous became synonymous with all transgenic food, called Frankenstein food in the tabloids. There are varieties of potatoes, bred to be eaten, engineered to resist insects, viruses and fungi. All these varieties have been fed to rats and have never harmed them. The opponents of transgenic food have no explanation for that - they are content to use one probably misinterpreted experiment with potatoes nobody will ever eat, to stir up doubts about food safety.

During safety testing of transgenic tomatoes, Dr. Belinda Martineau discovered that when rats are fed huge amounts of tomato paste they can develop stomach lesions. It doesn't matter whether the tomatoes are transgenic or not. But before there were transgenic tomatoes, nobody had ever fed rats such large doses. Martineau's results were reported to the FDA but opponents routinely call this a "cover-up" of a health hazard with transgenic tomatoes. As another example, I read an op-ed article in the Boston Globe in August of 1999, written by Paul Billings, a member of the board of the Council for Responsible Genetics. The gist of the article is that dangerous untested foods are being foisted upon an unsuspecting American public, by mad scientists. Every genetically engineered crop has been tested for safety. The testing has been much more extensive than that for any other foods, including foods developed by radiation induced mutation. Dr. Billings knows this. He knows about the government testing rules that establish the safety of each individual crop.

He wants genetically modified food to be tested as strictly as drugs are tested. Billings' op-ed article contains only one 'fact' that most people would not have known before - the rest is either his opinion or just wrong. He says that transgenic soybeans have been shown to be deficient in a certain unidentified nutrient. It is not easy to track down the source of this 'fact', but I did it. The nutrient in question is a phyto-estrogen (also known as a phytosterol). Although phyto-estrogens are not essential to human health, there is some indication that they help prevent cancer. The study that indicates that transgenic soybeans are deficient in phyto-estrogens comes from Dr. Marc Lappe, who wrote the book "Against The Grain", a polemic against genetic engineering and especially against the Monsanto Co., the leading company in the field, which developed the soybeans in question. Here is how Dr. Lappe established that genetic engineered soybeans are defective. Understand that there are dozens of varieties of soybeans with the herbicide tolerance trait and over a hundred varieties of conventional soybeans. Dr. Lappe compared one conventional variety with one transgenic variety. He found a 12 per cent difference. But individual soybean varieties vary by more than 100 per cent in their phyto-estrogen content.

Also phyto-estrogen content is not stable. It declines with storage, by much more than the 12 per cent difference. But Lappe at least reports his data along with his biased interpretation of it. Dr. Billings reports only the interpretation without any indication that it comes from a biased scientist whose own data show insignificant variations in a nutrient whose role in human health is not even firmly established. Dr. Billings seeks only to mislead people. Not one reader in a thousand would do what I did, track down the data. His purpose is to plant a little seed of doubt about food safety, hoping that it will fester in our minds, mingle with similar misinformation, and eventually become accepted fact. Not everything misleading is false. Propagandists are very skilled at making true statements that seem to imply something quite different. For example, they frequently say that the FDA's rules about testing genetically modified food are voluntary, implying that some testing doesn't get done. Under current law, FDA has no authority to require safety tests for any food, transgenic or conventional, although it can prevent the sale of foods it considers unsafe. But, in fact, each developer of a transgenic crop has consulted with FDA and performed every test FDA suggested. They also like to say that safety test results are trade secrets. True, it would be legal to keep test results secret.

But no developer has done it. Earlier I mentioned a problem with a soybean with a Brazil nut gene. People allergic to Brazil nuts should not expect to have to avoid soybeans, but the allergen was identified by testing and therefore the modified soybeans were never created. Also, the scientists who demonstrated this published their results in a scientific journal, and this led the FDA to not allow any gene to be transferred into a food from a species known to cause allergies. That should be seen as evidence that the genetic engineers are responsible people, and that testing is working well. But the unethical opponents of biotechnology routinely present this episode, carefully worded, as if there was a near disaster, revealing gross problems with the current regulatory system.

MORE FROM THE SUN

Plants derive energy from sunlight and use it to make sugar from carbon dioxide and water. This is photosynthesis. Scientists still do not have a complete understanding of how photosynthesis happens, although they know most of the steps. They know many of the genes which create the proteins needed for photosynthesis. They also know that there are differences in photosynthesis from one species to another. It happens that corn is an overachiever.

Corn plants make more sugar per unit of sunlight than any of the other grains. An international team of scientists from Japan and from Washington State has transferred three of corn's photosynthesis genes into a rice plant. Early indications are that the transformed rice is more productive than the original rice variety. A more important potential application may be the development of very fast growing trees. If global warming cannot be prevented by adding less carbon dioxide to the atmosphere, by burning less coal and oil, the only alternative is to depend on processes that remove it. Number one on that list is growing new trees. Anything that makes agriculture more efficient can make more land available for growing trees. Anything that makes those trees grow faster removes more carbon dioxide from the atmosphere. For many environmentalists, preventing global warming is the highest priority. But proposed measures to restrict burning fossil fuels have encountered fierce political resistance. Opponents claim that such restrictions would be cost too much money, and would cost people their jobs, their comfort and their prosperity. By contrast, nobody loses anything if carbon dioxide is removed from the atmosphere by growing new trees.

Toxic Soils

Some soils are poor for plant growth because their mineral content is toxic. A high aluminum content is the most frequent problem, especially in acidic soils. But it has been possible to identify a few genes which enable some plants to extract aluminum compounds from soil and sequester them harmlessly in their fibrous parts. Recently, Florida scientists discovered a type of fern which can extract arsenic from the soil, although they do not yet know how the fern does this. But other teams have identified genes that can enable plants to remove cadmium, zinc and mercury from soils. By transferring such genes to fast growing plants, it should be possible to clean up some toxic soils in much the same way as we can use bacteria to clean up oil spills. In the nearer term, there is the work of Mexican scientist Luis Herrera Estrella.

He transferred into corn a gene that allows the plant to overproduce a natural chemical, citric acid, which it then excretes through the roots. Citric acid binds to aluminum and prevents the plant from taking it up from the soil. Herrera's approach is not to extract aluminum from the soil but to prevent it from passing from the soil to the plant. A much larger problem is salt-contaminated soil caused by irrigation. Rainwater is very pure, but water borrowed from rivers contains some dissolved salt. Over many years of irrigation, the salt accumulates. But water cannot get from soil to roots if the soil water is saltier than the intracellular water. In fact, water goes the other way, from plant to soil, and the plant dies. A gene was identified in a relative of cabbage. This gene enables the plant to pump salt from the soil into an isolated part of a cell, called a vacuole, where it is stored without harm to the plant. When salt is thus removed from the soil around the roots the plant can then take up the less salty water. The salt-tolerance gene was experimentally transferred to a tomato plant, where a control gene keeps it turned on all the time. The resulting tomato plant is able to grow well in salty soils. Happily, the fruit is not high in salt, but the plant's stems, leaves and roots are loaded with salt, so after the growing season the plant parts could be shipped elsewhere, making the soil become less salty each year. It's one more case of an environmental problem that can be solved by gene transfer.

Biological Pest Controls

Farmers, for very obvious reasons, would prefer not to use pesticides. They cost money and they are dangerous to use. Farmers much prefer Integrated Pest Management (IPM), a system that combines many different methods of suppressing crop pests, including encouraging predatory insects. Farmers even buy them. Agricultural distributors can supply such insects as ladybugs, praying mantids, lacewings and parasitic wasps. Integrated pest management includes using pesticides when other means are insufficient. But when a crop is sprayed with conventional insecticide, the harmful insects are not the only victims. Predatory insects may also be wiped out. Without any predators available, the pest populations can recover quickly, so that a second application of pesticide is required, which also kills the insect predators. This vicious cycle could be broken if genetical engineers can develop predatory insects resistant to common pesticides.

Rust Resistance

To a plant scientist, rust has nothing to do with oxidized metal. It is a plant disease caused by a fungus. It blights all the cereal crops, barley, wheat, oats, corn, millet and sorghum, but not rice. The rust fungus reproduces itself by forming club shaped cells called basidia. Each basidium bears four spores. When the spores are ready, they are released and carried by the wind. The fungus infecting a single grain of wheat can easily produce millions of spores. The spores are so light that they can travel several times around the world before falling to the ground. Although some varieties are more resistant to rust than others, no variety is immune. But rice must contain some combination of genes that confers immunity to rust. If these genes can be identified and if their function can be deciphered, it should be possible to transfer them to other cereal crops and end, once and for all, this most important cause of famine.

Fast Growing Trees

Making paper requires large amounts of natural cellulose. Some can be derived from recycling, but most of our paper is made from freshly cut trees. The best trees for paper-making are fast growing softwoods with low resin content, like aspens. Genetic engineers have transferred genes for pest resistance and herbicide resistance into aspen and have tinkered with the genetic switches that promote growth to create a fast growing aspen that could supply our paper needs using considerably less land. The paper-making process must bleach out the brown colour of lignin, one of the components of wood. The bleaches used to be dumped in the nearest rivers, an important and highly visible kind of pollution. This is no longer allowed, but the disposal of chemicals from paper mills is still a major headache. At the Michigan Tech University, researchers have reduced the lignin content of aspen so that fewer chemicals are needed in the paper making process.

Fast Growing Fish

Most of the salmon we eat are caught wild, but some are grown in farm ponds. It takes about three years for a salmon to grow from fingerling size to optimum marketing size. In wild salmon a control gene turns on the gene for growth hormone, but only in the pituitary gland and primarily in warm water. So genetic engineers used a different control gene to turn on a growth hormone gene in cold water. That control gene was transferred from an ocean pout, and it originally turned on a gene for a protein that helped the pout tolerate very cold water.

The resulting creature looks and tastes just like the wild type salmon but it grows three times faster so it ought to be cheaper to produce. Wild salmon are now under environmental pressure from overfishing and because many of the streams where they lay their eggs are either polluted or inaccessible. If farmed salmon can economically replace more wild salmon, the pressure on this desirable species could be reduced dramatically. There's a need for fast growing fish in rice growing regions. Rice is planted in standing water, but it is harvested from dry ground. With a slow growing rice farmers often raised fish in the rice paddy alongside the young plants. But newer strains of rice mature almost twice as fast as traditional varieties. Although this lets farmers grow more crops per year, unfortunately the rice paddies are not flooded long enough to raise fish. If genetic engineers were able to make fish grow faster, the farmers could again exploit this valuable protein resource.

Consumer Traits

Most of the traits in the examples mentioned so far have been targeted at the producer. The no-till soybeans are cheaper to grow because ploughing costs money. The cotton is cheaper to grow because chemicals cost money. The chymosin from yeast is cheaper than chymosin from calves. The consumer doesn't know how much water or fertilizer the farmer used, or whether a salmon is one year old or three. The vitamin enriched rice is the only one of the examples where the final product is better, rather than cheaper, for the person who eats it. But in the future we can expect to see "consumer traits". One of the first to appear will be potatoes genetically engineered to have a higher percentage of solids.

If you love french fried potatoes but don't like the calories, you will love the new potatoes. They will absorb less oil but stay crispier longer. Another valuable trait coming soon is coffee beans without caffeine. Caffeine is now removed from coffee by a chemical treatment, invented by German chemist Ludwig Roselius. Decaf is the only coffee I drink so I am looking forward to cheaper decaf coffee. In addition, many common foods are not safe for everyone. For example, peanuts cause a life threatening allergy in some people, especially children. Allergens are unusual proteins which are digested very slowly. Someone who has an allergy to a widely used food needs to read the small print on product labels, quiz the waiter in a restaurant, etc. If a child has the allergy, the problem is that much harder to manage. But peanuts or other crops are now being genetically engineered to eliminate these allergens. USDA scientists have identified the principal allergen in soybeans and have successfully developed modified soybeans which do not produce that allergen. It is much easier to deactivate a gene, once its function is discovered, than it is to transfer a gene from one organism to another. The same antisense trick that was used to delay ripening can be used to suppress synthesis of an allergen. Therefore, once a gene has been identified which codes for an allergenic protein, the technology to eliminate that allergen from food crops is relatively easy, unless the allergenic protein is important to the life processes of the plant.

Lawns Don't Need Frequent Mowing

In addition to food and fibre, genetic engineers are working to modify grass. Farmland constitutes mankind's biggest footprint on the earth, but lawns, athletic fields, golf courses, etc. have the biggest footprint in some communities. Environmentalists have many criticisms of lawns. They must be mowed regularly, which uses gasoline, creates noise pollution, and takes up people's time. In addition, the growing grass uses a very significant amount of water, fertilizers and pesticides, to make longer leaves which are then cut off by the mower. But genetic engineers are trying to develop a variety of grass which reaches a desired length and then dramatically slows its growth. Combined with pest resistance genes, this new grass would be almost as simple to maintain as astroturf.

TRANSGENIC CROPS

The application of recombinant DNA technology has revolutionized plant breeding in recent years. Genetic transformation in plant breeding greatly increases the gene pool and broadens the scope of genetic changes that modern breeders may draw upon. The first generation of genetically modified (GM) crop plants hold great promise in carrying new or improved agronomical traits that require less intensive farming methods or which have improved quality traits. These GM crop plants have been produced to improve farming production systems, aiming, for instance, to reduce negative impacts on the environment, such as fertilizers, herbicides and pesticides.

In general they are of little direct interest to the consumer. However, the genetic manipulation of the primary and secondary metabolism of plants offers distinct possibilities in the development of nutritionally improved products, raw materials with added pharmaceutical value, and crops with real health benefits for the consumer. These are second-generation GM plants. Recently, several prototypes in molecular plant breeding with novel traits for (non-) food and industrial applications have been introduced into the market, and many more GM crops have the prospect of commercial exploitation in the near future.

The safety and wholesomeness of these GM crop plants raises questions beyond those posed by conventional foods. In this context, traditional plant breeding of most food crops has conformed to the standard 'generally recognized as safe' (GRAS). In some instances safety is assumed to be assured during performance and quality testing by breeders (*e.g.* glucosinolates and erucic acid in rape seed). However, the process is largely focused on the comparison of agronomic and phenotypic characteristics, such as colour, maturity, yield or disease resistance.

Products with an unusual taste or that have harmful effects following ingestion or skin contact have normally been rejected from the breeding programme. However, the safety of conventional plant varieties has usually been presumed from the assumption that prudent consumers will avoid those species once it is known that they cannot consume them without adverse health effects. When the commercial production of GM plants started, it sparked a debate centred on issues associated with concerns on environmental, human and animal safety matters. Many people questioned whether the 'simple' GRAS approach, based on historical and empirical experiences, is sufficient to guarantee food safety of the GM crop plants.

It is also asked whether the risks of chronic exposure of humans and animals to GM crops are adequately covered, and whether large-scale breeding of GM plants will have detrimental effects on ecosystems.

FOOD SAFETY ISSUES

Although the variety registration of existing crop plants has not resulted in adverse effects in humans, the conventional assessments carried out by plant breeders was not considered to be sufficient to ensure food safety of GM crop plants. As a result, Europe issued a regulation on novel foods and novel food ingredients that sets the legal framework for the market introduction of genetically modified organisms (GMOs). This came into force on 16 May 1997.

The Novel Foods Regulation establishes a system of mandatory pre-market approvals of novel foodstuffs, including GM plants. The accompanying EU guideline 97/618/EC gave a clear indication of the types of data that are needed to form the basis of an assessment. In particular, four central safety issues have been set forth related to food and feed safety:

- The nutritional and toxicological consequences of inserted gene products;
- The potential of pleiotropic (unintended) effects in the host organism due to the insertion event;
- The allergenicity of expressed proteins and novel foodstuffs;
- The potential of gene transfer to human and animal gut flora.

This means that in transgenic insect-resistant Bt tomatoes encoding the C-terminal truncated Bt2 gene derived from a *Bacillus thuringiensis* (Bt) strain, *IAb5*, the novel protein has to be evaluated on its own merit, while the safety and wholesomeness of the 'remaining' transgenic tomato is assessed separately. Whereas, for example, in antisense RNA exogalactanase tomatoes, which have improved rheologic characteristics due to downregulation of the endogenous exogalactanase activity, the safety assessment is primarily focused on the 'remaining' novel fruit. In general, there is basic agreement on the safety issues to be addressed.

One particular area of concern is the possibility of unexpected or unintended metabolic perturbations due to genetic modification that may alter, for instance, levels of nutrients and health-influencing components. The use of rDNA techniques does not necessarily result in fundamental changes in crop plants compared to the food produced by conventional breeding methods. However, it should be emphasized that a uniform international agreement for the evaluation of GMOs is urgently needed. For example, an agreement on how to establish in practical terms the similarities or differences between the 'remaining' GM crop plant and the traditionally bred plant seems to be much more difficult to achieve. This chapter is devoted to the description of advanced strategies for the evaluation of potential unintended effects in GM crop plants. Pleiotropic (unintended) alterations in agronomic traits or composition may arise from insertion mutagenesis or as a result of metabolic effects of the novel gene product(s). The challenge is to gain the ability to discriminate between metabolic alterations due to somaclonal variations, unintended effects or natural diversity. Thereto, emphasis is given to a platform of technologies designed to identify changes in the molecular machinery of crop plants. As is the case for detection methods for GM crops and derived products, there are needs and criteria for sampling strategies, statistical models and manufacturing standard reference matehers recently.

UNINTENDED EFFECTS

Strategies for assessing the food safety and wholesomeness of novel foodstuffs are currently in the exploration phase. Above all there is a need for uniform and harmonized quantitative methods for the testing of potential unintended effects in whole foods, such as GM crop plants are. The safety of, for example, synthetic chemicals and food additives is established by assessing separate elements of the compound in question.

Although very successful in case of single chemicals, the evaluation of a whole food product may not be a simple addition of individual assessments of the constituents. Traditional testing protocols involving labouratory animal 90-day feeding trials with (whole) food products are far from ideal. This type of animal feeding study with complex food matrices is complicated by the likelihood of nutritional imbalances leading to dietary problems, confounding factors, an insensitivity for specific endpoints, and the impossibility of using large safety margins, if any.

Therefore, in a global context several regulatory, research and industry-based bodies have proposed alternative concepts for the safety evaluation of genetically engineered foods and food ingredients. In particular the Organisation for Economic Co-operation and Development (OECD), the Food and Agriculture Organization (FAO) and the World Health Organization (WHO) have developed concepts and principles for the safety evaluation of GMOs and derived products. Their reports concluded that risk assessments should be directed at demonstrating that a GM crop plant or derived food product is as safe as a traditional or previously authorized product. Accordingly the EU Competent Authority incorporated these general principles into their Novel Foods Regulation and accompanying guidelines. In this context, novel foods and food ingredients that are considered to be 'no longer equivalent' to traditionally bred plants are subject to labelling as defined in the Council Regulation 1139/98.

Substantial Equivalence

The term 'substantial equivalence' appears in the EU Novel Foods Regulation as a major principle in the risk assessment. It emphasizes that, within the context of the Regulation the evaluation of the safety and nutritional value of a GM plant should be focused on a comparative analysis with conventionally bred products.

OECD's Group of National Experts on Safety in Biotechnology enunciated this comparative approach towards the safety assessment of transgenic food plants as the concept of substantial equivalence. In their report, it is assumed that existing food organisms possess a long history of safe use (*i.e*. GRAS). Consequently, the most practical approach to the determination of safety is to consider whether the GM plants are substantially equivalent to analogous food product(s), if such exist. Once substantial equivalence of a novel foodstuff has been established, it provides assurance of safety that is equal to or better than that of its comparators.

Further clarification of the concept of substantial equivalence originates from a WHO sponsored workshop in 1994. It has been concluded that the establishment of substantial equivalence is not a safety assessment *per se*, but a dynamic, analytical exercise. Thus, it may mean that analysis of, for instance, gene expression patterns, global changes in protein expression and/or differences in metabolic capabilities (the metabolome) of the novel product in comparison with conventionally bred products should be performed in order to examine equivalence.

This strategy has worked satisfactorily for the assessment of the first-generation crop plants for which sufficient background knowledge is available. However, relatively less experience has been gained with the safety and wholesomeness evaluation of novel food plants for which no substantial equivalence can be established. Even in the case of single gene modification, potential alterations in metabolic pathways are difficult to identify and, consequently, the implications of the genetic modification process on the metabolism of plants are poorly understood. Moreover, data on the mechanisms by which plants regulate, for instance, their response to environmental stress factors and other conditions are remarkably scanty.

Critical Nutrients and key Toxicants

Application of the concept of substantial equivalence appears in practice to lead to various interpretations. Different requirements for risk assessment with regards to the potential of unintended effects exist as a result of genetic modification. Actual approaches entail a consideration of the characteristics of host and donor organism, phenotypic properties and toxicological evaluation and include comparative determinations of any changes in critical (anti-)nutrients and key toxicants for the food source in question. In general, the analysis of an expanded spectrum of components is unnecessary, but should be considered if there is an indication from other traits that there may be an unintended secondary effect of genetic modification.

However, compositional analyses based on single parameters using a list of crop-specific critical (anti-)nutrients and key toxicants as a minimum requirement of screening for possible variations in the content has its limitations. Limited data are typically available for the less important food components, as in less well-known crops most critical (anti-)nutrients and natural toxicants will be unknown.

Moreover, the toxicants to be assessed may be partly influenced by the cellular function of the product encoded by the inserted gene. Furthermore, the natural variation of critical comparators may mask the evidence of secondary effects due to genetic modification. Thus, there is no criterion that determines whether a difference between a GM plant and its natural counterpart is outside the literature values for the ranges of parameters that express natural diversity.

Relevant information on indicators of unexpected effects may not have been observed during the development of an edible transgenic crop species. In the future other compounds could become of importance, such as anti-oxidative, oestrogenic and anti-carcinogenic compounds. In determining critical nutrients and/or key toxicants, differences in consumption patterns and practices of processing and consumption in various geographical settings and cultures must also be recognized. It is our opinion that conclusions about relative safety and wholesomeness based on a list of selected single components may not be equally valid in a global context and in all times. Therefore, there is an urgent need to set up generic approaches using platform technologies to evaluate undesirable metabolic perturbations in GM food crops.

A Post-genome Challenge

In our view the identification of unintended effects encompasses the ability to interpret and use innovations in molecular genetics research such as genomics, proteomics and metabolomics on a crop-by-crop basis. This will establish a whole data package directed towards a holistic view on possible unintended side effects due to genetic modification. In essence, the focus is shifting towards molecular characterization to understand functional activity. Vital to this approach will be informative profiles of molecules at different integration levels, *e.g.* mRNAs, proteins including post-translational modifications and molecules at the level of primary and secondary plant products and metabolites that might be useful benchmarks for the detection of unintended effects.

The relative importance of these various 'profiling' techniques in establishing data sets for assessing unintended effects is not indicated by their order and will vary from species to species. Alterations in, for example, expression levels, post-transitional modification, interactions or chemical composition do not *per se* imply that the product is less safe or even unsafe. Such a platform of technologies may determine whether unintended effects exist between the GM plant and its control(s) or whether there is substantial equivalence apart from certain well-defined expected differences.

Measures of the relative activities of various molecular constituents associated with metabolic capabilities in non- and modified plant cells under different conditions will provide new insights into how metabolism and genetic modification are orchestrated. But it should be emphasized that the fields of plant genomics and proteomics are still in their infancy.

CDNA MICROARRAYS

Over the past few years, the DNA microarray technology has emerged as a powerful high-throughput method for the analysis of gene expression. At present, microarrays for gene expression studies generally consist of cDNAs, which are physically deposited onto small glass surfaces. The major advantage of DNA microarray technology over conventional gene-profiling techniques is that it allows small-scale analysis of expression of a large number of genes in a sensitive and quantitative manner. Furthermore, it allows comparison of gene expression profiles under a large number of different conditions.

The cDNA microarray technology has already been successfully introduced in disciplines ranging from cell and developmental biology to drug development and pharmacogenomics. In the following a number of technical aspects of the technology as well as its potential value for the safety assessment of genetic modification of food plants will be discussed. The cDNA microarrays are produced by deposition onto a solid surface of purified PCR products corresponding to specific genes.

Typically a few nanolitres of DNA solution (100-500 μg/ml) are spotted using a micro-dispensing robot. Essential information on robot construction and protocols can be found on the Internet. In general, with respect to the spotting devices, a higher density of spots is accomplished when using a passive dispenser (> 2500 DNA spots/cm^2). Thereto, glass slides covered with a positively charged layer (*e.g.* poly-l-lysine) or carrying reactive groups are most commonly used.

To analyse the expression of genes in a plant cell or tissue sample, polyA+ RNA (typically 0.5-2.5 μg is required for each reaction) is purified and labelled by reverse transcription using an oligo-dT primer incorporating a fluorescently labelled nucleotide. The labelled cDNA pool is then hybridized to the microarray. The required hybridization conditions (*i.e.* sample concentration, stringency of hybridization) and parameters, such as detection level and the correlation between transcript concentration and hybridization have been determined for each experimental setting.

The intensity of a hybridization signal, which is a measure of the relative expression level of each gene, can be read by charge-coupled device (CCD) cameras or confocal/non-confocal laser scanners. RNAs extracted from two different samples can also be simultaneously analysed with a microarray by labelling each with a different fluorescent label using either Cy-3 or Cy-5. At RIKILT-DLO we are currently testing whether the analysis of differential gene expression using DNA microarrays is an informative strategy for the safety evaluation of GM plants.

To address this question, cDNA libraries are constructed and enriched by a subtractive cloning approach for cDNAs preferentially expressed in either green or red tomato fruit. These subtracted cDNAs, as well as control cDNAs representing known tomato genes (*i.e.* sequence information derived from public sequence databases), have been spotted on arrays and are used for comparison of gene expression profiling of control and genetically modified tomatoes. In the near future we will construct more defined arrays containing genes of toxicologically relevant pathways, such as those involved in the synthesis of natural plant toxins (*e.g.* the glycoalkaloids á-tomatine and á-solanine).

MRNA FINGERPRINTING

A very promising strategy to evaluate the equivalence of genetically modified crops is gene expression profiling. Conventional methods for the analysis of differential gene expression include Northern blotting, S1 nuclease protection, comparative expressed sequence tag (EST) sequencing, differential display and serial analysis of gene expression. More recently, Kok *et al.* have designed a method of detecting altered gene expression by means of mRNA fingerprinting or reverse transcription polymerase chain reaction (RT-PCR). Based on the original concept of Liang and Pardee specific subsets of the mRNA population of a tomato plant were amplified. This enables the expression of different genes in the GM crop and genes in the parental line to be compared and any empirical information about significant alterations to be established.

The method is still under development and requires further validation, but has been shown to be reproducible and able to detect important differences in gene expression. Some of the most significant remaining problems are that the method largely depends on individual skills based on experience and is time-consuming, even in its most simplified form.

PROTEOME PROFILING

The introduction of Bt genes of bacterial origin may theoretically influence the nature of post-translation modification (*e.g.* glycosylation) in the transgenic tomato fruit. This is because as well as the impact of the protein moiety, for example, the type of glycosylation pattern may alter due to inactivation of glycosidases and glycosyltransferases in the host cell. In addition, glycosylation of the inserted Bt gene product in the plant might result in altered protein activity and stability.

The importance of structurally elucidating these oligosaccharides, especially the *N*-glycans, can be attributed to their widespread functions in the cell. These functions include correct folding, biological activity and stability of proteins as well as involvement in plant development. Moreover, *N*-glycans represent antigenic epitopes by themselves. Immunological as well as biochemical studies revealed the allergenic potency for the á1,3-fucosylation of the proximal *N*-acetylglucosamine residue. Additionally, also the β1,2-xylosylation of the β core mannose is suspected to be an allergenic epitope. Recently, there is increasing evidence that these *N*-linked oligosaccharides may have a substantial impact on the immunogenicity of glycoproteins, as *N*-glycans are suspected to play a certain role in the adverse food reactions of hypersensitized patients. At the Institute for Agrobiotechnology, Austria, this issue was tackled by developing a generic technology for isolating, purifying and characterizing *N*-linked oligosaccharides in plant tissues.

Our low-temperature acetone powder method appeared to be most suitable for the isolation of (glyco-)proteins from plant tissue (*e.g.* tomato fruit). The resulting (glyco-)protein fraction was proteolytically digested and corresponding glycopeptides purified by cation-exchange chromatography. The oligosaccharides (*N*-glycans including á1,3-fucosylated oligosaccharides) were released from the peptide backbones by *N*-glycosidase A (PNGase A) and fluorescently labelled with 2-aminopyridine.

Subsequently, pyridylaminated glycans were separated and analysed structurally by repeated two-dimensional high-performance liquid chromatography (2D HPLC) using a size-fractionation and a reverse-phase column in combination with an exoglycosidase treatment. MALDI-TOF (matrix-assisted laser desorption ionization time-of-light) mass spectrometry was applied for further confirmation of the observed structures. Sixteen different *N*-glycosidic structures could be detected in tomato fruit. The two most abundant glycans in all the tomato samples showed identical properties to those of the major *N*-linked oligosaccharides of horseradish peroxidase (MMXF: Man(3)Xyl-FucGlcNAc(2)) and pineapple stem bromelain (MOXF: Man(2)Xyl-FucGlcNAc(2)), respectively, and accounted for about 65-78 per cent of the total glycan content. Oligomannosidic glycans occurred in only small quantities (3-9 per cent). The majority of the *N*-glycans found was β 1,2-xylosylated and carried an á1,3-fucose residue linked to terminal N-acetylglucosamine.

A structural element was identified and shown to be an IgE-reactive determinant, which contributes to cross-reactivities among non-related glycoproteins, and is present in a wide variety of plants extracts. It is proposed that this structural element might also be of importance in some cases of food and pollen allergy.

The carbohydrate profiling of, for example, populations of green and red-ripe fruit of non-modified and modified Bt tomato plants (expressing the *cryIA5* gene) revealed no differences in the nature of *N*-glycans, nor were there significant differences in the relative amounts of glycan. Interestingly, the relative amounts found were quite similar across seasons and comparable to the commercial variety Notoro used as an extended control. As far as the *N*-linked oligosaccharides were involved it was not possible to detect significant alterations in the á1,3-fucose content of Bt tomatoes due to genetic modification.

PROFILING OF THE METABOLOME

Among other factors, identity of a complex GM crop plant with its traditionally bred parent entails metabolism and the level of (un-)desirable substances contained therein. Therefore, we would like to build up an overall picture of components associated with different metabolic capabilities in the (non-)modified plant tissue. A multi-compositional analysis of the so-called metabolome may overcome the test limitations of, for example, single critical nutrient or key toxicant analyses and problems related to their limits of natural diversity.

An attractive alternative is the utility of high-resolution proton (^{1}H) nuclear magnetic resonance (NMR) spectroscopy (^{1}H-NMR) in combination with separation techniques (liquid chromatography (LC)). Recently, we have shown that the approach of chemical fingerprinting using off-line LC-NMR can obtain information on possible changes in complex plant matrices due to environmental effects as long as GLP-like conditions for sample handling, data acquisition and automation of data handling are maintained. In principle, it will be possible to screen all the different low molecular weight components (MW $<$10 kDa) present in the metabolome of the plant tissue.

TOXICOLOGICAL PROFILING

The significance of detected compositional alterations may need to be further explored by *in vitro* toxicity assays in order to screen for potential adverse effects and for mechanistic studies. It may even be that the latter approach leads to a redesign, refinement or even replacement of (sub-)chronic rodent feeding trials. *In vitro* systems derived from organs and tissues from animals and humans, and various types of cultured recombinant cell lines, are successfully used for screening for the toxic potential of single compounds. However, their ability to screen for the potential toxicity of whole foods or extracts thereof has been insufficiently explored up till now, and has been examined by us within the framework of a tiered safety evaluation of GM plants.

The validity of using *in vitro* test assays for comparative testing of whole food products, food ingredients and extracts thereof depends on the sensitivity and selectivity of the toxic response induced upon exposure. However, since these models are not yet validated, they should be considered as early warning or alert-systems to identify potential changes in toxicity of the engineered product only.

GENOTOXICITY ENDPOINTS

The mucosa of the gastrointestinal tract is clearly a primary, potential site of action for the effects of novel foods. Therefore effort has been concentrated on the development of a battery of intestinal cell systems (*e.g.* IEC-6, IEC-18, Caco-2 and INT407 cell lines) with different endpoints for initial cyto- and genotoxicity testing. The choice of a suitable method of sample preparation was considered of utmost importance in assaying vegetables.

Freeze-dried tomato fruit was extracted (10 per cent suspensions, w/v) in water, 0.9 per cent saline or chloroform:methanol (2:1, v/v) immediately before incorporation into the culture medium. It was encouraging that aqueous and chloroform/methanol extracts of, for example, red-ripe tomatoes up to a maximum of 10 per cent suspensions (w/v), had little or no general toxicity once the pH (> 6.0) and osmotic pressure had been taken into account. No toxicity to the intestinal cells Caco-2 and IEC-18 under short-term conditions was detected in red-ripe, non- and antisense RNA exogalactanase tomato using alpha-glutathione S-transferase (á-GST) lactate dehydroyrenase (LDH) leakage, reduction of 3-(4,5-dimethylthiazol-2-yl)-2,5 diphenyltetrazo-lium bromide (MTT conversion), neutral red (NR) uptake and total cellular protein as *in vitro* endpoints.

On the other hand, green tomato fruit displayed a pronounced toxic and cell-damaging effect. No statistical significant differences were found between transformants and isogenic control lines in terms of their activity in the assays. Treatment of tomato tissue with pepsin (ratio pepsin to tomato protein 1:250, w/w) did not result in an increase of or loss of inherent cytotoxicity. Supplementary investigations with naturally occurring plant components, for example, quercetin, β -carotene, á-lycopene, genistein showed that only steroid glycoalkaloids, such as á-tomatine, exhibited a rapid toxic effect, whereas á-tomatidine was ineffective. It was noted that the *in vitro* models did show general matrix-related toxic effects.

For example, á-tomatine diluted in a matrix of extracted red-ripe tomato tissue (final concentration of 5-10 per cent w/v) showed an increase of dose-response-dependent cytotoxicity in human Caco-2 cells (*e.g.* IC50 7.9 ± 2.1 mg/L versus IC50 33.9 ± 5.3 mg/L of á-tomatine only).

Using the COMET assay, suspensions of freeze-dried tomato fruit up to a maximum of 10 per cent (w/v) did not exhibit DNA-damaging effects in a range of gastrointestinal cells derived from rats (IEC-6 and IEC-18) and humans (Caco-2, INT407). Genetic modification did not result in increased genotoxicity, but fruit extracts were able to suppress the DNA-damaging effects of the known genotoxins H_2 O_2 and N-methyl-N2 -nitro-N-nitrosoguanidine (MNNG). Although these endpoints are important toxicological measures they do not indicate the molecular mechanisms of toxic damage. So far, pronounced adverse effects could only be observed with glycoalkaloids. The screening for subtle differences in toxic responses of other plant constituents appeared to be limited. However, therefore, the use of critical endpoints for such studies other than cytotoxicity should be clearly defined.

6

Biotechnology and Nutrition in Animals

Most developing countries are experiencing difficulties providing sufficient food (staples) for their resident population. In general in most third world countries food production is increasing at a slower rate than the population growth rate.

Developing countries on average are increasing food imports by 10% per year and this is likely to continue into the foreseeable future. Whilst ruminants need not compete with humans for food there are few economic, large scale systems for pig and poultry production established in developing countries which do not rely on grain based concentrate or on crops such as potatoes and other root crops which also take up land that could potentially be used for human food production. The recent developments of sugar cane as a source of feed (juice) for pigs, however, has a major advantage which resides in the high biomass production per hectare and a great potential to form the basis of an integrated farming system. There has been far too little emphasis on developing systems of production of pigs and poultry on agro-industrial by-products. The exception being the molasses and swill-feeding systems for pigs that have developed in Cuba.

By-products high in sugar such as molasses, palm oil sludge and fruit and vegetable pulps lend themselves to feeding to monogastrics. The constraints to production are a low protein content and in some cases an associated high level of fibre which is an intestinally indigestible carbohydrate. Pigs are able to utilise some fibrous feed, possibly up to 30% of their digestible feed intake because of caecal fermentation.

It appears desirable that in the future, pig production should utilise at least a proportion of fibrous carbohydrate polymers. Free-range pig production, where breeding stock obtain a fair proportion of their feed from pastures is a good beginning forced on producers in Europe by the high cost of grain feeding in intensive fattening systems.

If the microbial ecosystem in the caecum of the pig is well developed, basal diets high in sugar and low in fibre combine well with vegetable proteins such as cottonseed meal which are fibrous.

DEFINITION OF BIOTECHNOLOGY OR HIGH TECHNOLOGY

Biotechnology is a much used "in word" in research submissions since it conjures up images of research at the cutting edge of science. It also has an overtone of application, potentially patentable products or processes and therefore returns on investment. The promises have been much overrated in recent times and in general the promises for application from, for instance, recombinant DNA technology have failed to materialise in terms of animal production. In general, the results of recombinant DNA technology have been disappointing. For example transgenic animals have been produced but failure to control expression of the introduced genes has resulted in adverse effects often of a nature that horrifies animal welfare groups. This has been given 'bad press' but it has highlighted the need to study the basic aspects of control of gene expression and cell physiology.

This does not mean however, that the future for and importance of such work is debatable. In fact, the contrary is true, but it has indicated a major requirement for basic research to understand gene expression and that more time and funds will be needed before many new concepts can be applied.

Biotechnology in the context of this report cannot be restricted to recombinant DNA technology. Biotechnology is in reality, a science of great antiquity having its origins in processes such as brewing and wine making.

The broad definition used here is the application of biological organisms, systems or processes to production of animal products.

These include meat, milk, hides, wool or hair and draught power. In particular, biotechnology stresses the integration of microbiology, biochemistry and chemical and process engineering. It is always multi-disciplinary, and its strongest aspect is that it is directed to application. It usually requires disciplinary scientists working together and shepherded by person(s) with major integrative abilities. Young scientists (or disciplinarians) caught up in biotechnology and working in isolation often lose the direction for application or seek out compromise objectives to justify their approach which may or may not be rational.

On the tour of institutions prior to the writing of this report, it was an observation that a number of young, newly trained (overseas) scientists in the less well developed countries, were still working in the area of their overseas training without true recognition of the potential application of their work. They were, thus, in some ways competing with their previous supervisors and are therefore unlikely to produce novel innovations.

ROLE OF BIOTECHNOLOGY IN UTILISATION OF ALTERNATIVE FEED INGREDIENTS FOR MONOGASTRIC ANIMALS

Chronic undernourishment affects some 800 million people, i.e., 17% of the population of developing countries, and 34% in sub-Saharan Africa. Increased agricultural trade and efficient utilisation of available food resources will be a key to the future food security of the world. With an increasing level of affluence in some countries, such as China and India, the shift of diets towards livestock products has been significant.

For example, the average meat consumption in the developing countries has increased by 5-6% per annum over the past few decades (FAO, 2004). The driving force for the growth in meat consumption has been the monogastric livestock industry, in particular, the poultry sector, which has seen a threefold increase in per capita consumption since the 1960s. However, grain production is static while feed demand is rising each year. This disparity can only be addressed through significant gains in efficiency in utilising nonconventional ingredients and by-products.

This review discusses the type and availability of nonconventional or alternative feed resources, as well as technologies available to alleviate the nutritional constraints to these ingredients for monogastric animals.

NON-CONVENTIONAL INGREDIENTS AND THEIR AVAILABILITY

The so-called non-conventional ingredients here refer to the by-products of roots and tubers; of vegetable oil sources such as coconut, sunflower, safflower, groundnut (peanut), palm kernel; of alcohol and rubber processing. Compared with conventional feed ingredients such as cereal grains and vegetable protein sources, these ingredients are usually cheaper, contain higher levels of anti-nutrients, have poorer pelleting quality, and are more variable in quality. In addition, the quality of by-products is influenced by processing.

Two of the amino acids most adversely affected by processing conditions are lysine and cystine. Too high or prolonged heat during the extraction process will destroy essential amino acids in protein meals. Miller (1976) indicated that the proteins in oilseed meals are particularly susceptible to damage by heat treatments involved in oil extraction because of the high carbohydrate content of the oilseeds. However, the correct heat processing of oilseeds makes nutrients highly available and destroys anti-nutritional factors (Göhl, 1981; Kalinowski, 1993). Anti-nutritional factors contained in alternative protein meals will negatively effect the performance of animals, even though some efforts have been made to reduce or eliminate these factors (Cheeke and Shull, 1985). Most alternative ingredients contain high levels of non-starch polysaccharides (NSP), which are the most important limitation to their effective use in monogastric diets. Many other factors, including mould contamination and associated toxins, can impair the nutritional value of alternative protein sources. The detrimental effect of mycotoxins in feed has been thoroughly reviewed by Smith and his colleagues over the past 20 years (Smith et al., 2004). The energy and animo acid composition of alternative ingredients have been comprehensively reviewed by Hutagalung (2001).

POTENTIAL AREAS FOR BIOTECHNOLOGY IN PIG NUTRITION

FIBRE PROTEIN SUPPLEMENTS

The use of oil-seed protein residues for pigs and poultry are generally limited by the fibrous components and in some cases the presence of secondary plant compounds which may be toxic or simply lower production (*e.g.* gossypol in cottonseed meal). To overcome the constraint created by too much fibre it may be possible to develop pre-treatment to protect the protein and then find mechanisms for decreasing the fibre (*e.g.* growing fungi, such as the white rot species on the materials to decrease fibre and increase protein) or find methods to hydrolyse the cellulose to simple sugars before feeding.

Chemical treatment of the feed to hydrolyze fibre to monocarbohydrates may also be potentially possible (*e.g.* with pressurised steam and sulphur dioxide). Delignification with say *Coprinus* fungi and treatment by pressure/ steam of fibrous crop residues to produce simple sugars for inclusion in pig rations is a further possibility requiring research.

At least one group in the U.K. is attempting to introduce cellulase genes from *Clostridial* sources into embryos of pigs with the concept of producing pigs with a capacity to digest more cellulose. The enzyme, if expressed will be controlled by a promoter sequence encoding only in the pancreas this is given in a little more detail below as an example of:

- The possibilities,
- The problems and
- The requirements for basic research prior to any developments which might be applied.

Digestive Function Through Development of Transgenic Animals

By virtue of their hindgut fermentation, pigs have about the same capacity as humans for digesting plant structural polysaccharides.

In theory, substantially more energy would be available to pigs (and other simple animals) if cellulose and hemicellulose could be digested in the small intestine and the sugars absorbed. In addition the complex carbohydrates of grains (*e.g.* β-glucans of barley), fibrous protein sources (*e.g.* cotton seed meal) could be more efficiently used. In the latter case it would allow a much higher proportion of protein from such sources to be incorporated into the diets replacing more expensive less fibrous protein meals such as fish or meat meal. The ability to digest cellulose and hemicellulose by intestinal enzymes is not present in any animal. It is probable that in early evolution these abilities were not advantageous.

However, if encoding a gene for microbial cellulase or hemicellulase could be incorporated into an animal's genome and expressed in the pancreas it may allow these enzymes to be produced and secreted along with the other pancreatic secretions. This is being tested by Hazelwood and Gilbert by introducing into the early embryo of mice (and they suggest they will do this with pigs) the cel E gene of *Clostridium thermocellum* which encodes a thermostable endogluconase with xylan-hydrolysing activity. Expressions of this gene in transgenic mice will be regulated by the elastase I promoter/enhancer region (from rats) which is located "upstream" of the elastase gene and is responsible for restricting the synthesis of the digestive enzyme elastase to target sites of expression of other genes, to the exocrine enhancer cells in the pancreas of transgenic mice.

To modify porcine digestive enzyme secretion to include the enzymes, cellulase and hemicellulase, will require tissue specific expression of the cel E gene and secretion of mature active endogluconase enzyme into the intestinal lumen. For this reason gene constructs incorporating the elastase promoter/enhancer and cel E coding sequence have been designed. The concept of transgenic pigs is at the ideas stage, the research will undoubtedly require a return to basic investigations, undoubtedly over and under expression of the genes when transferred will be a major problem. Fibre breakdown in the pig by cellulase is likely to be slow and therefore gut capacity is likely to be a major constraint even when transgenic pigs are produced.

Porcine Growth Hormone (PSt.)

Undoubtedly porcine growth hormone injected in controlled amounts into pigs on grain based concentrates increase their growth rate, efficiency of growth and reduce fat deposition.

The place of PSt in pig production, as it pertains to developing countries, is unknown. The small production units, the likely high cost of injections and the "scavenger system" generally operated at the small farmer level in developing countries suggests that PSt is unlikely to be used.

The exception may be in grain-based feeding systems for pigs aimed at the markets provided by the higher income groups and tourists.

Transgenic Pigs—Porcine Growth Hormone

The technology for production of transgenic pigs has certainly been developed. Problems of over and under expression of the genes are still apparent. Overexpression results in considerable problems of leg weakness and reproductive inefficiencies whereas under expression gives no additional benefits.

The nutrition of pigs expressing for growth hormone again has been under-studied with the usually emphasis being placed on the thrust to produce transgenic animals. The need to develop special feeding and management systems for pigs seems to have been ignored.

Where these techniques have been applied in developed laboratories the absence of disease is mandatory. It is conceivable that in addition to a "special type of nutrition" there may be special requirements for disease and parasite control. Disease and parasitism are more likely to impact heavily on pigs and other animals with a greater potential for growth.

The problems to be faced in pigs transgenic for porcine growth hormone are well summed up in a paper by Wilmut et al.: "The pig exhibiting human growth hormone did not grow any faster than normal pigs but they had less back fat which might be of potential value to pig farmers.

Unfortunately, these transgenic pigs also suffered from a number of abnormalities, including arthritis, lack of co-ordination of their rear legs, susceptibility to stress, anoestrus in gilts and lack of libido in boars. So it is back to the bench to attempt to redesign the gene to avoid these problems."

BIOTECHNOLOGY AND ENVIRONMENT

Undoubtedly the environmental crisis arising from global warming or the commonly called "greenhouse effect" must develop as a major issue in the near future and it will inevitably influence aid programmes.

Global warming, attributable to the accumulation of gases in the atmosphere will have enormous repercussions on agriculture. Inevitable changes in sea levels, climate and temperature are forecasted with repercussion for both plant and animal agriculture.

The patterns of change are being modelled and predicted but our present level of knowledge suggests that these models will invariably be wide of the mark.

INTEGRATED FARMING

One of the major reasons for the development of livestock production in cropping systems in developing countries is to utilise crop residues, thus, eliminating waste and optimising the use of the total biomass produced within the small farm system for the benefit of the small farmer. The small farmer has often a requirement for draught power with animal production as a secondary consideration.

Integrated crop and livestock production systems are highly efficient; potentially crop residues are used as livestock feed; the waste products (*e.g.* faeces and urine) are fed into biogass digestors and the effluent used to fertilise ponds for aquatic plant/algae production, with fish farming as the terminal activity. These systems are very worthwhile pursuing as a means of providing nutrients/fuel for the family, minimising fossil fuel combustion and methane generation and, thus, reducing environmental pollution.

The array of integrated systems that could be developed is large. They all have as a central core a basic flow of nutrients through a number of systems.

At each of these steps research can be brought to bear to optimise the partitioning of the available biomass into food, fuel and residues. The environmental attributes of such systems are that methane emissions into the atmosphere and fossil fuel use are minimised.

In addition the efficient and almost total harnessing of the energy from high producing crops reduces the land areas required per unit of product. The problems are the necessary high level management that must be exerted by the small farmer, which can often be beyond his presently developed skills.

A complete discussion of these systems is beyond the scope of this document but two examples are:

1. The use of aquatic plants/algaes grown on biodigested effluent for protein production for feeding to pigs and ruminants particularly in the humid tropics and
2. The farming of suitable fish in biogass digesta fluids.

A system incorporating these aspects is currently being investigated and developed by T.R. Preston in Colombia.

NEW ON BIOFORUM: BIOTECHNOLOGY IN ANIMAL HEALTH

Vaccination is unquestionably one of the major success stories in modern biology and it is the main approach to protect animals from infectious diseases. Until today, no other technique has had such huge impact on animal diseases control. Practically all of available animal vaccines on the market were developed basing on conventional methods. Nevertheless, modern biotechnological tools have allowed to better understanding of molecular mechanisms and moreover have opened new opportunities to create better vaccines. BioForum has noted the crucial role of biotechnologies and therefore is creating special session dedicated to veterinary.

Animal health should be considered on at least two levels. Firstly over 70% of human diseases are acquired from animals therefore insufficient animal medicines would be resulted in higher risk from such illness. Secondly, humans depend on animals as food suppliers and thus without healthy livestock, number of food would be limited. In other words, healthy community strongly depends on healthy animals. In turn, the biggest impact on animal diseases control and thus animal welfare has had vaccinology.

Although classical approach of vaccinology has been sufficient and partially effective, it has not provided protection against many diseases. However, development of biotechnologies is providing crucial contributions to fight with pathogens by opening new ways to construct genetically recombinant non-pathogenic live vectored vaccines as well as vaccines attenuated by gene-deletion. It should be emphasized that these vaccines are very efficient in inducing long-lasting immunity through cell-mediated and humoral immunological responses. Despite these advantages, there are still no vaccines against a lot of diseases. A promising solution of this urgent problem seems to be reverse vaccinology. Generally this approach changed the way of thinking about novel vaccines developement. The real breakthrough was sequencing of the first bacterial genome in 1995. This achievement provided data about all proteins encoded by particular microorganism and thus allowed to identify potential surface-exposed proteins starting from the genome rather than the microorganism. Reverse vaccinology approach is powerful tool to manage with crucial issues such as noncultivable microorganisms or antigens that are not expressed in in vitro conditions. However, in spite of mentioned advantages, reverse vaccinology has some obstacles. The major limitation is undoubtedly the inability in identifying nonproteic antigens such as polysaccharides as well as glycolipids, a new promising group of vaccine candidate.

These obstacles together with recent disease outbreaks have emphasized the inevitability for not only producing new vaccines but also for improving existing ones. The European weapon against present animal diseases is joint initiative of industry and a wide range of stakeholders including the researchers and also regulators known as Discontools. In addition it is actively encouraged and funded by the European Commission. The issues of innovative solutions in animal health sector in the Central and Eastern Europe will be widely discussed during next edition of BioForum in Budapest.

ENVIRONMENTAL CONCERNS

Environmental concerns are often considered by far the most important, because of the difficulties of identifying them and of resolving them, once identified. Such concerns have intensified since the Rio Declaration on Environment and Development (adopted at the United Nations Conference on Environment and Development in 1992) whose Principle 15 states: "In order to protect the environment, the precautionary approach shall be widely applied by States according to their capabilities.

Where there are threats of serious or irreversible damage, lack of full scientific certainty shall not be used as a reason for postponing cost-effective measures to prevent environmental degradation."

The Cartagena Protocol was adopted in early 2000 with the following objective: "In accordance with the precautionary approach contained in Principle 15 of the Rio Declaration on Environment and Development, the objective of this Protocol is to contribute to ensuring an adequate level of protection in the field of the safe transfer, handling and use of living modified organisms resulting from modern biotechnology that may have adverse effects on the conservation and sustainable use of biological diversity, taking also into account risks to human health, and specifically focusing on transboundary movements... The Parties shall ensure that the development, handling, transport, use, transfer and release of any living modified organisms are undertaken in a manner that prevents or reduces the risks to biological diversity, taking also into account risks to human health."

A distinction is made between transgenic animals deliberately released into the environment and those released by accident. The problems of introducing transgenic animals into the environment concern their impact on the environment and possibilities of vertical and horizontal gene transfer.

GENERAL PRINCIPLES OF RISK ANALYSIS

Risk analysis has three elements: risk assessment, risk management and risk communication. The first two are most important as concerns risk analysis applied to the environment.

In risk assessment, the risk has two components:

1. A potential hazard
2. The likelihood of that event taking place.

Thus the risk is the *potential hazard* x *the probability of exposure.*

Risk assessment is a process based on scientific data and involving the following stages:

- Hazard identification;
- Hazard characterization;
- Exposure assessment;
- Risk characterization.

Potential hazards and the likelihood of their occurrence are studied in this manner and models are built to predict risk. Risk management, which is distinct from risk assessment, consists of examining policy alternatives with all interested parties, bearing in mind risk assessment and other factors relevant to protection of consumer health and promotion of fair trade, while, where needed, determining appropriate prevention and control options.

An environmental hazard is not so easy to quantify as a health hazard, and concerns a public good rather than a private good (health). When an appropriate risk management strategy is applied to environmental problems and not to safety problems, the problem must first be set out, as must the objectives and values that will serve to seek a solution.

All options are then examined to cover the maximum number of solutions, for weighing the advantages and drawbacks of a vast range of possible solutions helps get all parties involved. Risk communication is the interactive exchange of information and ideas between risk assessors, risk managers, consumers, industry, academia and other interested parties of the risk analysis process. The exchange of information concerns factors linked to the risk, as well as the ways in which the risk is perceived, and includes a presentation of the conclusions of the risk assessment and a justification for the risk management decisions. Risk assessment relating to the environment is a complex undertaking because it is very difficult to identify and quantify all the potential risks of releasing GM animals into the environment. The risk analysis procedure needs to be adapted and the animals need to be classified according to their capacity for dispersal in the environment.

Proposed stages are:

- Identification of potential harms;
- Identification of hazards that could cause these harms;
- Definition of what exposure means for gm animals and likelihood of exposure;
- Estimation of likelihood of risk;
- Classification of risks according to their priorities.

As not all potential risks can be identified, the process of risk analysis needs to be ongoing to take account of new information as it arises.

CLASSIFICATION OF TRANSGENIC ANIMALS ACCORDING TO RISK TO THE ENVIRONMENT

The harm can be partly envisaged: perturbation to ecosystems, reduction or even elimination of wild populations, etc. The hazard is the transgenic animal itself, whose presence in an environment can have direct and indirect consequences for natural ecosystems through the action of various factors. The risk is not so much that the animal finds itself in the environment but that it finds itself there AND establishes itself. It is therefore a matter of studying the impact of the establishment of transgenic animals on the wild populations. The establishment of a GM animal and thus of the transgene in the environment depends on three factors: its capacity for survival and breeding (which itself depends on five factors: viability of juveniles and adults, age at sexual maturity, fecundity, fertility and reproductive success); capacity of escape and dispersal; and the characteristics of the receiving community.

A reduced community of wild animals will be under greater threat from the introduction of transgenic animals than a large community that will be able to withstand the impact of the transgene. If a transgenic animal is better adapted to environmental conditions than its wild peers, it will establish itself among the wild population, which it might in fact replace. If not, the transgenic animal will disappear. If the fitness of the transgenic animal is the same as that of the wild animal, they might be able to coexist. Fitness can be determined prospectively or retrospectively. The prospective method studies four factors: heightened adaptability to environmental conditions, traits and characteristics of the GM animal, capacity to produce novel products and animal's aptitude to escape.

HEIGHTENED ADAPTABILITY TO ENVIRONMENTAL CONDITIONS

Heightened adaptability to environmental conditions, such as capacity to resist cold or disease, enables GM animals to establish themselves in an environment or to colonize ecosystems in which their wild counterparts cannot survive.

Enhanced Existing Traits

Enhanced traits often relate to rate of growth, feed efficiency or production performance (*e.g.* milk or eggs). An enhanced trait can sometimes be countered by physiological imbalance or increased food requirement, as has been observed among some transgenic farm animals. Genetic enhancement by means of genetic engineering differs from enhancement through the traditional selective breeding of animals with interesting traits, in the sense that the characteristics associated with the interesting trait do not follow the same evolutionary path.

A transgenic animal can have a vascular or skeletal system that is completely unsuited to its improved rate of growth, which produces a developmental imbalance of physiological systems and renders the animal less fit for survival. Transgenic animals developed for production and breeding often present an altered characteristic that counterbalances the impact of the improvement and therefore represent a low risk of establishment in the environment.

However, in some cases the improved trait could outweigh the altered characteristic; for example, it has been noted among certain transgenic fish that a 25% increase in size boosted their reproductive success by 400% as compared to the norm. Such success benefits the transgenic animal and rapidly disseminates the transgene among the wild population, even if the transgene reduces the lifespan of the transgenic animal. In such a case, the aggregate impact favours the establishment of the transgenic animal in the environment, an advantage that is further accentuated if the conspecific wild population is small. Fragile wild communities could be reduced or even disappear.

PRODUCTION OF NOVEL PRODUCTS

Transgenic animals producing more proteins and pharmaceutical substances in their milk, urine, or any biological by-product, have only limited capability of establishment in the environment, because they have not acquired any special advantage in fitness and their added functionality in fact requires higher energy expenditure which makes them less competitive with their wild conspecifics. Such transgenic animals therefore only represent a low risk for the environment.

PRODUCTION OF ANIMALS OR ANIMAL PRODUCTS FOR HUMAN HEALTH

Companion animals genetically engineered to reduce allergenicity and animals modified for xenotransplantation only present a low risk of establishment in the environment. Their fitness remains unchanged or perhaps compromised. Transgenic insects modified to control the spread of a vector-borne human disease such as malaria are meant to establish themselves in the environment.

RISKS TO THE ENVIRONMENT

Terrestrial Vertebrates

Escaped domestic or laboratory mammals with high feral capacity have disturbed the environment in many countries (*e.g.* rodents, rabbits, cats, pigs.). Any trait of a transgenic animal of these species that increases its fitness (resistance to climatic conditions, resistance to disease, improved feed conversion efficiency, etc.) and its dissemination among wild populations represents a high risk to the environment.

Fish, Molluscs and Crustaceans

In the fisheries sector, most transgenic animals have higher growth rates as these represent tangible benefits to fish farmers. But transgenic fish can also have a positive impact on the environment as they produce less contaminating waste. Improved feed conversion efficiency means lower feed requirements, reduced feed waste and effluent from fish farms, and thus lower eutrophication. Environmental concerns apply more to the "faster growth" characteristic, which can impact on predation, competition and genetic contamination: transgenic fish can eat larger quantities of prey species; their higher environmental tolerance enables them to invade new territories and possibly displace extant natural populations; finally, interbreeding could alter the genetic composition of the natural population. Four factors influence the ability of transgenic fish to escape into the environment: high mobility, the relative proximity of transgenic and wild species, the fact that ecosystems are often shared, and finally the fact that aquaculture animals are usually marketed live.

Advocates of transgenic fish argue that, as they are heavily domesticated, they will find it difficult to survive in natural conditions. However, studies on transgenic fish suggest otherwise. Transgenic Atlantic salmon have a higher energy and oxygen requirement than their wild counterparts. They lose their juvenile markings at a younger age and show early mobility, which reduces their status as prey, despite their larger size. Pacific salmon present more or less the same characteristics.

Their age at sexual maturity is lower but they are not so mobile, no doubt because of an imbalance or delay in the development of their muscular, vascular or nervous systems. Yet these transgenic species compete directly with wild species for food. As highlighted earlier, greater fish size combined with earlier sexual maturity but reduced juvenile viability can eventually lead to the reduction or disappearance of the receiving population.

The risks posed by transgenic fish culture should be assessed in the same way as for any other aquatic species coming from another ecosystem, under trial conditions that resemble the natural environment as closely as possible.

Dispersal in the ecosystem and the impact of each species need to be studied. Species should be kept in close confinement, but this is not easy for molluscs and crustaceans which have highly mobile larval stages, especially as they are cultured under extensive and thus dispersed systems.

Similarly, transgenic fish reared in ecosystems shared with wild populations can escape from the culture zones. Chromosome manipulation in molluscs and fish to render them infertile would be one way of solving the problem of dispersal in the environment, and is a technique used to reduce the risk to wild native populations. However, it presents a technical challenge and the low proportion of fertile individuals coupled with the so-called "mosaic" individuals resulting from sterilization techniques represent a risk of dissemination of the transgene among wild populations.

Not all the risks can be anticipated. Crucial information is still needed to assess the environmental risk from the presence of transgenic animals. For example, the growing body of empirical studies on risk from transgenic fish does not provide practical data to determine the level and impact of transgene establishment in a wild population, even though it draws attention to the need for serious risk analysis. Transgenic animals need to be monitored after their commercialization to discern unexpected impacts and attendant risks, and to adjust risk assessment and management.

Further information is required. In view of current uncertainties, many scientists call for the precautionary principle to be applied, especially for transgenic fish, molluscs and crustaceans. No transgenic fish has yet been exploited commercially, but countries involved in salmon fishing in the North Atlantic and adjacent areas have agreed to apply a number of precautionary measures. It is perhaps on account of the virtual disappearance of populations of fish species of economic importance to the region that decision-makers appear to be inclined to apply the precautionary principle, according to which *the absence of scientific certainty should not be used as an excuse to postpone management efforts*. Other important principles are that priority should be given to maintaining the productive capacity of a resource or ecosystem, that any impact should be reversible over a period of two to three decades, and that decisions should be taken on the basis of risk and benefit. Finally, benchmarks need to be established to identify areas of greatest uncertainty on which monitoring, research or study needs to concentrate.

GENETICALLY MODIFIED INSECTS

Transgenic Insects

Insect biotechnology is used to create genetically modified insects that are useful for pest control, especially in agriculture, and for the control of vector-borne animal or human disease. One of the major objectives of transgenesis is in fact to control the transmission of vector-borne diseases such as malaria. The vectors can be controlled by the usual methods, involving insecticides, actions against vector home environments or means of protection, such as mosquito nets, but these methods are only relatively effective because of growing resistance to insecticides and the physical limitations of the methods themselves. The only course of action is then to eliminate the vector itself. Biotechnology can be used to reduce or even eliminate vector insect populations, or to reduce the capacity of an insect to carry and disseminate a parasite or pathogen. Various methods have been proposed: the use of sterile males, the introduction of a lethal gene or the introduction into the insect vector population of a gene that gives resistance to infestation by human or animal pathogens.

Use of Sterile Males

Sterile males are used to reduce vector population density and sterilization can be done by chemical or physical irradiation. However, such methods reduce mosquito viability as the irradiation causes lesions in all insect tissues. Biotechnology can improve the sterility of males without damaging their viability and can therefore improve the effectiveness of control programmes and reduce the number of released insects. The use of sterile males requires sexing so that females are not released. Distinguishing male from female is easy in species that permit visual sexing on the basis of size of pupae, but can be a real challenge in others. Proposed methods include the use of a dominant gene associated with a female-specific promoter which would render the females dependent on a high concentration of tetracycline for survival and would therefore render them unviable in a natural environment.

The males would simply carry this gene which they would introduce into the wild population. Such a gene could however be a major problem if it were to establish itself among wild communities of non-targeted insects. The use of sterile males is effective if the females mate with them. However, the females of some species of mosquito are monogamous so this system would lose effectiveness under the massive inflow of already fertilized females. Depending on the target species, it would therefore seem best to combine methods, notably with another genetic technique.

Particular Characteristics

The production of Anopheles mosquitoes resistant to *Plasmodium* infection (malaria) is an interesting concept, but the gene in question has to be closely linked to the dissemination mechanism. However, the "resistance" trait would appear to depend on several genes. It has also been shown that 100% effective resistance is needed in areas with high prevalence and dissemination of the pathogen. The problem is that the creation of selection pressure based on

Plasmodium could eventually lead to the emergence of resistant *Plasmodium*, which has led to the idea of introducing several independent genes resistant to the pathogen. This is a major problem as biotechnologies are used for traits that are regulated by a single gene and cannot be used where a trait stems from a complex set of genes. Linking and transferring the set of genes is also complex. Another example is the sterilization of male ephemera larvae infesting cotton bolls using irradiation, which has not been 100% effective despite the high cost. Scientists have thought of inserting a gene capable of blocking the development of fertilized eggs, to help reduce the wild populations.

DIRECT PROBLEMS

Stability and Effectiveness of the Inserted Gene

Insertion needs to be stable and effective. Once the desired trait has been identified, the responsible gene needs to be defined and the inserted transgene needs to express itself at the right time and in the appropriate tissues. The resulting insect is then studied and can be reproduced in large volumes if observations are satisfactory. Dissemination models need to be prepared and trials conducted in confined areas.

Cost of Biotechnologies

The control of insect vectors of animal and human disease is a controversial subject. Traditional methods need to be seen to be ineffective if biotechnologies are to be introduced, and this for two basic reasons: the high cost of obtaining transgenic insects and the time and resources needed to conduct GMO risk analysis. A cost-benefit study is required.

The Problem in Human Health Terms

In the case of the control of certain vector-borne diseases, the elimination of the parasite within the vector population could result in an increase in population, which could foster the propagation of other vector-borne human or animal diseases. For example, mosquitoes rid of *Plasmodium* could become more susceptible to other human or animal pathogens. Tests are therefore required, especially for parasitic worms (Anopheles mosquitoes are vectors for filaria) and numerous viruses, including HIV. For the time being, mosquitoes do not seem to be carriers of the HIV virus, but it would be disastrous if they were to become so. Virologists see this as totally inconceivable but public opinion appears to see it as a major concern. Data need to be found to prove that this is not possible.

Problem of "Driver" Mechanisms

Three mechanisms are proposed:

1. *The use of transposable elements (which replicate and disseminate)*: Present know-how renders this mechanism too uncertain, and the

risk of horizontal transfer of an uncontrolled transgene means that it is forbidden.

2. *Wolbachia infection in mosquitoes*: This mechanism is still under development and thus untested.
3. The use of complex mechanisms using two complementary lethal genes that counteract one another. Hybrids only possessing one of these genes would be unviable which would benefit the animals carrying the two genes. Bonding the gene for resistance would enable it to implanted among the target population.

Without a driver mechanism for dissemination, the method would be seriously restricted by the migratory movement of vectors, notably female mosquitoes. Combining these methods with the use of sterile males would be more effective as they would eventually lead to a reduction in the wild population.

Problem of Bonding

This is a major problem, for it is impossible to envisage creating insect resistance to infestation by a pathogen if there is no guarantee of complete bonding between the driver mechanism and the transgene. There would be crossings and hybrids, without bonding, and those with the driver mechanism would not be effective for control.

RISKS TO THE ENVIRONMENT

Problems of Insect Confinement and Release

Insects have high mobility and considerable potential for colonization of new environments. They must therefore be confined to permit study without risk to the environment. This is important to avoid the propagation of a transgene in a wild population, especially in the case of biopharmaceutical insects. Transgenic insects should be confined in a laboratory, as there is no way they can be detected after release into the environment, especially if they proceed develop in an unexpected manner.

Two guides with differing emphases have been drawn up for insect confinement:

1. A classification of risks to the environment and of GMOs, with specific procedures for each. A distinction is made between different kinds of GM insect and corresponding levels of confinement.
2. Recommended practices of the US Animal and Plant Health Inspection Service (APHIS) which constitute a single standard procedure without distinction between transgenic insects, and highlighted by the OECD in 1998.

At present, each laboratory develops its own procedures for the confinement of GM animals, without uniformity. The thorny problem of insects created for biopharmaceuticals is not taken into account. Some observers argue that confinement requires time and investment at the expense of pure research, but the release of genetically modified insects could feasibly have damaging and more or less irreversible consequences for the environment.

HORIZONTAL TRANSFER OF A TRANSGENE

The risk of horizontal transfer of a transgene is regularly emphasized and, if this were to happen, could have significant consequences for the environment. The insertion of genetic material into wild populations, such as genes resistant to antibiotics (markers) or to insecticides, is highly undesirable. Such an occurrence is impossible to evaluate as we lack information on horizontal transfer mechanisms and related frequencies.

There has been very little study, but horizontal transfer would appear to be possible between conspecifics. The underlying mechanism is unknown, but this phenomenon has been known to occur with small fruit flies. High plasmid exchange is also possible between insects and intestinal bacteria, which are then found in faeces and thus the environment. The impact of such gene transfers is unknown, but close links exist between insects and other organisms (pollinizers, preys or predators, vectors, etc.) which means negative impact on an insect population could have a major impact on all ecosystems.

CONTROL AND REGULATORY BODY

National regulations often tend to forget transgenic insects. No regulations or international recommendations exist for the confinement of transgenic insects. However, their release is subject to authorization in most countries. There are sometimes several regulatory bodies at national level and control procedures can be complex. In the US, several different federal and national institutions need to be consulted before GM or non-GM insects can be released. Some commentators would also like to see the health agencies involved. So far, the only releases of transgenic insects that have taken place in the US have been for short duration trials. Their subsequent destruction, or at least total recovery, are prerequisites for authorization to conduct such trials.

NEED FOR RISK ANALYSIS

Risk analysis should be conducted prior to any release of transgenic insects and should include studies on:

- The stability of the transgene
- The possibility or not of horizontal transfer of the transgene
- The present and potential geographic distribution of the gm insect
- The different traits transferred to the gm insect and their impact on the ecosystem
- Possible unforeseen side effects on the environment.

The traits and fitness of a transgenic insect appear to be easily verifiable. On the other hand, the horizontal transfer of the transgene and short-, medium- or long-term side effects on the environment are more difficult to determine. A large number of scientific uncertainties need to be cleared before transgenic insects can be released into the natural environment.

DETERMINING THE LEVEL OF RISK

There are three possible scenarios:

1. A field or controlled-environment trial permitting the recovery or destruction of all the insects;
2. The release of sterile insects for biological control; this poses less risk than insects meant to establish themselves among the wild population as they cannot breed.
3. Insects for permanent establishment in the environment and crossbreeding between GM insects and wild insects to disseminate the transgene.

We can identify the risk but assessing the likelihood of risk is virtually impossible. Domesticated insects are virtually incapable of surviving alone in the environment (*e.g.* silk worms). The risk of dissemination of these insects is thus very low. Sterile insects represent the same low risk of dissemination because horizontal transfer is very rare. The case of bees would appear to be more uncertain as bees are half domesticated, so have higher potential for survival and dissemination in the environment. Risk analysis models for transgenic insects do exist, but have limitations as designed on the basis of possibly biased theoretical data. There are too many known and unknown interactions between the different factors (population dynamics, genetics, environment and space, etc.) Transgenic insects require proper risk analysis if reliable risk-benefit analysis is to be conducted. The possibility of finding alternatives to the costly and time-consuming production of GMOs needs to be looked into, as do the objectives of producing transgenic insects and the achievements to date. Negative impacts need to be studied and reduced, which will mean standardizing the risk analysis to be carried out before any release of GM insects. Retrospective studies following the introduction of these animals will also be needed.

NEW HEALTH PRODUCTS FOR FARM ANIMALS

The Animal Health Division at Bayer produces and markets a wide range of products to maintain the health of livestock and companion animals, as well as a range of farm hygiene products. 'Protect animals, benefit people' is the Bayer Animal Health mission statement and our philosophy – one in which we are among the leaders worldwide. A responsible relationship between people, companion animals and livestock requires that we safeguard the health of animals. Keeping the close relationship healthy is not just an obligation to our animal companions, it also protects people from the transmission of disease pathogens. To that end, the team at Bayer is constantly developing new products and forms of administration. Not just for companion animals, but for millions of farm animals as well, Bayer offers extensive protection in the form of effective and safe animal health products.

BETTER VACCINES TO PREVENT DISEASE

In addition to producing new diagnostic tests and therapeutic proteins, animal scientists are using biotechnology to develop vaccines to prevent disease. A safe, effective vaccine for swine pseudorabies, a fatal herpes-virus, is already in use. An advantage of the swine pseudorabies vaccine, and other vaccines made through recombinant DNA technology, is that they use only a small portion of the original microbe. Conventional vaccines use killed or weakened forms of the disease-causing microbe.

Sometimes, when the microbes are not completely killed or sufficiently weakened, a conventional vaccine may cause the disease it was supposed to prevent. Disease-causing genes are not included in the genetically engineered vaccines. Therefore, the recombinant vaccines build up the body's immunity without the risk of causing disease. Another advantage of recombinant vaccines is the speed with which they can be developed.

"Conventional vaccine development can take 20 or 30 years, maybe even 100 years," says Karen Jacobsen, D.V.M., University of Georgia College of Veterinary Medicine. "This is one reason why we don't have vaccines for a lot of animal diseases today. " Recombinant vaccines are being developed for foot-and-mouth disease, a highly contagious viral disease that infects cattle, sheep, and other animals. Although this disease has been eradicated in North America, it still causes substantial livestock productivity losses elsewhere, particularly in developing countries. Infected herds must be slaughtered, and contaminated ranches must be left idle for months to prevent new outbreaks of the disease. Conventional foot-and-mouth vaccines are made by weakening the virus that causes the disease. These vaccines sometimes revert to the virulent state, and they have caused outbreaks of the disease in Europe. The foot-and-mouth vaccine made with biotechnology cannot cause the disease because, as in other biotech vaccines, the disease-causing genes have been removed.

The biotech vaccine does not need to be kept cold and, therefore, can be used in developing countries. In addition to being used in treating swine pseudorabies and foot-and-mouth disease, recombinant DNA technology is being used to develop vaccines against dozens of livestock and poultry diseases, such as scours and a variety of respiratory disorders. Some vaccines may be injected into eggs so that chickens are immunized before they hatch. Most vaccines protect against viral or bacterial infections, but using genetic engineering, researchers in New Zealand and Australia have developed a vaccine against a parasitic disease. Farmers must destroy the meat from sheep infested with a species of tapeworm common in those two countries. The new vaccine, which protects sheep from the tapeworm, may save farmers millions of dollars annually.

HOW VACCINES PREVENT DISEASES

The diseases that vaccines prevent can be dangerous, or even deadly. Vaccines reduce the risk of infection by working with the body's natural defenses to help it safely develop immunity to disease.

When germs, such as bacteria or viruses, invade the body, they attack and multiply. This invasion is called an infection, and the infection is what causes illness. The immune system then has to fight the infection. Once it fights off the infection, the body is left with a supply of cells that help recognize and fight that disease in the future.

Vaccines help develop immunity by imitating an infection, but this "imitation" infection does not cause illness. It does, however, cause the immune system to develop the same response as it does to a real infection so the body can recognize and fight the vaccine-preventable disease in the future. Sometimes, after getting a vaccine, the imitation infection can cause minor symptoms, such as fever. Such minor symptoms are normal and should be expected as the body builds immunity.

FEEDING STANDARDS AND FEED EVALUATION

To ration concentrates effectively to the individual cow, the dairyman needs to know (a) how much TDN the cow needs and how much can be fed profitably, (b) the TDN value of the forage to be fed, and (c) how much of the particular forage will be ingested. Thus, effective rationing requires a satisfactory feeding standard and information on the over-all nutritive value of the particular forage (s) to be fed. A limitation in either the feeding guide employed or the appraisal of the nutritive value of the forage naturally is reflected accordingly in the amount of concentrates fed.

Recent attempts have been made to develop rapid and inexpensive, but indirect, methods for the assessment of the nutritive value of forage in programs which have become known as forage testing. The ultimate aim in the use of a forage-testing program is that of determining the kind and amount of concentrates that should be included in the ration of the individual cow as dictated by the feeding standard. Although forage-testing schemes in their present form and manner of application provide an estimate of the TDN value per unit weight of forage, they do not provide an estimate of the more influential determinant of the total energy intake, the expected forage intake. As a consequence of this limitation, and others which have been detailed in the text, it is concluded that forage-testing schemes contribute very little information of genuine value to the daily rationing and the improvement in milk yield of the individual cow. This is particularly true when the shortcomings of forage testing are considered in relation to the limitations of existing feeding standards and certain commonly encountered defects in practical feeding.

Most forages consumed by livestock in developing countries have a low digestibility which rarely exceeds 55% and is mostly in the range of 40–45%. The calculated metabolisable energy in the dry matter (ME/DM), thus, ranges from 7.5 MJ down to 4.8 MJ.

Feeding standards indicate that feeds with a metabolisable energy content, of 7.5 MJ will support growth rates of cattle of approximately 2 g/MJ of ME intake. On a forage at the lowest level of ME, cattle would be in negative energy balance.

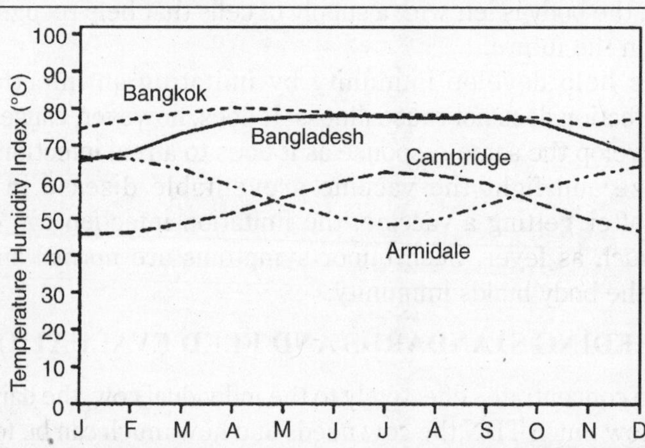

Fig. Temperature Humidity Index (THI) of Climates in Temperate Countries as Indicated by Cambridge (U.K.) and Armidale (Australia) as Compared to Tropical Countries as Indicated by Bangkok (Thailand) and Mymensingh (Bangladesh). The THI is Calculated on the Mean of the Maximum-minimum Temperatures and Humidities THI(°C) = Temp. (dry bulb) + 0.36 Temp. (dew point) + 41.2°C

Contrast this with results of supplementary feeding trials based on balancing the nutrition of animals with urea/minerals and bypass protein, where cattle growth rates equivalent to 18 g/MJ of ME intake have been achieved in cattle fed straw. Obviously the presently accepted feeding standards have been very misleading and can not be used as a means of predicting animal performance. Of vital importance however, is that the application of the concept of balanced nutrition can improve animal growth by 2–3 fold and the efficiency of animal growth by as much as six fold over previous estimates (a range of 2–10 fold). In addition it also shows that although growth rates of cattle are below those on grain based diets cattle on forage based diets can highly efficiently convert feed to product.

Implications of Low Productivity of Ruminants in the Tropics

Low productivity of ruminant livestock has been accepted in developing countries as an inevitable result of the poor feed base and a low feed conversion efficiency. The concept being that there is a large heat production (energy requirement) associated with the ingestion, movement of digesta along the tract in animals on forages as compared to concentrates. This conclusion is contrary to the conclusions of Leng and the concept of balanced nutrition presented here.

SOME BASIC EXPLANATIONS FOR THE INEFFICIENCY OF RUMINANTS ON FORAGE DIETS

Inefficiency of Acetate Utilisation

The original calorimetric studies of Graham and his colleagues indicated that infused acetate or butyrate were utilised by sheep with low efficiencies, *i.e.* there was a high heat increment when acetate was given compared with propionate. Considerable effort has since been expended on testing the hypothesis (or disproving it) that acetogenic substrate is used wastefully.

Blaxter and his colleagues used diets based on dried grass which was chopped and cubed. It had a metabolisable energy content (M/D) of about 8.5 MJ/kg dry matter. As most of the protein in the diet could have been highly soluble, the P/E ratio in absorbed nutrients would have been relatively low. The controversy concerning the efficiency with which acetogenic substrate is utilised may be rationalised at least to some extent by considering the balances of nutrients available to the ruminants in the various experiments and the ability or otherwise of the animals to synthesise fat, dissipate heat or to oxidise substrate to keep warm.

For example the presence of small amounts of fish meal, that has a considerable amount of protein that escapes the rumen, in a concentrate diet provides an explanation for the differences between these authors' results and those summarised by Blaxter where sheep were fed dried grass which may have contained a highly soluble source of protein.

Requirements for Glucose by Ruminants

The need to manipulate or supplement diets for ruminants in order to ensure an adequate supply of glucose and of glucogenic compounds was discussed fully by Preston and Leng, who made a comprehensive literature survey. In outline, the rationale that is used to justify the concept of glucose being a limiting nutrient is as follows: Little glucose is absorbed by ruminants but they synthesise considerable glucose from precursors such as propionate and certain amino acids, largely in the liver.

Glucose is certainly required, by ruminants, as a major substrate for cell synthesis, as an important oxidative energy supply in the brain and red cells and as an important nutrient for the growing foetus and for milk lactose and fat synthesis in lactating animals. Glucose needs to be oxidised in the adipose tissue, and to a lesser extent the mammary gland, to supply the reduced cofactors for fat synthesis from acetate (in this case of NADPH, generated in the pentose phosphate pathway).

It is possible therefore that fat synthesis in an animal may be limited by glucose availability for oxidation. If there is a block in acetic acid utilisation, it cannot be allowed to accumulate in the blood as this would lead to acidosis, the animal therefore, needs to oxidise the excess acetate in a futile cycle in which acetyl CoA is produced with a requirement for ATP and then hydrolysed. Alternatively an animal could increase its muscular activity by standing instead of sitting or walking long distances which would ensure increased acetate oxidation.

The animal that is most at risk to a deficiency of glucose is the animal with a big demand for glucose (late pregnant or early lactating) fed forage based diets (high acetogenic fermentation in the rumen) with little bypass nutrients in the diet (receives only amino acids from microbial sources) and is tethered (no requirements to oxidise acetogenic substrate to walk) living in a tropical country (no requirement to oxidise acetate to keep warm).

As discussed above, reduction in the need for acetate oxidation and a high glucose demand can establish a situation where a deficiency of glucose could lead to an imbalanced nutrition with a need to burn off excess acetate with a high thermogenic effect. Where such a thermogenic effect cannot be tolerated (high environmental temperatures, high humidity and perhaps heavy insulative coat) the animal needs to respond by reducing its feed intake. Conversely the thermogenic effect may be prevented by balancing the diet with nutrients that supply glucogenic substrates for absorption from the intestine.

Balancing Nutrition for Reproduction /Pregnancy and Lactation

Considerable research has indicated that the level of protein nutrition considerably influences reproduction of both male and female animals and subsequent pregnancy and lactation. Leng *et al.* indicated that on forage based diets each physiological function is affected adversely by a low P/E ratio in the nutrients absorbed by ruminants; this could be overcome, to some extent, by feeding protein meals which bypass the rumen.

In summary, feeding a supplement which improves the P/E ratio in the nutrients absorbed by ruminants on a low true protein, forage diet has the following potential effects, particularly in a hot climate, on reproductive efficiency of female ruminants:

- Stimulates liveweight gain of dam or reduces liveweight loss
- Improves ovulation rates

- Improves placental size
- Improves birth weight and so increases survival and because of the increased birth weight possibly lowers the incidence of retained placenta
- Increases milk yield and efficiencies of milk production.

Prevention of protein deficiency in early life also prevents stunting of final body size. Differences in size of animals of the same breed, in the same country, is almost always a result of differences in nutrition and not inherent differences. This has recently been emphasised with N'Dama breed which has always been considered to be a small breed weighing up to 250 kg liveweight. With good nutrition and adequate management the bulls have now been shown to grow to 500 kg liveweight.

Work from Nigeria and Australia has also shown that young and old bulls are also very susceptible to low P/E ratios in the nutrients absorbed. Young bulls that grow on diets that would have given a high P/E ratio in the nutrients absorbed, compared with animals fed a diet giving a low P/E ratio, had better testicular development and produced larger ejaculates with double the sperm content. Older bulls that go through a period of protein undernutrition have decreased testicular size and probably are less fertile.

IMPLICATION OF PARASITE/DISEASE AND NUTRITION

Undoubtedly any parasitic or disease condition that drains protein from the animal will increase the animals requirements for protein relative to energy. Similarly, infective agents that utilise glucose may also increase the demand for this critical nutrient.

For example, trypanosomes and epyrythrozoan parasites which invade red cells, increase protein requirements by increasing red cell turnover rate and also increase the animals requirements for glucose as this is the major substrate used in the parasite's metabolism.

It is suggested that improving the protein nutrition of ruminants through providing bypass protein directly to the animal (*i.e.* avoiding rumen fermentation) may considerably ameliorate the detrimental effects of intestinal and blood parasites and may assist in development of early immunity. (J. Steele personal communication)

IMPLICATIONS OF AN INCREASED NUTRIENT REQUIREMENTS FOR WORK

Light work requires acetogenic substrate probably acetate for muscle contraction but heavy work is probably dependent on long chain fatty acids mobilised from adipose tissue. All working muscles use some glucose and there is always a concomitant increase in glucose utilisation when free fatty acid metabolism is increased by the work load.

Imbalanced diets fed to draught animals which result in a low P/E ratio in the nutrients absorbed are not so disadvantageous as they are for a lactating or growing animal. The excess acetogenic substrate is used for work removing the necessity to otherwise enzymatically dispose of the substrate. However, the increment in requirements for glucose oxidation in muscles although small could be a limitation. In addition the need in heavy work to supply long chain fatty acids which can be oxidised to produce ATP at a faster rate than from acetate, indicates that there is an increased need for glucose not only in muscle metabolism but also to aid fat synthesis from acetate in the period of rest.

An imbalanced feed (low P/E ratio or low G/E ratio), which is associated with a high metabolic heat production, could also reduce the recovery of body temperature of draught animals when the animal is resting and, thus, reduce feed intake. A decreased fat deposition because of such an imbalanced diet during the non working season may also be a major constraint to draught capacity because of lack of body reserves for mobilisation in the working season.

The P/E ratio required to support fat deposition (in periods of low work load) or to reduce heat stress by reducing metabolic heat is likely to be much lower than for example for growth and milk production. A draught animal fed poor quality forage probably requires little more nutrients than can be supplied from an efficiently functioning rumen. This is attained for example by supplying the animal with a molasses/urea multi-nutrient block.

Implications for Areas of Research

This discussion highlights the manipulable aspects of ruminant nutrition, where substantial improvements in productive efficiency of animals fed largely on fibrous feeds could result.

The differences between temperate and tropical conditions also indicate opportunities which are often not apparent in temperate countries and mitigates against direct transfer of results from the former to the latter. In tropical conditions, protein nutrition of ruminants is more crucial than in temperate areas.

The general conclusions are that:

- The primary constraint to ruminant production from fibrous feeds (in the tropics) is the low efficiency of feed utilisation often coupled with an environmental/physiological effect which results in a reduction in potential feed intake.
- In practice the most effective mechanisms for improving productivity will be to improve the p/e ratio in the nutrients absorbed because of the large effect on efficiency of feed utilisation.
- Draught animals will also benefit from technologies that increase microbial growth efficiency in the rumen and at times improve digestibility, intake and p/e ratio in the nutrient absorbed.

- Contrary to many statements in authoritative reviews, only when p/e ratios have been optimised does the energy density of the feed become a primary constraint.
- As the primary constraint is an inefficiency due to a low p/e ratio in the nutrients absorbed it is unlikely that other inputs [*e.g.* agents to repartition nutrients or to stimulate growth (B-agonists, BSt)] could be successful without first applying strategic supplementation.

MAXIMISING DIGESTIBILITY OF FIBROUS FEEDS

SUPPLEMENTATION

A deficiency of any nutrient in the rumen must first affect the rumen ecosystem by reducing microbial growth efficiency which, as the deficiency becomes more acute, will eventually reduce microbial pool size and inevitably digestibility and intake of roughage by ruminants, Decreased digestibility will only be seen with time as the deficiency of the nutrient becomes progressively greater. Supplementation of a critically deficient animal with nutrients for the rumen will increase microbial growth efficiency, microbial pool size and digestibility of forage which will almost always increase feed intake. The feed intake effect, however, often depends on the degree to which the limiting nutrient is deficient. The most likely deficient nutrients, are non-protein nitrogen, P, S, Mg, Na and also trace elements such as Co, Cu, and Zn. Supplementation with multi nutrients of the type listed above is generally successful in increasing intake and digestibility but the large increases in productivity can be anticipated from combining such supplements with a source of bypass protein.

Sources of Mineral Nutrients and Molasses Urea Blocks

One of the best sources of multiple trace minerals in developing countries is a concentrated plant juice, such as molasses or palm oil sludge. Multi-nutrient blocks are no better than the mixtures of materials from which they are made. Blocks are useful because they form a package readily handled, marketed and used by the farmer with little impact on the time he has for other activities. The major problem with these multi-nutrient blocks occurs in the grasslands where there is a need for appropriate means of encouraging stock to consume them in sufficient amounts. In these areas molasses is often unavailable or is too expensive. More research is needed to produce suitable packages (*e.g.* blocks without molasses) for such localities. The importance of block supplementation relates more to the access it allows to livestock managed by small farmers. It may be possible in the future to apply innovations through inclusions in the blocks of chemicals and other materials that will manipulate the rumen (*e.g.* antiprotozoal chemicals) and manipulate the animal (*e.g.* protected chemicals that are absorbed from the intestines) or even control disease and parasitism (anthelminthics). In this respect the production and testing of blocks incorporating anthelminthic is appropriate and being undertaken already by the Commomwealth Scientific and Industrial Organization in co-operation with the Australian Centre for International Agricultural Research (ACIAR).

Improving the Enzymatic Ability of Rumen Microbes to Degrade Fibre

This is an area that has received emphasis and has aroused considerable expectations. The concept is simple, if rumen microbes can be developed with greater abilities to degrade fibre, or its components, the microbes will be able to extract a greater amount of the available nutrients from the feed consumed.

This often receives a high priority because of the concept that animals fed low quality pastures are primarily energy deficient. In fact, these animals more often are inefficiently using the available nutrients. It must be stressed that the application of balanced nutrition is the first priority and if other perturbations of the system are to succeed an animal must have an efficient rumen and also metabolism. However, in an animal fed a roughage diet, with an efficient rumen and balanced for protein, it is possible that for every 5 units increase in digestibility of the basal roughage resource, there will be 50% increases in liveweight gain on the same basal feed resource. Silva and Ørskov demonstrated that the extent of breakdown of a fibrous feed was dependent on the rumen medium. Untreated ground straw placed in nylon bags, in the rumen of sheep fed ammoniated straw, was fermented to a greater extent than the same untreated ground straw placed in the rumen of sheep fed untreated straw. This difference might be due to a higher concentration of enzymes in the rumen of sheep fed ammoniated compared to untreated straw.

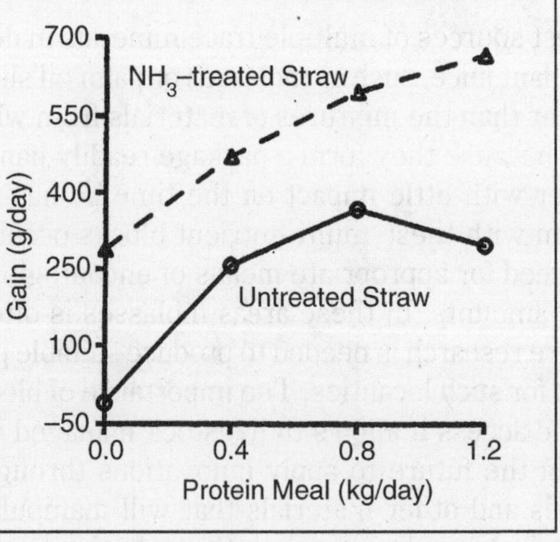

Fig. Growth Response to Different Levels of Protein Meal Supplementation of Cattle given Ammoniated (NH_3-Treated Straw) or Untreated Rice Straw (Untreated Straw)

The two observations discussed above when combined indicate the great potential for improvements in productivity from a successful establishment of a super bug in the rumen, *i.e.* a micro-organism engineered to produce more, or a greater variety, of enzymes for fibre digestion.

Potential Targets for Improving Fibre Digestion

There are potentially three ways to enhance the ability of rumen microbes to degrade fibrous feeds:

1. Selection of microbes, particularly with a high fibrolytic enzyme secretion or that secrete fibrolytic enzymes of high specific activity
2. Creation of microbes with greater spectrum of enzyme secretions by recombinant dna means, *e.g.* Create cellulolytic capacities in xylolytic organisms and vice versa
3. Create microbes with enhanced fibrolytic activity by recombinant DNA means.

In all cases, the prerequisite will be that the organisms are able to grow in and maintain their space in the rumen. These same prerequisites also apply to the production by genetic engineering techniques of high biological value proteins, other secretions of proteins, amino acids and peptides by rumen organisms (see earlier discussion).

State of the Art in Bioengineering of Rumen Organisms

There are number of well established ways of going about incorporating DNA encoding for specific enzymes into anaerobic bacteria by recombinant DNA technologies.

Progress has been relatively swift, with at least 4 groups having now achieved transformation of rumen microbes with foreign DNA. Virtually all transformations have been achieved with plasmid constructs and through electroporation methods. None of the engineered organisms have been reintroduced into the rumen to test their stability and in only one case was the engineered organism a recent isolate from the rumen. The approach of using laboratory strains of rumen organisms which have definitely changed in culture clearly demonstrates that the research is at the basic level in which the transformation was the major object and not the improved digestibility that might result.

A major constraint has been the lack of suitable monitoring systems for identifying the engineered organism within the milieu of the rumen. DNA probes are not proving very appropriate because of the genetic diversity of bacterial strains within a species.

The above discussion is intended to emphasise that their is a great need for considerable basic research before modified organisms that could survive in the rumen is developed. It should also emphasise the long time lag that is likely from the beginning, where virtually no information was available to application.

Selection of Anaerobic Fungi for Better Fibre Degradation in the Rumen

Filamentous structures of the rumen anaerobic fungi penetrate and substantially weaken the xylem cell walls of forages in the rumen, whereas the bacteria do not adhere to, nor degrade these highly resistant plant tissues to any significant extent. Rumen fungi, but not bacteria, penetrate the cuticle barrier that protects the leaf's surface. Research indicates that these fungi have unique enzymes, or enzymes with higher specific activities, that give them the ability to weaken and degrade the most limiting structural barriers to degradation.

On the basis of these observations it appears that research aimed at promoting the biomass of fungi in the rumen or the establishment of highly active rumen fungi (*e.g.* by selection) could be a major step forward in increasing the digestibility of poor quality forages by ruminants. The identification of strains of anaerobic fungi that most rapidly degrade the cell walls of forage plants is a necessary prerequisite to the successful inoculation of fungi into the rumen and which will nutritionally benefit the host.

Culture, selection of strains and the identification of new species of anaerobic fungi, testing of their ability to weaken plant fibre and improve the rate of digestion are worthwhile research areas for development.

The developing countries may be well placed to initiate and develop this research as they have access to animals under extreme dietary conditions that have been adapted over long periods of time and therefore could have already selected fungi for high fibrolytic activity.

The research must include developments which allow selected strains to multiply in the rumen and to prevent the competition from wild strains. As rumen fungi are spread by resistant spores which are passed in faeces and then, when taken in by the animal in feed, multiply by sporulation in the rumen, the support of specific species to ensure the survival is a research area with a high priority.

The selection criteria will include the measurement of the rate of solubilisation of fibrous carbohydrates by selected fungi and the weakening of the forage stems following a period of incubation with the forages.

Selection of Bacteria for Fibrolytic Activity

Selection of bacterial strains for improved cellulolytic activity appears to be less vital as it is the fungi that are relatively more important in the breakdown of the most resistant components of the forages, particularly the low quality forages.

Solubilisation of Lignin

Lignin is found within the cell walls of forage plants where it is closely associated with hemicellulose, forming a matrix surrounding the orderly cellulose microfibrils. Lignin is also found in high concentrations in the middle lamella between fibres where it functions as a binder for contiguous cells. Polymeric lignins are not degraded by any known anaerobic organisms. The aerobic white rot fungi *Basidiomycetes* degrade lignin more rapidly and extensively than any other studied microbial group. It is generally accepted that the delignification of a ligno-cellulose material will increase the rate of attack upon the cellulose/hemicellulose compounds when this enters the rumen.

Under aerobic conditions there are two mechanisms of biological delignification:

1. Mineralisation, with lignin being converted directly to carbon dioxide
2. Solubilisation when lignin is released (solubilised) from ligno-cellulose with a variable amount of hemicellulose material attached.

Solubilisation of lignins of grasses is particularly important for its degradation in soils. Something similar to solubilisation seems to occur in the rumen, as considerable lignin appears to be precipitated in the abomasum and therefore, is soluble in the rumen fluid (this is often referred to as acid-precipitatable polymeric lignin (APPL)). The enzymes for solubilisation of lignin, if present in rumen organisms, may be capable of being boosted by genetic means or by selection. In the event that they are not present in rumen organisms, the potential for inducing lignin-solubilising enzymes becomes a major possibility for improving digestibility of fibre in the anaerobic ecosystem of the rumen.

Detoxification of Anti-microbial, Toxic and Antiquality Factors in Plants

Plants in order to survive insect, fungal and bacterial attack, have developed secondary compounds which detract from these organisms colonising the leaf tissues. Some shrubs and trees respond to leaf damage as occurs by grazing and produce greater quantities of secondary compounds which often make them inedible. Recently it has been shown that the grazing of an Acacia tree caused it to release ethylene which within minutes, caused the adjacent acacias in the group to commence synthesising tannins which then caused the grazing deer to cease consumption of the leaves of these trees. Many secondary plant compounds are toxic to the rumen microbes or directly or indirectly toxic to the animal following microbial metabolism in the rumen. This restricts the feed base of ruminants and, at times, prevents stocking of land. Goats and probably other grazing ruminants are able to detoxify many of these secondary plant compounds in the liver.

FACTORS INFLUENCING EFFICIENCY OF FEED UTILISATION

Digested feed clearly has a major effect on nutrient supply, the most important factor governing the extent to which cows can meet their energy, protein and other nutrient needs is the amount of feed they consume. This is especially important in the early stages of lactation when the energy demand for production is higher than intake can support, creating an increasingly negative energy balance which cows have to meet from body reserves, milking off their backs.

The efficiency with which absorbed nutrients are converted to ruminant products (liveweight, milk etc.) is dependent on precisely meeting the animal's requirements above maintenance for individual nutrients required for the particular function. These, at times, are influenced by body condition as affected by previous health and nutritional history, the demands for body temperature control and the requirements for substrate oxidation in exercise (or work).

Graham *et al.* showed that the quantitative oxidation of individual nutrients (largely fat) depended on the degree of heat/cold stress of the animal. A cetogenic substrate was largely used to keep an animal warm when it was required to increase its metabolic rate in cold stress. In this report it is assumed that a cold stressed animal will oxidise acetogenic substrate for heat production until 'surplus' acetogenic substrate is totally utilised, after which fat mobilisation provides an extra and often the major source of metabolic fuel.

The apparently preferential oxidation of circulating acetate leaves a higher ratio of amino acids (and glucose) in the nutrients available for production than would be available to an animal in its zone of thermoneutrality. Conversely, an animal that is not cold has more acetogenic substrates available for anabolic purposes. The environment can, thus, alter the partitioning of nutrients into productive functions and therefore affect the efficiencies of feed utilisation. The design of supplements, to balance diets for ruminants, needs to account for the varying demands for nutrients brought about by the thermal environment of the animal.

It is recognised that cold stress in animals often increases voluntary feed intake and rumen turnover rate. And in this way on some diets it increases microbial cells moving to the lower tract and this has been suggested to increase the P/E ratio in the nutrients available for maintenance or production.

As an example of how environmental factors can change the nutrient balance available to ruminants for anabolism and maintenance, a model used previously to predict the relative availability of specific nutrients to a "standard steer" has been modified to use with sheep and includes the effects of cold stress. The model is based on the sheep (closely shorn) used in the studies of Graham *et al.* which were fed on a daily basis 600, 1200, or 1800 g of a dried grass pellets and subjected to short periods at environmental temperatures ranging from 8 to 40°C at a relative humidity of 50%.

The data in Table indicate that the need to maintain body temperature may require a considerable proportion of the available acetogenic nutrients to be oxidised. In the absence of a cold stress considerably more of digestible nutrients (and in particular more acetogenic substrate) is available for maintenance and synthesis. If in thermoneutral conditions the acetogenic substrate cannot be utilised for synthesis of tissue component because of a low availability of essential amino acids and/or glucose (*i.e.* imbalance in P/E or G/E) then acetate must be dissipated as heat. If the animal is able to oxidise the excess substrate but cannot dissipate the heat-generated because environmental temperature and humidity are high then it could allow its body temperature to increase to some extent but eventually it must reduce its feed intake. If the animal's body temperature rises, metabolic rate increases through the oxidation of protein which may have implications for protein requirements of ruminants in the tropics and differential responses to supplementation in the tropics as compared to temperate areas.

The temperature/humidity at which ruminants are cold stressed depends greatly on the level of feed intake, the insulation provided by the hair or wool coat and the environmental conditions prevailing, *e.g.* wind, rain and availability of shelter. Thus, the environmental temperatures at which minimum extra heat production to combat the cold stress occurs will probably move through a range of from around 10°C to 40°C.

CLIMATE, SUPPLEMENTATION AND INTAKE OF "LOW QUALITY FORAGES"

There has been vigourous debate on whether supplementation of sheep and cattle on low quality forage based diets with urea and/or bypass protein increases intake of the basal feed resource. The differences in results may be hypothesised to be a result of an interaction between climate and the balance of nutrients available from a diet. When research results (Australian) on the effects of supplementation to balance nutrition of cattle on low quality forages are grouped according to climatic zones a pattern emerges.

It appears to be in the tropics and subtropics where poor quality forage intake by cattle is low without supplementation and where significant responses in feed intake occurs when a non-protein nitrogen deficiency is corrected and extra protein that escapes rumen fermentation is provided in the diet.

It is strongly stressed that supplementation with urea and proteinmeals increases voluntary intake of poor quality forages by cattle under tropical conditions to approximately the same level of intake as unsupplemented cattle under temperate conditions. In this situation the supplement is only correcting a depressed intake back to normal intake.

The conclusion that can be drawn from this is that supplements which improve the P/E ratio in nutrients absorbed by cattle fed low quality forage reduces metabolic heat production.

Where metabolic heat production in unsupplemented cattle fed low quality forages would increase body temperature then the animal reduces its feed intake. This reduction in voluntary feed intake is ameliorated by the supplement which allows the acetogenic substrate which would otherwise have to be oxidised to be partitioned into synthetic reactions with a resultant decrease in heat production.

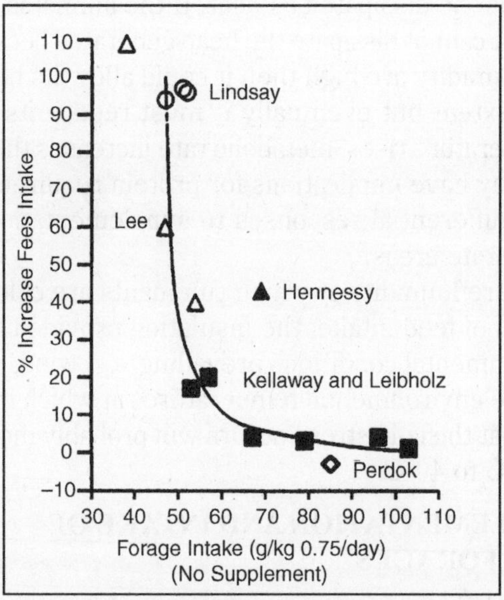

Fig. Intake of Low Digestibility Forages by Cattle Either Unsupplemented or Supplemented with Bypass Protein or Bypass Protein and Urea

The concept of small increases in P/E ratio being able to reduce metabolic heat and at times therefore allow an animal to consume more food might explain the effects of increasing levels of urea in a diet on forage intake (when digestibility is no longer increased) and also the occasional effects on feed intake of branched chain VFA supplements.

The concept is that it is a supplement that improves microbial growth efficiency which has an effect on feed intake and this is only seen in the hot conditions when feed intake is depressed.

The interaction of nutrition and climate may explain why there is a stubborn disbelief by some researchers from developed countries (largely in the temperate areas) of research carried out in developing countries in the tropics. Many of the results of supplementation indicate that a protein that escapes rumen fermentation stimulates both the level and efficiency of production of milk (or live-weight gain) in ruminants fed on crop residues.

The discussion above indicates that ruminants in hot countries have an advantage of not having to oxidise much acetogenic substrate (or body fat) to keep warm. By balancing the diet with supplements, this acetogenic substrate may be captured in products or oxidised to provide ATP for assimilation of the additional nutrients into products. In cold/cool countries supplementation with protein is less necessary, as the utilisation of surplus acetate for heat, decreases the need to balance nutrients. As long as feed intake is high (*i.e.* the diet is highly digestible and perhaps cold stress stimulates intake) production remains relatively high as the nutrients for heat production are extracted and the balance used in synthetic reactions.

Nevertheless, increases in the efficiency of utilisation are obtained when low protein diets are supplemented with a bypass protein even in temperate countries and at times feed intake is also stimulated but it is unknown whether the animals in such studies were actually subjected to hot conditions.

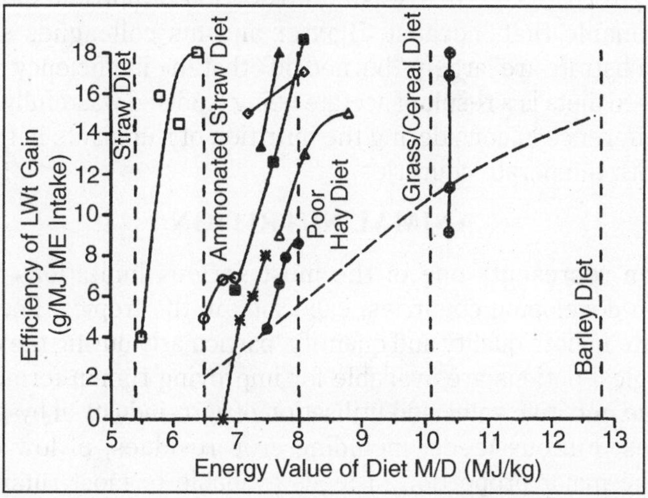

Fig. Schematic Relationship between Diet Quality (Metabolisable Energy mj/kg Dry Matter) and Food Conversion Efficiency (G Liveweight Gain/mj me) (---). The Relationships Found in Practice with Cattle Fed on Straw or Ammoniated Straw with Increasing Level of Supplementation. Australia (%, o, •), Thailand (") and Bangladesh (%). Recent Relationships Developed for Cattle Fed Silages Supplemented with Fish Proteins (o) and Tropical Pastures Supplemented with Cottonseed Meal (*) are also Shown. This Illustrates the Marked Differences that Result when Supplements High in Protein are given to Cattle on Diets of Low me/kg dm

It can be concluded that ruminants in the tropics that are adequately supplemented with small quantities of essential nutrients may produce at the same rate on a lower digestibility feed as an animal on higher digestibility feed in a cold environment.

To emphasise the differences in potential thermal stress of animals under different conditions the average temperature humidity index (THI) (which is an index of potential heat stress conditions for ruminants on a monthly basis for Cambridge (England), Chittagong (Bangladesh), Bangkok (Thailand) and

Armidale (Australia) are shown in Figure. The critical THI (72) for high milk producing cows as determined by Johnson (1987) is included in the figure. However, it must be emphasised that in addition to temperature/humidities, the critical THI will depend on the insulation provided by the animal's coat and its behaviour in seeking shelter, as well as the incidence of wind and rain in addition to level and quality of feed intake.

Many studies have shown that at the same forage intake by ruminants with an already efficient digestion that a supplement of protein that reaches the small intestine increases the efficiency of feed or metabolisable energy utilisation for growth.

This is positive proof that wasteful oxidation of nutrients can occurs. It seems reasonable that, because Blaxter and his colleagues showed that acetogenic substrate are largely "burned off" that the inefficiency of ruminants on forage based diets is a result of acetate being oxidised wastefully. This points to a major difference in considering the nutrition of ruminants in the tropics as compared with temperate countries.

ANIMAL NUTRITION

Nutrition represents one of the most serious limitations to livestock production in developing countries, especially in the tropics. Feed resources are inadequate in both quality and quantity, particularly during the dry seasons. Biotechnological options are available for improving rumen fermentation and enhancing the nutritive value and utilisation of agro-industrial by-products and other forages. Fibrous feeds, including crop residues, of low digestibility constitute the major proportion of feeds available to most ruminants under smallholder situations in developing countries. The associated low productivity can be overcome to some extent by several means, among which are: balancing of nutrients for the growth of rumen microflora thereby facilitating efficient fermentative digestion and providing small quantities of by-pass nutrients to balance the products of fermentative digestion, enhancing digestibility of fibrous feeds through treatment with alkali or by manipulating the balance of organisms in the rumen and genetic manipulation of rumen micro-organisms, currently acknowledged as potentially the most powerful tool for enhancing the rate and extent of digestion of low quality feeds. Rumen micro-organisms can also be manipulated by adding antibiotics as feed additives, fats to eliminate or reduce rumen ciliate protozoa (defaunation), protein degradation protectors, methane inhibitors, buffer substances, bacteria or rumen content and/or branched chain volatile fatty acids.

Increasing Digestibility of Low-quality Forages

Low-quality forages are a major component of ruminant diets in the tropics. Thus, much progress can be made by improving the forage component of the ration. The characteristic feature of tropical forages is their slow rate of microbial breakdown in the rumen with the result that much of the nutrients of the feed are voided in the faeces. The slow rate of breakdown also results in reduced outflow rate of feed residues from the rumen which consequently depresses feed intake. At present, the main treatment methods for forages such as cereal straws are either mechanical (*e.g.* grinding), physical (*e.g.* temperature and pressure treatment) or a range of chemical treatments of which sodium hydroxide or ammonia are among the more successful.

The lignification of the cell walls prevents degradation by cellulase or hemicellulase enzymes. Fortunately, it is possible to use lignase enzyme produced by the soft-rot fungus (*Phanerochaete chrysosporium*) which causes a high degree of depolymerisation of lignin. The enzyme acts like a peroxidase and causes cleavage of carbon-carbon bonds. At present the levels of the lignase enzyme produced by the basidiomycete fungi are insufficient for the treatment of straw on a commercial scale. However, it is conceivable that the use of recombinant DNA engineering techniques will allow the modification of the lignase genes and associate proteins to increase their efficiency and stability. The lignin gene has to date been cloned and sequenced from *P. chrysosporium*.

Improving Nutritive Value of Cereals

Moderate protein content and low amounts of specific amino acids limit the nutritive value of cereals and cereal by-products (*e.g.* barley is low in Iysine and threonine). This is a major limitation in the ration formulation for non-ruminant livestock which necessitates addition of expensive protein supplements. There are on-going studies to enhance the low level of Iysine in barley by genetically engineering the grain genome. Genetic modification through insertion of genes into rice protoplasts and generation of transformed plants has already been achieved.

Removing Anti-nutritive Factors from Feeds

Anti-nutritive factors in plant tissues include protease inhibitors, tannins, phytohaemagglutinins and cyanogens in legumes, and glucosinolates, tannins and sanapine in oilseed rape (*Brassica napus*) and other compounds in feeds belonging to the *Brassica* group. As with amino acid deficiencies, the adverse effects of these compounds are more marked in non-ruminants than in ruminants.

Conventional plant breeding has been used to reduce and, in some cases, eliminate such anti-nutritive factors. An example is the introduction of cultivars of oilseed rape which are low in, or free from erucic acid and glucosinolates. A combination of genetic engineering and conventional plant breeding should lead to substantial reduction or removal of the major anti-nutritive factors in plant species of importance as animal feeds.

Transgenic rumen microbes could also play a role in the detoxification of plant poisons or inactivation of antinutritional factors. Successful introduction of a caprine rumen inoculum obtained in Hawaii into the bovine rumen in Australia to detoxify 3-hydroxy 4(IH) pyridine (3,4 DHP), a breakdown product of the non-protein amino acid mimosine found in *Leucaena* forage demonstrates the possibilities.

Improving Nutritive Value of Conserved Feed

The conservation of plant material as silage depends upon anaerobic fermentation of sugars in the material which in turn is influenced by the ability of naturally occurring lactic acid bacteria to grow rapidly on the available nutrients under the existing physical environment. Unless the ensiled material is sterilised, lactic acid bacteria are always present. However, the ensiling conditions may not always be ideal for their development. In addition to the number and type of bacteria, other interrelated factors may affect quality of silage, including availability of water-soluble carbohydrates, the dry-matter content, the pH and extent of air exclusion. For example, lack of water-soluble carbohydrates may be overcome by wilting the material to raise the dry matter to a level at which less acid is required to stabilise the fermentation. The availability of sugars in the material and the rate at which the different micro-organisms multiply also influences the ensilage process.

Throughout this century, research workers have investigated ways through which the fermentation process in silage making can be controlled in order to improve the feeding quality of the resulting silage. Use of additives, to restrict the activity of the microorganisms, to stimulate the fermentation by the lactic acid bacteria or simply as nutrients has been one of the approaches. Additives used in the early studies included chloroform, toluene and cresol (to inhibit bacterial growth) and sulphuric acid and hydrochloric acid (to reduce the pH). Indeed, over the last 40 to 50 years, corrosive, acid-containing additives have been widely used in silage making. Other fermentation inhibitors which have been studied include organic acids, salts of acids, formaldehyde and other aldehydes, sodium hydroxide, and antibiotics. Of these, formic acid is probably the most widely studied and has been reported to have a beneficial effect on the fermentation process and on the nutritive value of silage. Sulphuric acid is cheaper than formic acid and is popular in some countries. However, acids are a hazard on the farm and can be particularly dangerous if recommended to uninformed farmers. Salts of acids are safer to handle but are less effective than the acids from which they are derived.

The hazardous nature of some of the chemical additives has necessitated a search for alternative compounds for improving the ensilage process. A group of compounds classified as fermentation stimulants have been widely studied. These include sugar sources (*e.g.* molasses and whey), enzymes and inocula of lactic acid bacteria. Molasses is of particular relevance to smallholder farmers in developing countries in the tropics where sugar-cane is produced and processed. Enzymes are essential for the breakdown of cell-wall carbohydrates to release the sugars necessary for the growth of the lactic acid bacteria. Although resident plant-enzymes and acid hydrolysis produce simple sugars from these carbohydrates, addition of enzymes derived from certain bacteria, *e.g. Aspergillus niger* or *Trichoderma* increases the amount of available sugars. Commercial hemicellulase and cellulase enzyme cocktails are now available and improve the fermentation process considerably. However, prices of these products preclude their viability for farm level application, especially in developing countries.

There are two forms of indigenous lactic acid bacteria: the homofermentative type which converts hexose sugars to lactic acid with no loss of dry matter and the heterofermentative type which produces a range of compounds accompanied by loss of dry matter as carbon dioxide. Thus, the native bacteria are not the most efficient. Considerable research in the USA and Europe has been directed towards the development of microbial silage additives (inoculants). Commercial bacterial inoculants designed to add sufficient homofermentative lactic acid bacteria to dominate the fermentation are now available The objective of using such additives is to ensure the rapid production of the required amount of lactic acid from the carbohydrates present to preserve the ensiled material. Most such inoculants contain *Lactobacillus plantarum* with or without other bacteria such as *L. acidophilus, Pediococcus acidilactive* and *Streptococcus thermophilus*. In general, the results with bacterial inoculants have been quite variable. However, with an effective product, it is possible to improve the fermentation of low dry-matter silages and to enhance the efficiency of their utilisation.

In order to improve the effectiveness of microbial inoculants in breaking down structural carbohydrates to glucose, detailed knowledge of the lactobacilli bacteria is essential. Work already undertaken on the molecular biology of *Lactobacillus plantarum* and other species suggest that the rapid progress in this area will make it possible to construct novel genes encoding highly active fibre-degrading enzymes. Such genes could then be inserted into strains of *L. plantarum.* Successful silage making incorporating these technologies can only be achieved with strict adherence to recommended application procedures, including rates of additives, inoculants etc. This technology is available in most developing regions including Africa. However, it is not fully exploited. Indeed, in Africa silage making is still generally restricted to large-scale commercial farms.

Improving Rumen Function

Armstrong and Gilbert and Forsberg *et al*. have reviewed the major areas of rumen function which might benefit from transgenic technology. These include development of transgenic bacteria with enhanced cellulotic activity, capability to cleave lignohemicellulose complexes, reduced methane production capability decreased proteolytic and/or deaminase activities, increased capability for nitrogen "fixation" and increased ability for microbial production of specific amino acids. The first successful transfer of foreign genes into rumen bacteria (*Bacteriodes ruminicola*) was reported by Thomson and Flint. However, we are still a long way from commercial production of genetically engineered rumen bacteria.

Although several workers have isolated genes encoding plant structural carbohydrate-degrading enzymes from rumen bacteria, there are limited reports on the genetic engineering of these microorganisms. In contrast to conditions in which single species of organisms are grown in controlled environments and where the energy supply is usually in excess of demand, the rumen environment is very complex, competition between different microbial species is intense and energy is usually the limiting growth factor. This is probably the main reason why reintroduction of genetically modified rumen bacteria into their natural habitat has met with variable success. Advances being made in transformation methods for obligate anaerobic bacteria will certainly result in successful genetic engineering of a range of rumen bacteria. However, it is not possible to predict if any of these bacteria will be capable of colonising the rumen. It can be concluded that there are several potential opportunities for improving the efficiency of ruminant digestion and possibilities for utilising a wider range of feeds than is currently possible. Modification of rumen microbial population is one such opportunity. However, technical difficulties associated with making genetic modifications to individual species of rumen bacteria hinder progress in this area.

NUTRIENT ADDITIONS TO FOOD PRODUCTS

The following is not an exhaustive review of all food products which are fortified. Food products have been selected for discussion on the basis of their nutritional impact or their applicability to fortification programmes in developing countries.

CEREALS AND CEREAL BASED PRODUCTS

This is one of the most important areas for consideration of food fortification technologies. Foods from this category form the major component of diets around the world. This is especially true in the case of developing countries where dietary diversity is limited. On average, cereals provide 52 per cent of caloric intake globally. For Africans and Asians they represent 60-75 per cent of the caloric intake; for Latin Americans, 50 per cent; and in the United States of America, 26 per cent. In developing countries 95 per cent of the population consume cereals as a dietary staple which also provide about 47 per cent of the per capita protein intake. Milling of cereal grains prior to their consumption is a common practice. During the milling process a substantial proportion of the nutrients are lost from the refined product. The fortification/enrichment of cereal grains can therefore be rationalised in more than one way. One valid reason is to restore to refined products, nutrients which have been removed during the milling process. Another reason is to improve the nutrient intake levels of target populations which are at risk of micronutrient deficiency.

RICE AND OTHER WHOLE CEREAL GRAINS

Among the cereals, rice presents unique problems in fortification. These are due to the fact that it is most commonly consumed as a whole grain and also in many countries, extensive washing of the grain prior to cooking is the normal practice. The earliest methods of rice enrichment involved the production of parboiled and converted rice. By this means nutrients from the bran layer were transferred to the starchy endosperm. The parboiling process involved the soaking of the rough rice, the application of heat followed by drying and milling.

It was demonstrated that in this way 50-90 per cent of the thiamine was retained. The process for converted rice developed by Huzenlaub was similar to that for parboiling, but also employed pressure differences to facilitate transfer of nutrients. Acid parboiling, described by Kondo was similar to the parboiling except that it was carried out in the presence of acetic acid. After parboiling and converting, the next methods of enrichment involved the actual addition of nutrients to the milled products. Techniques used for this have been classified into two main groups 'powder type' and 'grain type' enrichment.

In powder type enrichment, a powdered pre-blended mixture of vitamins and minerals has been added at a rate of 1, 0.5. or 0.25 oz. per 100 lbs of rice (a w/w ratio of 1:1600, 1:3200 or 1:6400). For white parboiled rice, the normal practice has been to add the premix soon after milling as the heat and moisture at the grain surface at this point facilitates adherence of the powder. A major disadvantage of this method of nutrient addition has been that 20-100 per cent of the nutrients are lost on washing. In the USA, rice enriched in this way must bear a label stating 'to retain vitamins do not rinse before or drain after cooking'.

In the second major type of enrichment, a powdered nutrient mixture has been applied to the milled rice grains followed by coating with a water insoluble substance. A fortified rice premix produced in this way has then been added to milled rice at a rate of 0.5 per cent, to yield an enriched product conforming to the required standard of identity. In the United States of America, this standard requires between 2.0 - 4.0 mg of thiamine, 1.2 - 2.4 mg of riboflavin, 16 - 32 mg of niacin or niacinamide and 13 - 26 mg of iron per 100 lbs. of rice.

It may contain 250 -1000 USP units of vitamin D and 500 -1000 mg of calcium. The requirement regarding riboflavin has been stayed pending final action. The objections regarding the use of this vitamin in the premix have been due to its role in the colour deterioration of the enriched grain. 'Grain type' or 'coated grain' enrichment have also been carried out by spraying the premix solution onto the rice which is contained in a rotary cylinder, followed by hot air drying of the rice, application of a water insoluble sealant, addition of the iron compound and finally a second application of the water insoluble compound. Water insoluble coatings which have been reported include an ethanol or isopropanol solution of zein, palmitic or stearic acid and abeitic acid. Other coating materials have contained ethyl cellulose. They dissolve at the elevated temperatures employed during cooking. Using this 'coated grain' procedure, Cort et al and Rubin et al. were able to successfully enrich rice with niacin, thiamine, pyridoxine, vitamin A, vitamin E, folic acid, iron, and zinc using expanded versions of the procedures already described. The water insoluble vitamins and minerals were added in different layers with intermittent coatings of shellac. Stability to rinsing treatments were reportedly high, with vitamin losses in the range of 0.2-1.1 per cent.

Ricegrowers Co-operative Ltd. (RCL) in Australia revised the HLR rice fortification procedure for the enrichment of rice with thiamine, niacin and iron. The HLR process involved dissolution of vitamins in dilute sulphuric acid, application of vitamin solution to rice, application of a water insoluble substance, application of ferric pyrophosphate and talc followed by reapplication of a water insoluble layer and iron mixture before screening and packaging.

Due to increased cost of raw materials, low factory output, use of hazardous materials, reported problems with browning of enriched grains, RCL modified the procedure to include the application of the nutrient mixture to the rice followed by drying, screening and packing. The nutrient mixture was composed of ferric pyrophosphate suspended in an acid solution of the vitamins.

They reported that the browning found in the HLR premix on storage was due to the formation of ferric sulphates and that this problem was eliminated with the use of an alternative acid. Acid hydrolysis at the surface of the rice grain provided a sufficiently sticky surface to ensure adhesion of the iron compound. Cost of the process was reduced and the factory output was doubled.

There was no mention, however, of the stability of premix to nutrient loss through rinsing. In the Philippines, attempts at fortifying rice by application of vitamin A followed by a water insoluble layer were abandoned since washing losses of between 10 - 20 per cent were recorded. Rice fortification with iron has also been tried in the Philippines using the coating method. The fortificant used in this programme was anhydrous ferrous sulphate. Some discolouration was found after 20 weeks of storage at room temperature and washing losses were 9 per cent.

Ferric orthophosphate (white iron) is a recommended form of iron for use in the fortification of rice. This iron compound is almost water insoluble and has been preferred for mixing with milled rice due to its white colour. When it is oxidised or contains excessive moisture it may become tan, yellow, purple and or black. Hurrel reported that the bioavailability of ferric orthophosphate varied widely from batch to batch and was highly negatively correlated with particle size as was also found with elemental iron. This fortificant compound is more expensive than anhydrous ferrous sulphate. According to Hurrel the cost of ferric orthophosphate was about six times that of anhydrous ferrous sulphate, for the same level of total iron. In Japan, a multinutrient enriched rice has been on the market since 1981. The first step in this procedure was acid parboiling in the presence of thiamine, riboflavin, niacin, pantothenic acid and pyridoxine. The second step involved coating the grain with separate layers of vitamin E, calcium and iron in separate layers and finally a protective coating material. Modified atmosphere packaging, utilising aluminium laminate and carbon dioxide, ensured the stability of vitamin E during storage. The yellow or brown colour of the premix was not a problem in this case as it was used in the fortification of 'brown' rice.

Apart from these well established procedures there have been other innovations regarding alternative methods for rice fortification. Joseph et al. described an enrichment procedure using a premix containing thiamine, riboflavin, niacin and pyridoxine. This procedure involved soaking the milled rice in an acid medium containing the water soluble vitamins followed by the cross linking of starch granules in the enriched grains. The cross linking procedure itself was demonstrated to have caused significant vitamin loss, but the added vitamins were highly cook and wash stable. It is possible that this method could have some utility in the future.

Use of fortified simulated grains has featured prominently in attempts at rice fortification in the developing countries. Murphy et al. described the production of such a fortified rice product for use in the Philippines. The synthetic rice grains were produced by extrusion of rice flour in a pasta machine. The best formulation contained vitamin A stabilised by a mixture of tocopherol, ascorbate and lipids with a low level of unsaturation.

Retinyl palmitate stabilised in an acacia matrix, type 250 SD was the fortificant used. Retention of vitamin A after washing was reported to be 100 per cent. Vitamin retention after cooking, however, ranged from 60 - 94 per cent. The formulations which demonstrated the better storage stability, particularly at high humidity, suffered greater cooking losses. Drawbacks with this technology have been reported to exist with respect to blending with the natural product and in the consistency of the simulated grains after cooking. Field trials in Brazil showed no problems with acceptability of rice enriched in this way.

The enrichment of whole wheat grains with vitamin A has been attempted. The fortificant used was a premix comprised of concentrated vitamin A attached to wheat grains, for mixing at a level of 0.25 per cent with wheat grains. The feasibility of this procedure was not determined. The fortification of whole grain cereals with soluble iron compounds is difficult because they promote oxidation of the lipid component of the grain, thus reducing the shelf life.

FLOURS, CORNMEAL AND BREAD AND PASTA

In the fortification of flour the required nutrient mixture is mixed with an appropriate diluent to produce a premix, which is then accurately metered into the flour. The addition of vitamins B_1 and B_2, niacin, iron and calcium to wheat flour is a common practice in many developed countries. It is technologically feasible to add other vitamins and minerals as well. Cort et al successfully used two premixes to fortify wheat flour; the first comprised vitamin A, pyridoxine, folic acid, tocopherol acetate, thiamine, riboflavin, niacin and iron while the second contained calcium, magnesium and zinc. The vitamin/iron premix demonstrated excellent stability on storage. The flour enriched with both premixes also demonstrated excellent stability on storage at room temperature. Under conditions of accelerated storage at elevated temperature (45 °C), however, there was substantial loss of vitamin A beyond 4 weeks of storage. Parrish et al. also reported good stability of enriched wheat flour stored at room temperature, but about 50 per cent losses in flour stored at 40 °C for 6 months.

The form of vitamin A most commonly used in the fortification of flour was dry stabilised vitamin A palmitate (type 250-sd) powder form. The water soluble vitamins (thiamin, riboflavin, niacin, pyridoxine, folate and calcium pantothenate) are used in pure crystalline form. The mononitrate salt of thiamine is preferred for this use. Iron is normally used in its reduced elemental form.

In the work of Parrish et al. ferrous sulphate was also used as the source of iron and there was no significant difference demonstrated between the flour enriched with ferrous sulphate and that enriched with elemental iron. It has been recommended that ferrous fumarate be used in the fortification of wheat flour within the P.L. 480 programme, due to its greater bioavailability. When mixing dry fortificants with dry foods, careful selection of the physical characteristics of the fortificant compound is important so as to ensure adequate mixing and to minimise segregation on storage.

Rubin et al. investigated the stability of bread made from flour enriched by the 6 vitamins and 4 minerals described by Cort et al. They found that the inclusion of calcium and magnesium adversely affected vitamin A retention during the baking process. There was also an off-flavour detected in bread which included magnesium after 5 days of storage. Crumb colour and grain were also negatively affected by the nutrient multimixes tested by Rubin et al., with the greater affect occurring with the nutrient mix which included magnesium.

Fortification of bread with zinc salts exerted no adverse effect on loaf volume, flavour or any other index of bread quality. Iodisation of bread has been carried out in the Netherlands and Tasmania by the addition of 2-4 ppm KIO_3 to the bread improver which was already in general use. KIO_3 has been used in bread production in the past, not as a fortificant but as an oxidising agent to improve dough quality. The fortification of corn meal with 6+4 vitamin/iron mixture was shown to be technically feasible. Losses of vitamin A on storage at room temperature for up to six months were greater than with wheat flour but still remained below 20 per cent. In many countries pasta or noodles are commonly eaten and these can therefore be important vehicles for fortification. The manufacture of such products involves the production of a dough which is then extruded and dried.

Enrichment can be through the use of enriched flour or alternatively, wet addition of a dispersion of the required vitamins can be carried out at the dough-making stage. Vitamin losses during production depend largely on the drying conditions employed. Dexter et al. enriched durum wheat flour with a vitamin mixture containing riboflavin, thiamine mononitrate and niacin for the production of spaghetti. They reported that high temperature drying treatments resulted in significant losses of riboflavin, whereas the other vitamins were stable to the processing conditions. In all cases cooking losses, estimated between 40 - 50 per cent, exceeded those experienced during processing.

BREAKFAST CEREALS

The fortification of ready to eat breakfast cereals is a wide-spread practice. In the United States the general practice has been to provide 25 per cent of the RDA of thiamine, riboflavin, niacin, pyridoxine, folate, ascorbic acid and vitamin A per 1 ounce. 10 or 25 per cent iron, up to 20 per cent RDA for calcium have also been added; vitamin D has also sometimes been included. Surveys have indicated that fortified breakfast cereals play an important role in ensuring nutritional adequacy of the diets of the U.S. population.

The minerals and the more heat stable vitamins like niacin and riboflavin have been added to the basic formula mix prior to processing. The heat labile vitamins such as vitamins A, C and thiamine are usually sprayed onto the cereals after the high temperature processes such as oven or extruder. According to Johnson et al. two important considerations in the design of the spray system are:

- To ensure uniformity in spray coverage
- To optimise protection of both vitamins A and D.

As the hydrophobic vitamins cannot form true aqueous solutions, they have been sprayed on in the form of an emulsion. The composition of the emulsion has been shown to greatly affect the vitamin stability. The inclusion of 15-20 per cent sucrose into the spraying emulsion stabilised the vitamin A. It acted by forming a protective coating around the vitamin on drying. This fact explained the greater stability of vitamin A included in raisin bran as compared with cornflakes. In the latter case a no-sugar spray was used. Vitamin E can be added to the spray emulsion, it was at least as stable when added under these conditions.

Good storage stability of the fortified breakfast cereals was reported. This was partly due to the fact that packaging materials with appropriate barrier properties have to be selected to ensure the maintenance of equilibrium moisture contents below about 5 per cent. Above this level there was textural deterioration of the product which was perceived as 'staleness'. Lund reported almost 60 per cent loss of vitamin C from raisin bran on storage for a 12 month period.

This was ascribed to the fact that the moisture levels in this cereal were maintained at a higher level to prevent hardening of the raisins. Improved procedures for the coating of raisins with a moisture barrier material prior to mixing with the material can minimise this problem. The addition of ascorbic acid was found to stabilise vitamin A to a greater extent than did phenolic antioxidants. The addition of tocopherols did not provide any additional stability, possibly because they were already present at optimal levels.

THE RUMEN AND ITS MICRO-ORGANISMS

As the utilisation of forages by ruminants depends on microbial fermentative digestion, the principles of digestion in the rumen are discussed as a framework to view the requirements for biotechnology innovations in nutrition.

The rumen is the dominant feature of the digestive tract of cattle. This maintains a medium that supports a dense and varied population of microorganisms. These organisms ferment feed materials to produce mainly shortchain organic acids or volatile fatty acids (VFAs), methane and carbon dioxide and the process provides substrate (the feed) and ATP (energy) for the growth of micro-organisms. The microbial mix in the rumen is complex and highly dependent on diet. The main agents that break down fibre, sugars, starches and proteins in the rumen are all anaerobic and include bacteria, protozoa and fungi. The bacteria are the principal organisms that ferment plant cell-wall carbohydrates but the anaerobic phycomycetous fungi may at times be extremely important.

Protozoa are now recognised as having an overall negative effect in the rumen, particularly where ruminants are fed forage diets low in true-protein. Protozoa ingest and digest bacteria and reduce the bacterial biomass in the rumen and consequently the protein supply to the animal. Thus, they decrease the protein to energy ratio in the nutrients absorbed (see later) and increase the requirement of animals for true protein. The net result of the presence of protozoa is an increased requirement for dietary bypass protein and on low protein diets a decreased efficiency of utilisation of feed for growth and milk production. The presence of protozoa in the rumen may also reduce the rate at which bacteria colonise and degrade the ingested feed particles. In studies with sheep fed straw based diets, it has been found that the apparent digestibility of dry matter was increased by 18% after protozoa had been removed from the rumen (*i.e.* defaunated). This research indicates that large increases in productivity may be achieved with ruminants fed fibrous diets, particularly those low in true protein by controlling or removing protozoa from the rumen. Other workers have not seen the differences in digestibility and in some instances removal of protozoa from the rumen has led to decreased digestibility of mixed, starch containing diets.

FERMENTATIVE EFFICIENCY IN THE RUMEN

A deficiency of a nutrient needed by rumen micro-organisms reduces microbial growth efficiency which reduces microbial biomass and eventually reduces digestibility and feed intake, particularly of fibrous feeds.

The first priority in feeding ruminants is to ensure no deficiencies in the diet of nutrients for microbial growth in the rumen. Of major importance is that the efficiency of microbial growth (that is, the amount of microbial biomass available for digestion in the intestines per unit of digestible carbohydrate entering the rumen) also determines the proportion of digested feed that is converted to methane and VFA. Methane production accompanies the formation of acetate or butyrate, whereas methane and VFA production are inversely related to microbial cell production.

MEETING THE REQUIREMENTS FOR EFFICIENT MICROBIAL GROWTH IN THE RUMEN

On most diets based on crop residues and low-digestibility forages, the primary limitation to the growth of rumen micro-organisms is probably the concentration of ammonia in rumen fluid The second consideration is deficiencies of minerals, particularly sulphur, phosphorus, magnesium and certain trace minerals. Ammonia in the rumen must be above a critical level for a considerable period of the day to ensure a high rate of microbial growth and digestion and therefore feed intake. The level of ammonia that supports the optimal population of micro-organisms in the rumen the highest protein to energy ration in the nutrients absorbed, and therefore maximum digestion, will vary among diets. In general on forage based diets the ammonia level should be above 200 mg nitrogen/litre. It must be stressed, however, that any nutrient, (including many minerals required in the growth of micro-organisms), that is deficient in a diet will result in low microbial cell yield relative to VFA and lead to a low protein (from microbes) to energy (from VFA) in the nutrients absorbed. The ratio of protein digested and absorbed from the intestines to the VFA produced in and absorbed from the rumen is termed the P/E ratio.

CONSEQUENCES OF THE RUMINANT MODE OF DIGESTION

One of the consequences of the ruminant mode of digestion is that fermentation results in up to 20% of the digestible energy intake being lost as heat and methane. A second major disadvantage is that proteins that are fermented in the rumen are not then sources of amino acids for the animal because they are hydrolysed and their constituent amino acids deaminated by microbes. In general, where ruminants are fed forage based diets typical of that available in tropical developing countries, small amounts of extra nutrients are needed to balance nutrient availability to requirements. Proteins which are directly available to the animals and are protected from degradation increase the efficiency of anabolism of the absorbed nutrients in growth, pregnancy, lactation or work.

QUANTITATIVE ASPECTS OF FERMENTATIVE DIGESTION IN THE RUMEN

The end products of rumen fermentative digestion are governed by the feed, the rate of consumption of feed, the balance of nutrients in the feed for microbial growth and the balance of micro-organisms that develop in the rumen (bacteria, protozoa and fungi). In general, a proportion of the digestible feed dry matter is converted to VFA, methane and carbon dioxide and the balance is assimilated into microbial cells. The pathways of these reactions are well known and a schematic outline is shown in Figure.

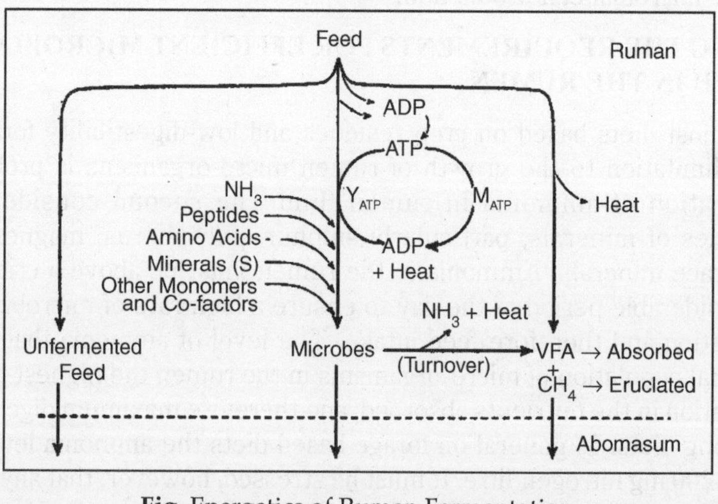

Fig. Energetics of Rumen Fermentation

Microbial cells, that are synthesised from the feed resource use the ATP that is generated in the formation of VFA from the feed to provide the energy for synthesis. The microbes are lost from the rumen pool either by passage out of the rumen to be partially digested in the intestine or by death and breakdown within the rumen (with formation of VFA, CO_2 and methane). Lysis and degradation in the rumen is inefficient as it makes the protein of microbes unavailable as such to the animal. Because microbial cells are more reduced than the substrate fermented, the quantity of microbial cells leaving the rumen per unit of carbohydrate consumed is related to methane production. The efficiency of microbial growth is then a primary determinate of the quantity of methane produced.

A Model of Fermentation in the Rumen

For the purposes of the present discussion and to demonstrate the underlying principles of the concepts developed, a model for a 200 kg steer will be used to illustrate the quantitative availability of nutrients from rumen fermentation. The steer consumes 4 kg which represents 25 Mole anhydroglucose or organic matter which is completely fermented in the rumen.

It is assumed:

- That the fermentation of 1 mole of carbohydrate gives rise to either 2 mole acetate, 2 mole of propionate or 1 mole of butyrate, according to the following stoichiometry:

Hexose	→	Pyruvate + 2ATP + H_2
2Pyruvate + $2H_2O$	→	2HAc + $2CO_2$ + 2H2 + 2ATP
2Pyruvate + $8H_2O$	→	2HPro + $2H_2O$ + 2ATP
2Pyruvate + $4H_2O$	→	H Bu + $2H_2$ + $2CO_2$ + 2ATP
CO_2 + H_2	→	CH_4 + $2H_2O$ + 2ATP

- In the stoichiometry, H_2 indicates reduced co-enzymes, HAc is acetic acid, HPro is propionic acid and HBu is butyric acid
- That the animal's rumen is functioning at "normal" level of fermentative efficiency in which one-third of the organic matter fermented is converted to microbial cells and the rest to VFA, CO_2 and CH_4.
- That the moles ATP generated per mole of end-product are for acetate 2, butyrate 3, propionate 3, and methane 1.

On chemical principles, the equation of substrate use and end products from fermentation of 4 kg of carbohydrate is:

- *Carbohydrate to* VFA: 16.7 CHO → 21 HAc + 6HPro + 3HBu + 7.5CH_4 + 78ATP
- *Carbohydrate to microbial cell precursors*: 8.3CHO → 1.4 polysaccharide + 13.8 pyruvate + 2CH_4 + 17ATP
- *Overall*: 25CHO → 21HAc + 6HPro + 3HBu + 9.5CH_4 + 1300 g dry cells.

In the example, one-third of the carbohydrate provides the substrate for microbial cell synthesis 1300 g dry microbial cells are produced at a Y_{ATP} of about 14.5 (Y_{ATP} is a measure of the efficiency of utilisation of ATP generated in fermentation of carbohydrates to VFA; it is defined as the g dry cells produced per mole ATP available.) The upper level of efficiency has a Y_{ATP} of 26. On the other hand the lowest efficiency of a microbial growth in the rumen that is deficient in, say, ammonia, is possibly below a Y_{ATP} of 4. The relative efficiency of the system (indicated as Y_{ATP}) is governed largely by the availability of essential nutrients for microorganisms.

The ranges of Y_{ATP} are shown for:

A. A relatively inefficient rumen (*i.e.* Ammonia deficient)
B. A 'normal' rumen with no deficient nutrient for microbial growth
C. A rumen free of protozoa with no deficient nutrient for microbial growth
D. The theoretical optimum microbial growth efficiency.

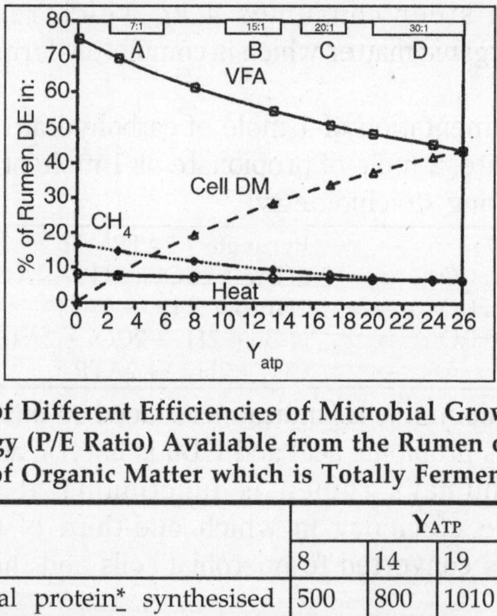

Table. The Effect of Different Efficiencies of Microbial Growth on the Ratio of Protein to VFA Energy (P/E Ratio) Available from the Rumen of a Steer Consuming 4 kg of Organic Matter which is Totally Fermentable.

	Y_{ATP}			
	8	14	19	25
Microbial protein* synthesised (g/d)	500	800	1010	1212
VFA produced (MJ/d)	41	34	30	26
Methane produced (MJ/d)	9.4	8.5	8.0	7.6
Heat (MJ/d)	6.4	5.1	4.3	3.1
P/E ratio (g protein/MJ)	12	25	34	47

Note:

* Microbial protein may be only 75–85% digestible and this will change the P/E ratio markedly in the animal.

Based on this model, but assuming a varying efficiency, the microbial cells produced relative to VFA and methane production change are shown in Table. The main point to emphasise is that, depending on the efficiency of utilization of ATP, the amount of carbohydrate converted to microbial cells can be highly variable. It is the efficiency of microbial growth that largely controls the amount of methane produced by an animal.

PROTEIN UTILISATION BY RUMINANTS

Protein that is fermented in the rumen is largely wasted as a source of amino acids to the animal because:

- Dietary protein is degraded and essential amino acids are deaminated to form ammonia and VFA
- Fermentation of 1 g of protein generates only half the ATP that would be produced from 1 g of carbohydrate and therefore anaerobic microbial growth on protein is approximately half that on carbohydrate.

In combination these effects result in only 30 to 60 g of microbial protein becoming available to the animal for digestion for every kilogram of dietary protein that is fermented in the rumen. The fermentation of protein is, however, associated with relatively small amounts of methane production. On the other hand methane is not generated when protein bypasses the rumen.

Protein that is insoluble, or has a high component of disulphide bonds or is associated with tannins tends to bypass rumen fermentation but is digested in the intestines and in this way it augments the microbial protein and alters the ratio of protein to energy (P/E) in the nutrients absorbed. The better the balance of nutrients for microbial growth the higher the ratio of P/E in the nutrients produced in fermentation. The higher the content of bypass protein in the diet the higher the P/E ratio in the absorbed nutrients.

Ensuring a Balanced Nutrition For Ruminants on Forage Based Diets

From the above discussion the first priority for improving the utilisation of a low digestibility forage by ruminants is to optimise the availability of nutrients from fermentative digestion by:

- Ensuring that there are no deficiencies of microbial nutrients in the rumen and therefore the microbes in the rumen grow efficiently and, through fermentative activity, extract the maximum possible amounts of nutrients from the forage (*i.e.* The ratio of microbial cells to vfa produced is high as are production rates of these end products)
- Ensuring that the microbial cells (which provide most of the protein to the animal) synthesised in the rumen are not lysed and fermented in the rumen but are available for digestion and absorption as amino acids from the intestines.

The second objective of a feeding strategy should be to optimise the efficiency of partitioning of absorbed nutrients into product by:

- Supplementation with critical nutrients that escape or bypass rumen fermentation to augment and balance the nutrients absorbed to those required for maintenance of homeostasis, maintenance of body temperature, exercise (or work), and the particular physiological or productive function.

As the nutrients needed for different functions differ in priority, supplementation strategies will need to vary according to climate, environment, management and production targets in any one location.

DEVELOPING DETOXIFICATION MECHANISMS IN RUMEN ORGANISMS

Research into the possibility of detoxification of secondary plant compounds was given impetus by the discovery that organisms in the rumen of goats in Hawaii that was able to degrade the breakdown product of mimosine (*i.e.* dihydroxypyridine) in *Leucaena leucocephala,* but ruminants in Australia were unable to do so as they had lost the specific rumen microbe with the necessary enzymes. The microbe isolated from the rumen of the Hawaiian goats degraded dihydroxy-pyridine, the toxic breakdown product of mimosine. Introduction of the microbe from goats in Hawaii into ruminants in Australia prevented the symptoms of mimosine toxicity in previously susceptible ruminant livestock. This has raised the possibility of genetically engineering into rumen organisms the ability to degrade such toxic compounds. Research at the University of New England is in progress to attempt to introduce genes into bacteria for defluorination of fluoroacetate, a lethal compound present in *Acacia gidyea,* which often kills cattle in northern Australia when consumed in the dry season. The successful transformation of rumen bacteria with a single gene for defluorination of fluoroacetate would indicate the possibility of the utilization of a range of other plants, which may be weeds (*e.g.* lantana), or potential high biomass producers (some legumes and legume trees) and also high protein crops (*e.g.* canavalia).

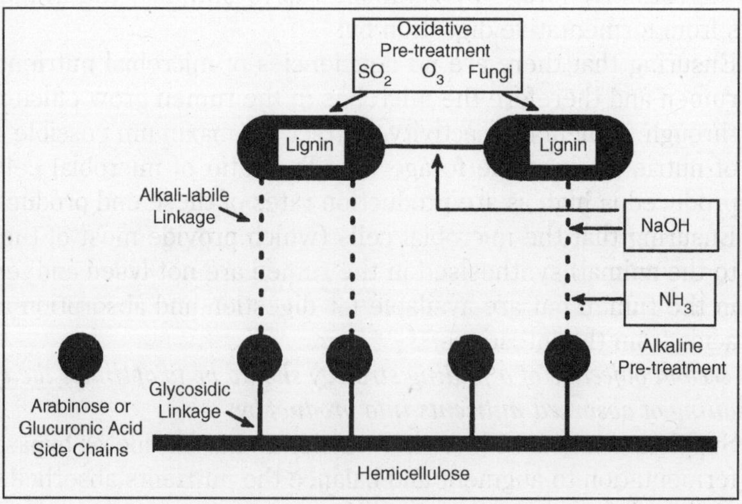

Fig. Sites of Action of Chemicals that, when Added to Straw, Improve its Digestibility

The problems are universally the same. The first is identification of bacterial DNA encoding for the suitable enzyme followed by development of methods to introduce this into an appropriate bacteria, stabilisation of the DNA in the cell and the last problem is whether the modified organisms will survive in the rumen and be capable of expressing the gene. The major research area would be in finding, isolating and packaging the DNA for the appropriate enzymes which must be capable of expression of being switched on easily and function in an organisms in the rumen ecosystem.

There is a danger in developing the ability of ruminants to degrade toxic compounds, in that often these are the last defence of plants being eaten and in the arid areas could lead to deforestation and soil erosion.

TREATMENT TO INCREASE DIGESTIBILITY

Improving the Digestibility of Crop Residues by Chemical Treatments

There are a number of acid and alkaline hydrolytic processes that will solubilise lignin, cause disruption of fibre by swelling or improve the potential digestibility of fibrous roughages in other ways. It seems that there is adequate information available on the technology of treatment and on the benefits of treatment of low quality forages. There is therefore little stimulus in pursing research of the methods. A new and novel approach will at times, however, require research but at present it can be safely assumed that alkaline hydrolysis of wet straw by ammonia generated by urea (usually 4% of the dry matter) by microbes in the straw, is probably the most feasible treatment for application in developing countries. The use of anhydrous ammonia application is also potentially applicable.

The ammonia in the straw has two effects:

1. It improves the fermentability of the straw in the rumen by breaking some of the ligno-hemicellulose bonds and
2. It provides non-protein-nitrogen for growth of microbes in the rumen when the forage is consumed.

Although urea ensiling of straw can be economic and effective the technology has not been widely taken up by small farmers in developing countries in general, although there are a few reports of individual farmers regularly using the technology. It is necessary for researchers to understand the reasons for non-acceptance of such technology which is almost always associated with a slow return on funds expended and sociologic constraints.

Microbial Treatment of Straw and Other Crop Residues to Improve Digestibility

In recent years considerable emphasis has been placed on the improvement of fibrous crops by the strategy of growing non-toxic fungi on the straw. In particular, the white rot fungi have been used because of their ability to delignify the plant material. The major problem that has arisen is that the technology, although relatively simple, is more complex than that of urea treatment of roughages and although protein is produced and digestibility of the residue increased, a considerable proportion of the total biomass is lost, particularly when white rot fungi is allowed to proceed to the mushroom formation stage. To some extent the loss of biomass can be prevented by a reduction in the incubation time of the straw and forages, such that biomass losses are reduced to less than 10%. Dr. B.N. Gupta and his colleagues at the National Dairy Research Institute (NDRI) in Karnal, India have developed a two stage process using an altophilic fungus (*Coprinus fimetorius*).

The technology is two-stage and relies on:

1. Alkali treatment (urea ensiling)
2. A second incubation period where a fungi is cultivated on the straw.

The technology is relatively simple, although the two stages make it more complex than the urea treatment alone for application by the small farmers. Both chemical and microbiological treatments may, however, be very effective where large amounts of feed must be stored and moved to areas where forages are scarce because of drought. However, again caution is needed in that to obtain maximum benefits from such feed, cattle will require supplements of bypass protein.

Potential for Increasing the Digestibility of Poor Quality Roughage by Manipulating the Digestive Physiology of the Animal

Goats compared with sheep, and buffalo compared to cattle, tend to retain fibrous materials in the rumen for longer periods and extract a few more units of digestibility at the same time. Within these comparisons, goats and buffaloes recycle a higher amount of urea-N back to the rumen. In the same way, *Bos indicus* cattle recycle more urea-N back to the rumen that *Bos taurus* cattle. Mechanisms that slow rumen turnover, increase rumen volume and allow greater recycling of urea-N may therefore improve the utilisation of poor quality forages. Immunisation against somatotropin release inhibiting factor (SRIF) has recently been demonstrated to increase flow rate through the gut, leading to increased digestibility and greater availability of protein to sheep on concentrate/forage diets. The effects of immunisation against SRIF on cattle on forage/bypass protein diets would be extremely interesting. However, the technology will not be easily applied to village farmers because of the numbers of injections required at the present time to effect immunity.

ALTERING THE PARTITIONING OF NUTRIENTS WITHIN THE ANIMAL

A whole range of injectable compounds will lead to more efficient utilisation of nutrients by livestock. In particular, repartitioning of nutrients into lean meat production or into milk production. These include anabolic steroids, growth hormone, and other hormonal growth factors, beta agonists and techniques such as the suppression of hormones by immunisation. These compounds in general, act through manipulation of hormone status and they lead to improved lean meat in the carcass and a greater efficiency of feed conversion into product. For a recent review of this area, the reader is referred to the publication by Buttery and Dawson.

Anabolic Steroids

Natural sex steroids are used to promote growth and reduce the fat/protein ratio in meat in many countries of the world but their use has been recently banned in Europe. The injected hormone alters the influences of endogenously secreted hormones (oestradiol injections into cattle usually lift growth hormone levels in blood) and growth hormone is more tightly bound to sites in the liver. As is usually the case there is a marked interaction with nutrition and on imbalanced diets in general the effect of exogenous hormones is less obvious than for instance when a diet of silage is balanced by a bypass protein (fish meal). Again this stresses the differences and therefore the need for a new approach to biotechnology in developing countries where ruminant diets are usually low in true protein.

Growth Hormone

Recent developments in fermentation technology together with recombinant DNA technologies have made it possible to produce large quantities of peptide proteins which appear to be the same as bovine or porcine growth hormone (BSt. or PSt.). These growth hormone injected into animals stimulates growth and dramatically decreases fat content of the animals carcass. They also stimulate milk production by some 20–25% in dairy animals.

In the initial studies with growth hormone injections milk production increases appeared to result from repartitioning nutrients and mobilisation of nutrients from tissue deposits in recent times a long term feed-intake effect has been reported. In addition the stimulation of milk production was dependent on the level of management and was not apparent in herds with poor management. In Italy and India in well managed buffaloes, milk yield has been increased by exogenous growth hormone by up to 25% but the cost of the injections may limit this application.

β-Agonists

β-agonists given by mouth, increase lean meat deposition and reduce fat content of the carcass in pigs and ruminants but with only small increases in liveweight gain. Again it is more than likely that the responses are dependent on the nutrition of the animal.

Injected Growth Promotants Versus Supplementation to Balance Nutrition

There are strong similarities between the response observed to growth hormone/β-agonists etc. and the response to correcting the P/E ratio in nutrients absorbed. At least one paper reports an interaction of bypass protein and response to exogenous hormones. It is thus possible that the optimisation of P/E ratio in the nutrients absorbed elicits responses in hormonal release that mimic the injection of growth hormone etc. Conversely and if true, this means that effects of exogenous hormones may not be so obvious where the P/E ratio in nutrients absorbed is optimised. In addition the requirements for protein relative to energy in the transgenic animals expressing growth hormone will be much increased as will requirements for minerals and vitamins.

The important conclusion from the paper by Buttery and Dawson is that: "the current technology has given many exciting possibilities but there are, however, still major gaps in the knowledge of the major metabolic control processes associated with growth which will have to be filled before a universally accepted method of growth promotion is developed"

This is a reservation which is repeatedly used by biotechnologists and indicates that there will be an ongoing requirement for basic research at the level of metabolic control.

OPTIMISING MICROBIAL GROWTH IN THE RUMEN

MINERAL REQUIREMENTS OF RUMEN MICROBES

The rumen microbes have specific requirements for both macro and micro minerals to meet the needs of structural components of cells and for components of enzymes and co-factors. Little is known about the requirements of the microbial milieu for trace elements and as a 'rule of thumb' it is accepted that if the animal is not deficient then it is unlikely that the rumen microbes will be deficient. As Suttle has so aptly put the situation, "it will rarely be possible to approach a suspected mineral deficiency situation with a table of minor nutrient requirements or biochemical criteria in the hand, and define a scale of the animal health (microbial health) problems". In practice, either no mineral supplements are used or a "shot gun" mixture is given as a salt licks or as molasses (which is concentrated plant juice rich in minerals) suitably fortified with minerals. As with any deficiency of a nutrient the likely scenario of a mineral deficiency for rumen organisms is first a reduced growth efficiency of microbes (lowered ratio of cells to VFA produced) with or without a decrease in digestibility. As the deficiencies become more extreme the digestibility of forage must decrease along with the decrease in microbial pool size and it is only then that feed intake will decrease. Feed intake however, will be decreased as P/E ratio decreases if the animal is heat stressed (see later). Correction of deficiency will obviously reverse these effects.

Requirements for Ammonia

Ensuring adequate ammonia N in the rumen to supply the majority of nitrogen for microbial growth is the first priority in optimising fermentative digestion of forage. Satter and Slyter suggested that 50–80 mg NH_3-N/1 of rumen fluid was the optimum for maximising microbial growth yield and this has been widely accepted. However, recent studies from two laboratories in Australia have clearly indicated that the minimum level of ruminal fluid ammonia for optimum voluntary intake of low N, low digestibility forage by cattle is about 200 mg N/1, even though the digestibility of the forage (in nylon bags) was optimised below 100 mg NH_3-N/1. All these studies were carried out in hot environments and the effects on feed intake are possibly explained by an improving P/E ratio in the nutrients absorbed, which reduces the metabolic heat load.

The effects of increasing ruminal fluid ammonia by infusion of urea into the rumen of steers on the intake of rice straw and its digestibility in nylon bags in the rumen are shown in Figure.

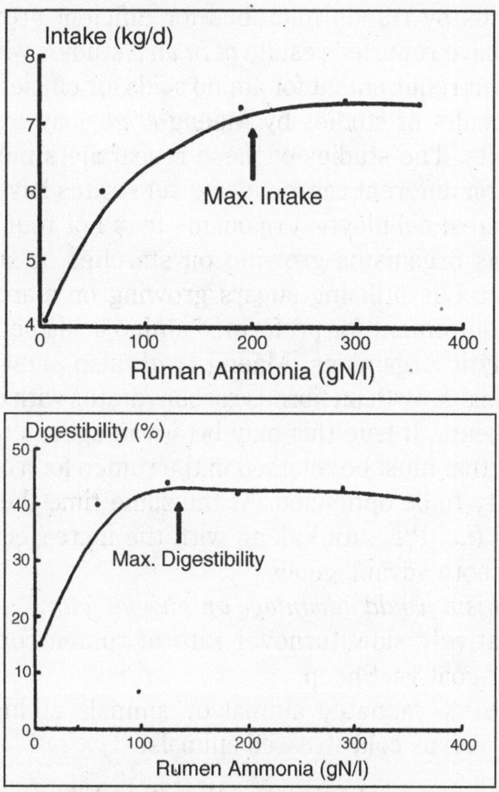

Fig. The Effects of the Level of Ammonia in the Rumen on the Intake and *in Sacco* Digestibility of Straw by Cattle. The Ammonia Levels were Adjusted by Infusing Urea in the Rumen.

Timing of Urea Supplements and the Ratio of Sugars and Starches to Fibre in a Diet

Supplements must provide adequate levels of ammonia in the rumen for continuous growth of both fibrolytic and saccharolytic organisms. The only satisfactory approach to meeting these changing requirements for ammonia is to provide ammonia continuously. One way of doing this is to provide salt/urea or molasses/urea licks and allow the animal to take these as needed. There are indications that cattle and buffaloes given continuous access to multi-nutrient blocks based on molasses/urea are able to control fairly closely their intake of urea. Once buffaloes were accustomed to molasses/urea blocks they adjusted their intake according to the N content of the basal diet (NDDB—unpublished data). Lambs given wheat straw and molasses/urea blocks also had similar abilities and consistently maintained their rumen ammonia levels above 200 mg N/1.

Requirements for Amino Acids/Peptides by Rumen Organisms

There has been considerable controversy concerning the requirements for peptides/amino acids by rumen microbes for efficient growth. A number of studies, however, have reported results of *in vivo* studies which appear to have indicated no apparent requirement for amino acids for efficient growth of rumen organisms. The results of studies by Maeng *et al.* may explain some of the contradictory results. The studies of these researchers indicated that rumen microbes growing on different carbohydrate substrates have requirements for different N-substrates; celluloytic organisms may not require amino acids to the same extent as organisms growing on starches or sugar as the major substrate. For microbes utilising sugars growing on starches there was an apparently high requirement for preformed amino acids/peptides but this was not so for cellulolytic organisms. Maeng *et al.* also showed an increase in efficiency of microbial growth on fibrous carbohydrates with decreasing dilution rate of rumen contents. If true this may be advantageous to ruminants given low quality forages that must be retained in the rumen for a considerable period to allow digestibility to be optimised. At the same time the improved ratio of cells: VFA yielded (*i.e.* P/E ratio) along with the increased availability of the total nutrients are both advantageous.

Such a mechanism would advantage an animal with:

- A comparatively slow turnover rate of rumen contents (*i.e.* Buffalo vs. Cow or goat vs. Sheep
- Fauna-free vs. faunated animal or animals at high environmental conditions versus cold stressed animals.

Amino Acid Requirements of Microbes Digesting Fibre

The organisms in the rumen that are largely responsible for the fermentation of cellulose (*Ruminococcus albus, Ruminococcus flavefaciens* and

Fibrobacter succinogenes (previously called *Bacteroides succinogenes*) appear to have minimal requirements for amino acids and grow on ammonia. Conversely, organisms important in starch hydrolysis (*Butyrivibrio fibrisolvens, Bacteroides ruminicola, Selenomonas ruminantium, Streptococcus bovis* and *Ruminobacter (Bacteroides) amylophilus* readily incorporate amino acid N and in many cases peptides.

Supplementation of sheep fed a poor quality forage with branched chain VFA has been reported to increase the apparent flow of microbial-N to the duodenum. The apparent stimulation of microbial growth with branched chain VFA has also been shown to increase feed intake on occasions. This together with the suggested requirements for peptides/amino acids by rumen organisms (which on the basis of the results of Maeng *et al.* must now be questioned) has tempted many scientists to explain the increased feed intake of ruminants, on poor quality forages that are supplemented with bypass protein, to the slow release of amino acids, peptides and branched chain fatty acids to the rumen milieu from the protected protein, even though in most studies there was no evidence of increased digestibility with such supplements in predominately fibre based diets.

The above discussion indicates that the cellulolytic organisms in the rumen even of cattle on straw based diets, are rarely if ever deficient in amino acids, peptides or branched chain VFA in the rumen. This is not to deny that these organisms may still need these nutrients in "catalytic" amounts but that they are rarely if ever at such low concentrations in rumen fluid to bring about a deficiency.

The Roles of Small Amounts of Fresh Forage in Straw Based Diets

Farmers in developing countries have generally recognised the benefits to cattle of adding a small amount of fresh green herbage to straw based diets. These practices, which have evolved through trial and error, may have a number of beneficial effects which include the supply of vitamin A, essential minerals, ammonia, peptides/amino acids in an otherwise unsupplemented diet.

Recently it has been shown that where the supplemental forage in a straw based diet given to sheep is of high digestibility a boost to digestibility of the basal diet occurs even at relatively small levels of supplementation. The rate of digestibility of straw depends on the rate and extent of colonisation of fibre and the biomass of adherent organisms and the high digestibility forage supplement may act to seed microbes onto the less digestible straw. On the other hand, other influences cannot be ruled out. For example, in the studies of Silva and Ørskov in the absence of an effect of supplemental forage on digestibility, the rumen ammonia levels were often not significantly below 200 mg NH_3-N/1. Where increases in digestibility of the basal forage occurred to supplemental forage the ammonia levels in the rumen were significantly below 200 mg N/1 and the supplement apparently improved the concentration to above the critical level.

Elimination of Rumen Protozoa and Preservation of the Fauna-Free State

The physiological research which has shown that there is an increased availability of microbial protein for digestion in fauna-free as compared to faunated ruminant has supported the feeding trials with large numbers of animals which has demonstrated significantly high feed conversion efficiencies and wool growth rate in sheep without rumen protozoa as compared to control animals.

Fauna free cattle on the same intake of a low protein, molasses/urea based diet grew at a 43% greater rate than faunated cattle on the same feed intake. The improved production was therefore an effect of a higher efficiency of feed utilisation. The discussion to follow, on the implications of environmental temperature/humidity for the nutrition of ruminants, will indicate why a major change in P/E ratio in the nutrients absorbed (*i.e.* the major difference between faunated and fauna-free ruminants) will be more effective in improving ruminant production in the tropics as compared to temperate/cold countries. In the tropical areas the available forages used to feed to ruminants are generally lower in digestibility, lower in true protein and animals are rarely cold, but heat stress at times may be severe. It should also be noted that animals brought into animals houses from fairly could environments may at times suffer severe heat stress through a combination of a well insulated fleece or coat and an imbalanced diet.

7

Productivity of Livestock in Developing Countries

Few developing countries have surplus tracts of land that can be regarded as fertile and therefore produce high quality fodder.

In general, human population pressures and the primary need to supply human food ensures that livestock are restricted to:

- Poor quality (often overgrazed) pasture lands, either infertile or with terrain that makes them impossible to utilise for crops or where erosion has made them unusable.
- Standing crop residues or weeds following cropping when environmental conditions are poor for plant growth and a further crop in that year cannot be taken.
- Crop residues (cut and stored)
- Agro-industrial by-products
- Extensive pastures of poor quality and otherwise sparse (*e.g*. The Llanos of Colombia and Venezuela).

Table. Average Carcass Weight (kg) Per Animal Slaughtered (1986 Statistics)

	Cattle	Buffalo
Europe	185	206
Asia/Pacific	120	161

Table. The Change in the Average Milk Yield Per Cow in Industrialised and Third World Countries

Country/Region	Average Yield/Cow (Kg/Year)		Percentage Increase (%)
	1976	1986	(1976 to 1986)
North America	3,250	5,200	60
EEC Countries	2,900	4,100	35
Asia	620	700	13
Africa	322	354	7

The net result is in general an extremely low rate of productivity with animals often at 5 years of age at maturity and productivity at 0.1–0.25 of that of ruminants in temperate countries grazing fertilised pastures or fed high quality feeds based on grain and immature pasture plants. A comparison of the levels of production of cattle between developed and developing countries is shown in Tables.

THE IMPLICATION OF LOW PRODUCTIVITY

Biotechnology research in industrialised countries is not generally aimed at low yielding animals. In addition to poor nutrition, adverse climate and disease, other stresses are also generally high, particularly in those countries situated in the tropics. These additional constraints are likely to affect and often remove any advantage of biotechnology transferred from developed countries. The considerable differences in feed resources available in developing countries compared to developed countries (where most biotechnology research is presently centered) and the low cost of production systems employed in the developing world as compared with the industrialised world indicates that direct technology transfer is unlikely to be successful.

FEED RESOURCES AVAILABLE TO RUMINANTS

The feeds that are available to ruminants in developing countries are fibrous and relatively high in ligno-cellulose. They are usually of low digestibility and they are often deficient in critical nutrients, including protein, non-protein nitrogen and minerals. As a generalisation, the forages consumed by ruminants in developing countries are almost always below 55% (usually 40–45%) digestibility and are often less than 8% crude protein and this protein level is more often around 3–5%, *e.g.* cereal straws. The only exception to this is in the early growth phase of pasture and when stocking rates are extremely low and the animals are able to select for leaf material. The low levels of production per head and per acre of most grazing systems are indicated by the summary of data given by Walker (1987). Without supplements, these low levels of production lead to a highly inefficient use of the available feed, with a possibility that up to 30% of the feed consumed is dissipated as heat. This heat has at times great effects on feed intake particularly in the tropics.

It is necessary, therefore in discussing the nutrition of ruminants, to appreciate the digestion of forage based diets and the constraints to the utilisation of nutrients that arise largely from a fermentative digestion system, since this knowledge must considerably influence research priorities.

CHEMICAL COMPOSITION OF LOW QUALITY FORAGES

In the literature concerned with the nutritional value of forages, considerable emphasis is placed on the crude chemical composition of the forage.

Although analyses that indicate cell solubles and cell wall materials are highly useful for studying the fermentative characteristics of a forage, they bear little relationship to its feeding value to the animal. Considerable efforts is often put into feed analysis which is often totally unwarranted particularly in the developing countries. In this report the overriding effects of a balanced nutrient approach to feeding are emphasised.

It is suggested that in most production systems for ruminants based on a poor quality forage, it is the balance of those nutrients providing the major building blocks for tissue synthesis and milk production, that should be the primary concern. With most forages and contrary to the assertions in the past that energy deficiency is the primary constraint to ruminant production on low quality feeds, the efficiency of feed utilisation is the major determinant of the production levels achieved. Therefore in reviewing the need for biotechnology research, this is the area for most consideration.

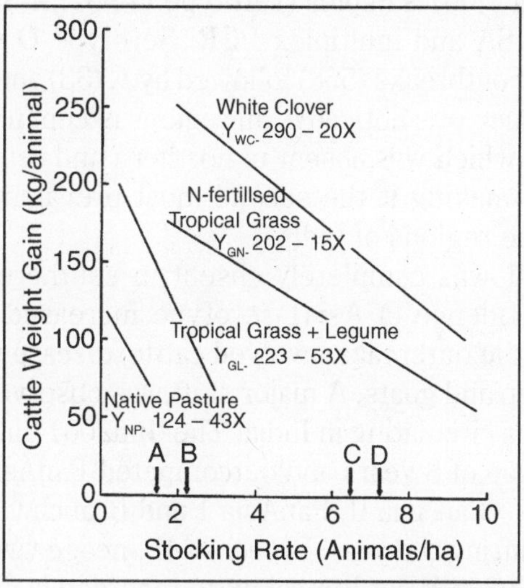

Fig. Cattle Growth on Pasture is a Function of Pasture Type, Fertilizer Applications and Legume Content. Productivity Per Unit Area is Maximised for the Different Pastures at Different Stocking Rates: 89 kg/ha for Native Pastures, 223 kg/ha for Tropical Grass with Legume, 682 kg/ha for Tropical Grass with Fertilizer and on Temperate Pasture (Clover) 105 kg/ha.

Recent evaluation of data from studies in tropical countries has indicated that medium to high levels of production at very high feed conversion efficiencies can be achieved by ruminants on poor quality forages adequately supplemented with critical nutrients. Of more importance is that the efficiency of utilisation of the metabolizable energy of a straw based diet appropriately supplemented can be higher than that of grain based diets. This suggests that the efficiency of utilisation of metabolisable energy of a forage can be markedly improved simply by supplementation (by up to 10 fold). It also discounts theories that the digestible nutrients from such feeds are less efficiently utilised than from grain based diets.

LIVESTOCK PROTECTION AND FOOT-AND-MOUTH DISEASE

During the period under the report, 799 outbreaks were recorded throughout the country. Maximum outbreaks were recorded in the eastern region where there was increase in the number of outbreaks compared to last year. Disease incidence in West Bengal directly affects the situation in NE region due to animal migration. Drastic reduction in outbreaks was noticed in southern region. FMD cases were not reported in Tamil Nadu, whereas Himachal and Punjab recorded a single case of FMD each. Maximum incidence of disease was reported in March and from August to November.

Out of 1,624 clinical specimens collected from 799 outbreaks, the virus could be detected in 1,067 samples (serotype O-991, A-38 and Asia 1-38) by using sandwich ELISA and multiplex PCR. Serotype O was responsible for maximum number of outbreaks (568) followed by A (33) and Asia 1 (9). Increase in serotype O incidence was noticed in all regions except northern and western region. Serotype A, which was absent in northern and central region last year appeared this year, making it the second most prevalent serotype affecting geographically all the regions of India.

Serotype Asia 1 was completely absent in southern, central and north eastern regions. Incidence of Asia1 serotype increased in western region. Though majority of the outbreaks involved cattle, disease was also reported in buffaloes, pigs, sheep and goats. A major shift was observed in genetic lineage of serotype O viruses circulating in India. The 'Ind2001' lineage viruses gained upper hand after a gap of 8 years and outcompeted PanAsia II lineage.

The cases due to PanAsia II, PanAsia 1 and Branch C-II viruses were not completely absent during the year. Branch-C II lineage viruses responsible for outbreaks during 2008 in West Bengal were recorded in Tripura, Asom, West Bengal and Odisha during 2009–10. In serotype A, it was observed that both VP359 deletion and non deletion mutants belonging to genotype VII are cocirculating in the field in recent times.

Precisely, the viruses from Punjab and Uttarakhand shared ancestry and clustered in the same VP359 deletion lineage, VIIf, whereas viruses from two different outbreaks from Odisha and those from Andhra Pradesh clustered together and revealed no deletion in the VP3 coding region. During the period, outbreak due to Asia1 serotype was recorded in Gujarat, Madhya Pradesh, Maharashtra, Uttar Pradesh and West Bengal.

The Asia 1 field isolates were grouped with lineage C reiterating the supremacy of this lineage in the field since 2005. In vaccine matching exercise 33 virus isolates were subjected to one-way antigenic relationship study. All the isolates showed close antigenic match with respective in-use vaccine strains indicating appropriate antigenic coverage. Virus isolates (133) including 97 type O, 12 type Asia 1 and 24 type A field isolates were added.

In National FMD Virus Repository 1,687 (1,096-O, 317-Asia 1,259-A, 15-C) well characterized field isolates are available. Under National FMD sero-surveillance testing of 11,560 serum samples from 13 states revealed the presence of protective antibody titre in 42.3%, 40% and 30.5% animal population against serotypes O, A and Asia1 respectively. This baseline response has improved due to application of vaccine. The current in-use vaccine strain IND R2/75 is found still the best and covers all the type O circulating outbreak strains in the country.

Classical Swine Fever

Classical swine fever (CSF), the topmost diseases in pigs, is one of the OIE listed diseases. The CSF or hog cholera is associated with high mortality, abortion, mummification of fetuses. It is of great concern in areas, where the pig population is in high density. CSF virus isolates maintained at the repository of PD-ADMAS, Bengaluru, were revived and processed for RT-PCR of NS5B genomic region. Phylogenetic tree based on the alignment of 408 nts of NS5B region was constructed. It showed three major branches, each representing three previously described genetic groups with further sub-branches.

The majority of the Indian CSFV isolates belong to subgroup 1.1. Clustering of the Indian field isolates away from the vaccine strain and other related strains revealed that the field isolates did not have any vaccine origin. Phylogenetic analysis based on 52 UTR classified subgroup 1.1 Indian isolates into two minor subbranches, however, no such classification could be seen from the analysis based on NS5b region; nevertheless in the latter analysis grouping of isolates was much more resolved. Lone isolate, IND294/08, is clustered within subgroup 2.2, ak into its grouping based on 52 UTR sequence.

Brucellosis

Serum samples from goat and sheep received from Gujarat, Rajasthan, Uttar Pradesh and Maharashtra revealed that among them 6.35% were positive from brucellosis. Similarly, 1.8% sheep and goat sera samples for Karnataka were positive for brucellosis. Porcine brucellosis is one of the most important emerging zoonoses and currently, no control programme for porcine brucellosis exists and the epidemiological situation is still unknown.

Indirect-ELISA for screening brucellosis in swine was standardized using smooth lipopolysaccahride antigen from a standard strain *Brucella abortus* S99; 22.6% were found positive, which is much higher than that found in sheep and goat. An indirect ELISA was developed to detect *Brucella* in any of the susceptible species. Expression of four recombinant *Brucella* proteins, *viz.* BP26,

BP39, superoxide dismutase (SOD) and *Brucella lumazine* synthase (BLS) is being envisaged to overcome the problem of cross reactivity in diagnosis of this disease.

Leptospirosis

PCR based techniques were developed for differentiation of pathogenic and nonpathogenic *Leptospira* as well as for typing. Initially, *Leptospira* genus specific PCR was carried out to differentiate the *Leptospira* from other spirochete organisms, *viz. Borrelia*, *Treponema* and *Leptonema* etc., which amplified 331 bp products from the *Leptospira* organism only.

Sero Prevalence of IBR

Serum samples collected from Andhra Pradesh, Gujarat, Jammu and Kashmir, Karnataka, Madhya Pradesh Maharashtra, Meghalaya, Manipur, Orissa, Punjab, Rajasthan, and Tamil Nadu, were subjected to Avidin Biotin ELISA. An overall apparent percentage positive was 34. The highest prevalence of 67% was in Tamil Nadu, and the lowest in Meghalaya. Region-wise positive percentage were as follows: East zone 17%, south zone 24%, west zone 48%, north zone 37%, north-east 39%, and central zone 25%.

Herbal Anticoccidial and Antidiarrhoel Drugs

The extracts CIRG-5 and CIRG-6 from two different plants were evaluated *in vitro* against sporulation of coccidian oocysts and found 96 to 100% effective to check the sporulation up to 96 h. Eight potential plant formulations of herbal extracts were subjected for clinical trials to control diarrhoea caused by *Escherichia coli* in kids of less than 1 month of age. Final product was developed from plant extract prototype CIRG-I and CIRG-2A and final products (formulation) @ of 10 mg/kg bwt for 1–2 days orally was highly efficacious to control EPEC diarrhoea in kid.

VACCINES AND THERAPEUTICS

- Live attenuated orf, buffalopox and camelpox vaccines met the thermostability and potency requirements of the biologicals.
- Bi-valent DNA vaccine containing HN and F genes of Newcastle disease virus was evaluated along with genetic adjuvant comprising IL-2, IL-4, IFNγ, and GMCSF gene constructs.
- An aroA gene knockout mutant of *Pasteurella multocida* serotype B:2 was constructed for the development of HS mutant vaccine.
- Chitosan coated cationic PLG nanoparticles were prepared and when tested for the transfection efficiency with commercial transfecting agent, indicated that these nanoparticles can be safely used for delivery of DNA vaccines.

- *Trypanosoma evansi* beta tubulin protein was expressed in both prokaryotic and eukaryotic expression systems.
- Two new target proteins of *Fasciola gigantica, viz.* leucine aminopeptidase and peroxiredoxin coding genes were cloned and successfully expressed in their native conformation in a prokaryotic expression system.
- Immunomodulators of microbial and herbal origin, *i.e. Mycobacterium phlei, Lactobacillus acidophilus* and *Tinospora cordifolia* were effective under field conditions to improve productive performance and mitigate the adverse effect of sub-clinical caecal coccidiosis and thereby improvement of body weight gain in broiler birds.
- Pharmacokinetics of gatifloxacin revealed that it has therapeutic potential for use in sheep.
- Medetomidine-butorphanol, thiopental and halothane combination provided better clinicophysiological and haemodynamic stability in general anaesthesia in buffaloes.
- Three TLR genes of yak (TLR 5, 7 and 9) and 2 TLR genes of mithun (TLR 4 and 7) were sequenced completely.

DIAGNOSTICS

- A recombinant lipoprotein 'P48' of *Mycoplasma agalactiae* was evaluated in an ELISA as a rapid immunodiagnostic tool in sheep and goats.
- A genus-specific PCR assay was standardized for rapid detection of *Arcobacter* sp. targeting 16S rRNA for confirmation of the isolates of this emerging pathogen.
- A 36 kDa protein was identified in different serovars of *Salmonella* during biofilm formation.
- Magnetic nanoparticles (MNPs) were synthesized by two different processes and were characterized for structural, morphological, size and magnetic properties. The MNPs were functionalized for immunoassay applications and evaluated for cellular toxicity in *in-vitro* experiments.
- A specific and sensitive diagnostic marker was identified for early detection of pancreatitis in dog.

MOLECULAR CHARACTERIZATION OF PATHOGENS/RECEPTORS

- Indian H5N1 isolates were characterized genetically and grouped with EMA *3/2.2.3 subclade,* which branched from two distinct nodes in the phylogenetic tree.
- The alignment of deduced amino acid sequences showed that all the Indian H5N1 viruses characterized contained the multiple basic amino

acid motif at their HA cleavage site, which is characteristic of highly pathogenic avian influenza viruses.

- Two viruses from 2010 outbreaks in West Bengal contained amino acid asparagine at position 31 of the M2 protein, suggesting that both viruses are resistant to drug amantadine.
- An isolate of buffalo pox from man and animal recovered from an outbreak from Pune was characterized for B5R, H3L, D8L and A27L genes.
- An orf isolate from camel in a sporadic infection was characterized for B2L, GIF and VLTF-4 genes.
- Genotyping and partial sequencing of rotaviruses from man and pigs revealed human-pig reassortants as well as presence of novel combinations in India.
- Complete HN gene of velogenic Newcastle disease virus containing kozak sequence was amplified and cloned into pcDNA3. 1 eukaryotic expression for enhanced expression.
- Phylogenetic analysis by AP-PCR, PFGE and ribotyping showed considerable similarity between human and fish isolates of pathogenic and pandemic *Vibrio parahaemolyticus,* which reflected the public health risk potentials of fish.
- Molecular characterization by PCR-SSCP analysis of *Oesophagostomum* spp. revealed two distinct conformers with six nucleotide polymorphisms (SNPs) among male and female parasites.
- Probability of evolutionary changes among prevailing genotypes of *Oesophagostomum* spp. was predicted by analysing secondary structure of RNA.

QUALITY CONTROL AND PRODUCTION OF VETERINARY BIOLOGICALS

- 223,500 doses of RD 'F' strain vaccine; 166,000 RD 'Mukteswar' strain vaccine; 132,500 doses of fowl pox vaccine; 184,195 doses of lapinized swine fever vaccine; 423,800 doses of tissue culture sheep pox vaccine; 160,600 doses of PPR vaccine; 57,925 doses of *Brucella abortus* strain-19 (live) vaccine; 3,640 ml HS adjuvant vaccine; 73,500 doses of tuberculin PPD; 50,000 doses of Johnin PPD; 25,250 doses of mallein PPD; 55,250 ml of *Brucella* agglutination test antigen; 6,900 ml of *Brucella abortus* Bang Ring antigen; 19,860 ml of Rose Bengal Plate Test antigen; 219 ml of *Brucella* positive serum; 9,000 ml of *S.* Pullorum coloured antigen; 45 ml *S.* Pullorum positive serum; 87 ml *Salmonella* 'O' serum and 3,750 ml of *S.* Abortus equi 'H' antigen were produced and quality tested.
- 7.32 million monovalent doses of FMD vaccine were produced at Bengaluru campus.

- 435,100 doses of PPR vaccine were produced; similarly, 33 PPR c-ELISA, 22 PPR s-ELISA and 100 doses of goat pox vaccine were supplied by Mukteswar campus.

Sero-monitoring of Important Equine Diseases

During 2009–10, equine disease sero-survey was conducted in Rajasthan, Haryana, Punjab, Uttar Pradesh, Uttarakhand, Madhya Pradesh, Chhattisgarh and Sikkim. A total of 176 out of 1,045 equines (16.84%) had antibodies to equine influenza virus while 28 (2.67%) were positive for EHV-1 and 107 (10.43%) were positive for antibodies to Japanese encephalitis virus. Antibodies to *Babesia equi* were detected in 320 (38.09%) of 840 equines, while *T. evansi* antibodies were detected in 49 (5.4%) out of 906 equines tested. One case of equine infectious anemia was detected in Uttarakhand. EIA-positive equine was reported after a gap of 11 years from India. None of the 1,045 equines tested showed antibodies to *Brucella* and *Salmonella* Abortus equi 'H' antigen.

Japanese Encephalitis (JE) Virus Isolation

JE, a mosquito-transmitted disease caused by JE virus, is causing disease in equines, pigs and humans. It is prevalent in eastern and southern Asia. Serum samples from clinically affected horses and pooled mosquito vector samples from the surrounding field were found positive for JEV RNA by RT-PCR. Virus was isolated from one of the horses. This is the first report of isolation of JEV from a horse in India. The phylogenetic analysis of the envelope gene (1500 bp) sequence of the isolated virus, JE/eq/India/H225/2009 indicated that the isolate belonged to genogroup III of JEV, and that the equine isolate clustered together with Vellore group of JE isolates from India.

Camelpox Virus Zoonosis

Camelpox virus (CMLV) causes skin eruptions in addition to severe generalized exanthema mostly in camels and outbreaks occur frequently in camel-rearing north-western region of the country. However, their zoonotic potential has not been proven conclusively. The disease was confirmed on the basis of isolation of the virus from diseased camels, detection of the CMLV antibodies in the serum of diseased camels and human patients by serum neutralization test and amplification of CMLV-specific three full-length envelope protein genes (A27L, H3L and D8L) and one partial gene (C18L) of the isolated viruses from camels and human specimens.

The phylogenetic analysis indicated that CMLV isolates cluster closely with variola virus (VARV) and vaccinia virus (VACV). The CMLV zoonosis indicates the changing pattern of host specificity of poxviruses and certainly points towards a declining cohort immunity against Orthopoxviruses in humans, which could be a public health concern. This is the first confirmed report of camelpox virus zoonosis.

Cysteine Proteinases from Trypanosoma Evansi

Parasite derived cysteine proteinases were proposed as potential targets for immunotherapeutic, chemotherapeutic and diagnostic reagents for parasitic diseases. The proteinases of camel isolate of *T. evansi* were analysed by gelatin substrate SDS-PAGE containing collagen antigen confirming the presence of cysteine proteinases in *T. evansi*. Some of these cysteine proteinases were reactive with experimental/field serum sample from equines with *T. evansi* infection. This indicated the potential of these enzymes in serodiagnosis of this economically important disease of livestock.

AVAILABLE BIOTECHNOLOGIES IN REPRODUCTIVE PHYSIOLOGY

One of the challenges for genetic improvement is to increase reproduction rates. Several reproduction techniques are available. The commonest of these are artificial insemination (AI), embryo transfer and associated technologies. Measurement of progesterone in milk or blood which is a widely used technique for monitoring ovarian function and for pregnancy tests is also an important technology for managing the reproductive function of the animal.

ARTIFICIAL INSEMINATION

No other technology in agriculture, except hybrid seed and fertilizer use, has been so widely adopted globally as AI. Progress in semen collection and dilution, and cryopreservation techniques now enables a single bull to be used simultaneously in several countries for up to 100,000 inseminations a year. This implies that a very small number of top bulls can be used to serve a large cattle population. In addition, each bull is able to produce a large number of daughters in a given time thus enhancing the efficiency of progeny testing of bulls. The high intensity and accuracy of selection arising from AI can lead to a four-fold increase in the rate of genetic improvement in dairy cattle relative to that from natural mating.

A wider and rapid use of selected males through AI will accelerate the rate of gender improvement. Also, use of AI can reduce transmission of venereal diseases in a population and the need for farmers to maintain their own breeding males, facilitate more accurate recording of pedigree and minimise the cost of introducing improved stock. However, success of AI technology depends on accurate heat detection and timely insemination. The former requires a certain level of experience among farmers while the latter is dependent on good infrastructure, including transport network, and availability of reliable means of transport.

Though AI is widely available in developing countries it is used far less, particularly in Africa, than in developed countries. Its use has been limited largely to "exploratory" purposes mainly by research institutions. A few countries including Botswana, Ethiopia, Ghana, Malawi, Mali, Nigeria, Senegal and Sudan have taken the technology to the field, mostly for programmes of "upgrading" indigenous stock and as a service to a limited number of commercial farmers keeping exotic dairy cattle breeds. A few others have used the technology more widely. Kenya and Zimbabwe, for example, have elaborate AI systems which include national insemination services incorporating progeny testing schemes. However, even these have gone through periods of collapse or serious degeneration and have had to go through "rehabilitation" phases.

The Republic of South Africa is probably the biggest user of AI technology in terms of number of inseminations. This country also has what is perhaps the best organised progeny testing scheme on the continent.

AI technology use is still more generally associated with dairy cattle than other domestic livestock species. The limitations of AI use in beef cattle include the difficulty in detecting heat in large beef herds kept on ranches and the less frequent handling of individual cows. In sheep and goats the failure to develop a simple, non-surgical insemination procedure has prevented extensive exploitation of the technology in sheep. However, the technical success of laparoscopic intrauterine insemination has prompted research into less invasive transcervical procedures. Also, in Africa, research to improve the freezing-and-thawing properties of sheep semen is underway in the Republic of South Africa. In pigs use of AI is hampered by the inability to successfully cryopreserve boar semen.

AI is credited for providing the impetus for many other developments which have had a profound impact on reproductive biotechnology. Foote (1982) noted that studies of oestrus detection and ovulation control which evolved out of a need to correctly time inseminations, led to the development of embryo-transfer technology.

Embryo Transfer (ET)

Although not economically feasible for commercial use on small farms at present, embryo technology can greatly contribute to research and genetic improvement in local breeds. There are two procedures presently available for production of embryos from donor females. One consists of superovulation, followed by AI and then flushing of the uterus to gather the embryos. The other, called *in vitro* fertilisation (IVF), consists of recovery of eggs from the ovaries of the female then maturing and fertilising them outside the body until they are ready for implantation into foster females. IVF facilitates recovery of a large number of embryos from a single female at a reduced cost thus making ET techniques economically feasible on a larger scale. Additionally, IVF makes available embryos suitable for cloning.

The principal benefit of embryo transfer is the possibility to produce several progeny from a female, just as AI can produce many offspring from one male.

For example the average lifetime production of a cow can be increased from 4 to 25 calves. Increasing the reproductive rate of selected females has the following benefits: genetically outstanding animals can contribute more to the breeding programme, particularly if their sons are being selected for use in AI; the rate of genetic change can be enhanced with specially designed breeding schemes which take advantage of increased intensity of female selection combined with increased generation turnover; transport of embryos is much cheaper than that of live animals; risk of importing diseases is avoided; facilitates rapid expansion of rare but economically important genetic stocks; and the stress to exotic genotypes can be avoided by having them born to dams of local breeds rather than importing them as live animals.

Embryo transfer is still not widely used despite its potential benefits. In developing countries this is mainly due to absence of the necessary facilities and infrastructure. Even in developed countries, cost considerations still limit the use of commercial embryo transfer in specialised niches or for a small proportion of best cows in the best herds. Thus, in North America and Europe, only about one out of 500 calves born in the last decade was from ET. Commercial embryo transfer is more popular with cattle than other species. This is mainly because ET is relatively easier in cattle than the other species and also because it is more economical in cattle (*i.e.* cattle are worth more). Additionally, the low reproductive rate and the long generation interval of cattle make ET much more advantageous in cattle. Production of several closely related, and hence genetically similar, individuals through ET techniques can make critical contributions to research. For example a project at the International Laboratory for Research on Animal Diseases (ILRAD) to locate the genes responsible for tolerance of some cattle populations to trypanosomiasis required large numbers of closely related crosses of trypanotolerant and trypanosusceptible cattle. Use of ET has made it possible to generate such families thereby facilitating the search for genetic markers of trypanotolerance. Additionally, ET could be useful in studying the extent to which a trait is influenced by the embryo (direct component) or the reproductive tract (maternal component).

Embryo Sexing and Cloning

Although embryo sexing may not have dramatic effects on rates of genetic gain it can considerably increase efficiency. Taylor *et al.* concluded from a study that an all-female heifer system using ET was 50% more efficient than the highest achievable in a traditional system. It has been suggested that, if multiple sexed-embryo transfer became as routine an operation as AI is, beef operations based on this system could become competitive with pig and poultry production in terms of efficiency of food utilisation.

Clones may be produced by embryo splitting and nuclear transfer. These offer the possibility for creating large clone families from selected superior genotypes which, in turn, can be used to produce commercial clone lines. However, some studies have concluded that cloning of embryos will not increase rates of genetic progress in the nucleus, but that it offers considerable advantages in increasing the rate of dissemination of tested superior genotypes in commercial populations. Other potential applications of cloning include efficient evaluation of genotype x environment interactions and testing and/or dissemination of transgenics. From a research standpoint, production of identical siblings should, by eliminating variability among animals, greatly reduce the size and hence the cost of experiments.

HORMONE USE

Use of hormonal assays to *monitor* reproductive function can be rewarding for both research purposes and commercial livestock operations. Reproduction can also be *manipulated* using hormonal treatments. But while hormonal treatments have produced desirable results in some studies in Africa, lack of awareness about their use and the fact that they are not economically viable under most prevailing production circumstances limit their use. Progesterone and PMSG treatment and immunisation against androstenedione increased ovulation rate in Ossimi sheep. Also, exogenous melatonin treatment of barren Rahmani ewes resulted in increased proportion of ovulating ewes and a higher ovulation rate. These responses, however, did not result in increased litter size because of increased ova wastage. Thus, in addition to the impracticability arising from prohibitive prices of hormonal preparations and the problems with hormonal administration at farm level, there are other technical problems associated with the use of these technologies. For example, technologies aimed at increasing litter size in traditional small ruminant production systems should not be applied unless management, including nutrition, can be improved in concert to ensure the survival of the additional progeny. Reproduction can also be manipulated without application of exogenous hormones. Aboul-Ela *et al.* reported that exposing ewes to rams one week prior to mating ("the ram effect") increased the percentage of ewes in oestrus (and hence the per cent mated) by 27%. Such management approaches offer practical options for increasing annual lamb (or kid) production in situations where other technologies are either not available or not appropriate. "Accelerated lambing"-increasing the number of lambings per year-can also be used to increase annual productivity. However, in tropical and subtropical situations where animals depend on seasonally available natural pastures, this practice may not be sustainable. Under such circumstances the reproductive cycle tends to be dictated by availability of feeds.

Animal Genetics and Breeding

Genetic improvement of livestock depends on access to genetic variation and effective methods for exploiting this variation. Genetic diversity constitutes a buffer against changes in the environment and is a key in selection and breeding for adaptability and production on a range of environments.

In developed countries, breeding programmes are based upon performance recording and this has led to substantial improvements in animal production. Developing countries have distinct disadvantages for setting up successful breeding programmes: infrastructure needed for performance testing is normally lacking because herd sizes are normally small and variability between farms, farming systems and seasons are large; reproductive efficiency is low, due mainly to poor nutrition, especially in cattle; and communal grazing precludes implementation of systematic breeding and animal health programmes.

Multiple Ovulation Embryo Transfer and Open Nucleus Breeding System

Multiple ovulation embryo transfer (MOET) is a composite technology which includes superovulation, fertilisation, embryo recovery, short-term *in vitro* culture of embryos, embryo freezing and embryo transfer. Benefits from MOET include increasing the number of offspring produced by valuable females, increasing the population base of rare or endangered breeds or species, *ex situ* preservation of endangered populations, progeny testing of females and increasing rates of genetic improvement in breeding programmes. Genetic improvement of ruminants in developed countries has made much progress in the last 35 or so years through the use of large-scale progeny testing of males. As has been pointed out, the general failure of extensive use of AI in developing countries has implied that progeny testing schemes cannot be operated with much success. In any case the generally small herds/flocks and uncontrolled breeding in communal grazing situations preclude implementation of progeny testing. Smith suggested that the Open Nucleus Breeding System (ONBS) may be especially valuable for developing countries where the use of AI has been a failure due to the reasons given above.

The ONBS concept is based on a scheme with a nucleus herd/flock established under controlled conditions to facilitate selection. The nucleus is established from the "best" animals obtained by screening the base (farmers') population for outstanding females. These are then recorded individually and the best individuals chosen to form the elite herd/flock of the nucleus. If ET is possible, the elite female herd is used through MOET with superior sires to produce embryos which are carried by recipient females from the base population. The resulting offspring are reared and recorded and the males among them are evaluated using, as appropriate, the performance of their sibs and paternal half sibs and their own performance.

From these, an elite group of males with high breeding values for the specific trait is selected and used in the base population for genetic improvement through natural service or AI. It should be noted that, while MOET improves the rate of progress substantially, it is possible to operate an ONBS without ET technology, especially in species, such as small ruminants, with high reproductive rates. Such schemes are being tried for sheep in West Asia by FAO and in Africa. However, availability of AI and ET, in addition to increasing rates of genetic gain, enhance the flexibility of the system. For example, germplasm from other populations can be introduced easily through semen and/ or embryos. One of the advantages of a nucleus herd is that it provides opportunity to record information on more traits than is possible in a decentralised progeny testing scheme.

The ONBS can be used for the improvement of an indigenous or exotic breed. It can also be used to improve a stabilised crossbred population. The level of the genetic response depends on the size of the scheme (that is, number of participating herds/flocks and total number of animals) and the selection intensity.

An ONBS can initially be developed to form a focus for national sire breeding and selection activities. In time, and with experience, the capacity can be expanded and ET introduced to increase the rate of genetic progress.

At one time it was suggested that application of MOET in nucleus breeding schemes could increase animal genetic gains by 30-80%. More recently it has been concluded that the earlier figures were over-predictions. The over-predictions arose partly because the assumed average number of progeny (eight) per donor female was unrealistically high and partly because of wrong assumptions made about genetic parameters. The realistic average number of live progeny per donor flushed is in the range of 2-3 in sheep and cattle and 6-8 in goats. Consideration of these figures suggest that MOET could increase annual genetic gains by 10-20% in large nucleus breeding schemes. However, costs of operating such schemes in developing countries need to be evaluated before they can be recommended.

INDICATOR TRAITS

Indicator traits are characteristics which are genetically correlated to traits of economic importance and are easier to measure than the latter. Such traits are usually not the target of genetic improvements but provide an indirect means of improving a targeted trait. Blair *et al.* reviewed some physiological and/or metabolic characteristics which might be considered as potential indicator traits. Traits such as testicular size in rams or bulls or FSH in ewe lambs have potential as indirect predictors of fertility. Indicator traits can improve genetic response by increasing accuracy of selection and reducing generation interval. The value of an indicator trait will depend largely on the magnitude of co-heritability (square-root of the product of heritability of the indicator and of the target trait) and the genetic correlation between the two traits. Woolliams and Smith concluded that, with high co-heritability, selection for the indicator trait alone can result in greater rates of response than is possible with progeny testing, especially when breeding values are not accurately measured by progeny testing. Packed cell volume (PCV), an indication of the extent of anaemia, is widely used as an indicator trait for pathological conditions associated with anaemia. For example, PCV is currently used at ILRAD and ILCA (International Livestock Centre for Africa) as an indicator of the effect of trypanosomiasis and hence of trypanotolerance, and at ILCA as an indicator of effect of the endoparasite *Haemonchus contortus* and hence as an indicator of resistance to the parasite.

Genetic Markers and Marker-assisted Selection

A genetic marker for a trait is a DNA segment which is associated with, and hence segregates in a predictable pattern as, the trait. Genetic markers facilitate the "tagging" of individual genes or small chromosome segments containing genes which influence the trait of interest. Availability of large numbers of such markers has enhanced the likelihood of detection of major genes influencing quantitative traits. The method involves screening the genome for genes with a large effect on traits of economic importance through a procedure known as linkage analysis. The chances of major genes existing for most traits of interest, and of finding them are considered to be high.

The process of selection for a particular trait using genetic markers is called marker assisted selection (MAS). MAS can accelerate the rate of genetic progress by increasing accuracy-of selection and by reducing the generation interval. However, the benefit of MAS is greatest for traits with low heritability and when the marker explains a larger proportion of the genetic variance than does the economic trait. Lande and Thompson suggest that about 50% additional genetic gain can be obtained if the marker explains 20% of the additive genetic variance and the economic trait has a heritability of 0.2. MAS also facilitates increased rate of genetic gain by allowing measurement in young stock thereby reducing generation interval.

Marker identification and use should enhance future prospects for breeding for such traits as tolerance or resistance to environmental stresses, including diseases. Already, identification of carriers of genes for resistance and introduction of such genes into a population seems feasible for resistance against *Trichostrongylus colubriformis* and *Haemonchus* contortus. It should also be possible to eliminate factors predisposing sheep to Listeriosis or Salmonellosis. There is also evidence for a major gene for resistance to the cattle tick (*Boophilus microplus*) in a Hereford x Shorthorn cattle line called Belmont Adaptaur. Research is currently underway at ILRAD to identify genetic markers for tolerance to African trypanosomiasis in N'Dama cattle and at ILCA for resistance to endoparasites in Red Maasai sheep.

LIVESTOCK MANAGEMENT PRACTICES

To maximize forage productivity, it is best to combine or alternate various livestock breeds on a range. Their differing food preferences can help to keep plants productive by minimizing overgrazing of a particular favoured area and allowing less preferred plant species time to mature. It is prudent to make superior forage available to those animals with the highest needs. When forage is limited, livestock managers may decide that young and milk-producing animals must have first access to new pastures and ranges with a wide variety of abundant forage.

Managers should investigate the value of various systems of rotating livestock. Rotation allows land to be grazed continuously throughout the year. Livestock can be rotated between fields or ranges to prevent the buildup of disease and to vary grazing pressures. Through either fencing or herding, they can be relocated into croplands to consume crop residues. Assessment of seasonal grazing patterns should include potential impact on soils.

Dry-season grazing can benefit the land by breaking up crusted soil and working seeds into the ground. By contrast, grazing on moist soil can cause considerable soil compaction, which reduces the soils ability to absorb moisture and can result in increased erosion from run-off during the rainy season. Many of the environmental impacts from livestock production are associated with particular practices of livestock managers.

Thus it is critical to understand current practices and how the proposed project might alter these practices or promote new ones:

- Who are the local community's livestock managers?
- What practices do a family or community use to control the size and composition of livestock herds?
- How do livestock managers currently control livestock movement? Will the proposed project change these movements in a way that might harm the environment?
- Does the proposed project require the construction of fences? If so, will they interfere with wildlife migration or transit of livestock belonging to other communities? Could the fences lead to overgrazing and land degradation? Will the fences be built with local materials? Would living fencing be practical? Would solar-powered electric fencing be technically and economically feasible?
- Are streams and riverbanks currently protected from livestock damage? If the proposed project will open new areas to grazing, will water supplies need to be protected?
- Must steps be taken to prevent new livestock and associated animals (*e.g.*, dogs) from transmitting disease to wildlife? Is there a vaccination/animal disease control programme available for this purpose?

- Will the project involve construction of improvements (*e.g.*, boreholes or other infrastructure)? Could these lead to unplanned changes in herding patterns and overgrazing?

ASSESS DEMAND AND USE OF LIVESTOCK PRODUCTS

- Who is marketing livestock and livestock products?
- Is the demand for livestock products coming from local or outside populations? How rapidly is it increasing or decreasing? How stable is the demand?
- In preparing livestock products, are people using technologies which reduce impacts on the environment, open additional markets, or improve health and nutrition (of people and animals.)

ASSESS LIVESTOCK ECTOPARASITE MANAGEMENT

Epidemic and endemic diseases continue to be a major constraint to livestock productivity in large parts of the developing world. Although vaccines have controlled many of the epidemic diseases, they continue to cause severe economic losses through morbidity and mortality. These diseases include the infections caused by vector-borne haemoparasites and helminths. Existing technologies, such as chemotherapeutic agents and live vaccines that were previously successful in controlling these diseases, are no longer effective—because of acquired resistance or weakened delivery services. Appropriately designed alternatives are often lacking.

CONSIDER POPULATION PRESSURE AND DISEASE BURDEN

Two factors that may affect the outcome and impact of projects population growth and fatal or debilitating epidemic diseases. Population growth may lead to conflicting uses of grazing lands or reduce individual farms or rangeland to sizes that cannot sustain livestock. Fatal or debilitating epidemic diseases may weaken effective dissemination or replication of proper livestock management techniques.

- What is the current and projected population growth rate in the project area? How might this affect project sustainability in the future?
- What is the current extent of the HIV/AIDS epidemic in the region? How might this affect the composition of the population (size, ethnic makeup, age/gender distribution) and family structures necessary for project sustainability? How will development and livestock technical support services be affected?
- Are there other epidemic diseases in the region, such as sleeping sickness, that might adversely affect project implementation?

LARGE AREAS OF LAND DEGRADED

Overgrazing of rangeland reduces the density of vegetation and the amount of organic matter generated. This, in turn, increases soil erosion from wind and water and decreases soil fertility through loss of nutrients. In arid and semi-arid areas these impacts may also contribute to desertification. Fortunately, ecosystems in these areas demonstrate considerable resilience and often recover when grazing pressure is reduced, either through traditional methods or through modern management practices.

Policy and Legal Problems

In areas that traditionally rely on grazing, the health of rangeland is generally best maintained by traditional pastoralist practices, which regulate grazing location and herd size in accordance with drought cycles and the supply of fodder. In fact, government policies or donor interventions that disrupt or discourage these practices may be a root cause of degradation. A variety of government policies may restrict the movement of livestock within a range area and prevent livestock managers from moving stock from areas that have been depleted of fodder to better supplied areas.

Two particular policy-based problems are:

1. *Land Tenure Insecurity:* Lack of confidence in secure title to rangeland (especially on communal lands) has been shown to reduce the incentive to manage the land sustainably. Many national governments have either implicitly or explicitly claimed ownership of range and wildlands and ignored traditional or customary claims.
2. *Privatization of Communal Resources:* Where national governments have privatized, or are privatizing, formerly state-owned or communal lands, new owners may erect fencing or prevent herds from crossing or grazing on their property.

Wells and Boreholes

Traditionally, access to water on critical grazing lands has been controlled to limit livestock populations and prevent herds from outgrowing the forage supply in dry areas. Thus, new wells or boreholes, generally sponsored by donors or governments, may undermine traditional livestock management systems practices by allowing herds to grow beyond sustainable levels for surrounding areas. Overgrazing and degradation are most noticeable in the immediate vicinity of the boreholes or wells, but their effects can extend (in gradually decreasing severity) over a considerable radius. Boreholes also reduce pressure on livestock owners to decrease herd size during drought and may discourage movement of herds to other rangelands, disrupting historic wet season/dry season grazing patterns. Larger herd sizes and reductions in pastoral movement may prove to be a recipe for severe degradation of soil and vegetation.

Wet-season Grazing

Poor timing in the use of rangeland can also damage the soil. Wet-season grazing can compact the moist earth, reducing its ability to absorb moisture This increases erosion from water run-off.

Poor Balance of Livestock Species

Each species or breed of livestock has foraging preferences and will graze favoured areas and plants while neglecting others. Browsing animals, such as goats and camels, prefer the leafy tops of shrubs. By contrast, grazers tend to consume ground-level grasses and leafy plants. A poor balance between browsers and grazers can change the mix of plants in ways that degrade the area. For example, too many grazers can diminish the number and populations of herbaceous plant species and allow woody plants to become dominant, changing, possibly irreversibly, the character and utility of the ecosystem.

DAMAGED HABITAT AND REDUCED BIODIVERSITY

Livestock production can damage habitats and reduce biodiversity in wildlife and domestic stock, in vegetation, and in aquatic and wetland ecosystems.

Harm to Wildlife and Domestic Stock and Loss of Wildlife Habitat

The loss of habitat caused by livestock production in grazing and mixed farming systems may be one of the greatest threats to wildlife. Human population growth and density, and the accompanying increase in livestock, often leads producers to expand livestock grazing ranges into wild lands and the conversion of wild lands to mixed farming use.

Slaughter of Wildlife by Livestock Managers

Another danger to wildlife is intentional slaughter by livestock managers. Fear that the wildlife will prey on livestock and damage crops is a common motivation, as is the belief that the wildlife are competing with livestock for fodder, the desire to prevent spread of disease to livestock, and concern for human safety.

Extinction of Local Livestock Breeds

Systematic livestock production may result in loss of genetic diversity in livestock species. This is unfortunate because genetic diversity is a measure of a species' robustness. Local breeds may have traits conferring resistance to emergent or future pathogens, or have other favourable adaptations to local environments. The consistent replacement of local breeds with more productive imported ones can contribute to the extinction of that breed and of all the genetic diversity harbored within its population.

HARM TO VEGETATION

Clearing of Forest and Wild Lands

Vegetation is typically altered or destroyed when forests/wild lands are cleared or are burned to promote new growth. This changes local ecosystems and may contribute to global warming. Fires to burn vegetation are dangerous and degrade air quality.

Loss of Rangeland Fertility

Ironically, mixed farming systems may reduce the fertility of rangeland while helping to solve a farmland problem. Traditional farming practices cause a net loss of nutrients in farm soils; that is, when crops are harvested and sold (or even when human waste is deposited in a latrine) nutrients that make the soil fertile may be lost.

Mixed farming reduces the extent of this loss, by transferring nutrients from the range to the farm in the form of manure. The gain in fertility for the farm is, of course, a net loss for rangeland. Over time, the altered nutrient balance can reduce the productive capacity of the range and/or lead to changes in the composition and density of plant species.

Loss of Farm Fertility

Damage from mixed farming often occurs when poverty and population growth pressure change the crop/grazing land ratio, where other nutrient sources are not available. As the land area available for each grazing animal shrinks, overgrazing becomes prevalent leading to soil erosion and nutrient loss. Harvesting livestock further reduces nutrients available for crop production.

Damage to Riparian Soil and Vegetation

Livestock in grazing and mixed farming systems often graze very heavily in riparian areas along streams and lakes. Results include trampling, loss of vegetation, soil disturbance, and soil compaction, erosion and/or sedimentation which can severely damage riparian habitats, increase siltation and adversely affect watersheds. Pesticide contamination from treatments to protect livestock from insectborne infections (*e.g.*, livestock "dipping") may ultimately reach the aquatic environment. Here it can be toxic to aquatic organisms, as well as people or animals who depend on these sources for drinking water.

Introduction of Invasive Plants

New breeds or fodder crops can introduce invasive non-native plants into a region. The manure, coats and hooves of newly introduced breeds can carry plant seeds. Most non-native plants are not invasive, but when they are, the results can be devastating.

Contamination from Manure

Livestock manure contains relatively high concentrations of nutrients, solids, enteric bacteria and other microorganisms, and organic material. The manure from industrial livestock operations is often discharged or "leaked" into lakes or streams, because it cannot be economically transported to replenish crop fields. When this occurs, the nutrients can cause eutrophication (rapid plant growth in water bodies), solids can create sedimentation, and organic material leads to oxygen depletion (BOD) of the water. Manure from mixed farming, if applied in a concentrated fashion, can lead to similar problems.

Degrade Water Quality and Reduce Water Supplies

Where water is scarce, either chronically or seasonally, the diversion of water to sustain livestock potentially limits its availability for other purposes. This is of particular concern in arid and semi-arid regions, where the construction of boreholes to supply livestock can lead to unsustainable withdrawal rates and the dangerous depletion of aquifer reserves. Stockpiled manure can contaminate bodies of water, causing myriad adverse effects. These include eutrophication, oxygen depletion, sedimentation, contamination with enteric bacteria and possibly other pathogenic organisms, toxic pollution from pesticides, and contamination of groundwater and aquifers with both nitrates and pesticides. Moreover, high concentrations of nitrate in potable water supplies represent a potential health hazard, especially for children.

Harm to Human Health

Using water for farm animals may make it less available for the many uses that influence health, such as bathing, washing, cooking, and drinking. Moreover, excessive contamination by enteric microorganisms, toxic pesticides or nitrates in may render water unfit for human consumption and may be especially dangerous to children. Pesticides or other vector control treatments used on livestock represent threats to the health of livestock managers, their families, and others exposed directly or through water use. These substances may be toxic, cause birth defects, alter children's proper development, promote cancer, or slowly poison one or more organ systems.

Climate Change/Global Warming

Approximately 17% of global methane is produced by livestock digestion and manure. Methane is a greenhouse gas that has 56 times more global warming potential than carbon dioxide over a 20-year time horizon. Furthermore, clearing wild or forest land for new fields reduces the land's ability to act as a carbon sink. Setting fire to vegetation to clear land or promote new growth for livestock to graze on may also contribute to global warming.

Odour

Concentrated manure stored at industrial livestock facilities can generate strong and unpleasant odours, damaging the quality of life of nearby residents. This problem is most evident when facilities are located in densely populated areas.

TECHNOLOGIES AND SPECIFIC IMPACT IN DEVELOPING COUNTRIES

There are a large number of technologies that have been developed for or adapted to the livestock of both developed and developing countries. However, the major technologies that are used effectively in livestock production in the developing world include conserving animal genetic resources, augmenting reproduction, embryo transfer (ET) and related technologies, diagnosing disease and controlling and improving nutrient availability.

TRANSGENICS

Although gene-based technologies have the potential to improve the efficiency of livestock production, thereby ensuring better returns for the farmers, the economic impact of transgenics in the livestock sector will be much less than in the crop sector. However, the global adoption of genetically modified (GM) crops, which were grown on 67.7 million hectares in 2003 compared with 2.8 million hectares in 1996, has had a substantial impact on livestock feed. It is estimated that the United States of America (USA), Argentina, Canada, Brazil and the People's Republic of China have 63%, 21%, 6%, 4% and 4%, respectively, of the global transgenic acreage and that the most frequently grown crops are GM soybean (61%), maize (23%), cotton (11%) and canola (5%).

Although few developing countries have released GM crop varieties, a preliminary analysis reveals that more than 20 developing countries are conducting research into the applications of GM crops. Although transgenic animals (especially mice) are used routinely in research (particularly in the medical field), no GM animals have yet been released on farms. A wide range of traits of potential interest to livestock producers have, however, been the subject of research; for example, the gene responsible for the production of growth hormone (which could be manipulated to increase growth rates), the phytase gene (which could reduce phosphorous emissions from pigs) and keratin genes (which could improve the wool of sheep).

The genetic modification of livestock has proceeded much more slowly than the genetic modification of crops for a variety of reasons, including the high costs, the inefficiency of the gene transfer techniques and the low reproductive rates of animals. Recombinant deoxyribonucleic acid (DNA) approaches have been used to promote the expression of desirable genes, to hinder the expression of undesirable genes, to alter specific genes and to inactivate genes so as to block specific pathways. It is estimated that at least 30 enzymes produced by GM bacteria, yeasts and moulds are currently commercially available worldwide; many of these enzymes are used in the food industry. Genetic engineering has been used to introduce foreign genes into the animal genome or, alternatively, to knock out selected genes. Genes controlling growth were introduced into pigs to increase growth and improve carcass quality. Currently, research is underway to engineer resistance to diseases that affect the animals or that pose an indirect risk to human health, such as Marek's disease and salmonellosis in poultry, scrapie in sheep and mastitis in cattle.

Other studies have tried to increase the casein content of milk or to engineer animals that produce pharmaceutical or industrial chemicals in their milk or semen. No agricultural applications have yet proved commercially successful. Nuclear transfer (NT) technology now provides an alternative route for cell-based transgenesis in domestic species, offering new opportunities for genetic modification. Livestock that produce human therapeutic proteins in their milk, that have organs suitable for xenotransplantation and that are resistant to diseases such as spongiform encephalopathies have been produced by NT from engineered cultured somatic cells.

CHARACTERISING GENETIC VARIABILITY

There is considerable genetic diversity in the livestock of developing countries, much of which controls traits that influence adaptability to harsh environments, productivity and susceptibility to disease and parasitism. However, little if any data on these genetic resources are available. Economic analysis can play an important role in ensuring that conservation efforts are appropriately focused. The primary challenge facing conservationists is to identify sound reasons why society should preserve animals that livestock keepers have abandoned. Jabbar and Diedhiou show that the breeding practices and breed preferences of livestock keepers can successfully be determined by using research techniques such as the revealed preference hedonic approach. On the one hand, conservation cannot be achieved through a conventional breeding programme because the animals carrying the most advantageous traits cannot be easily identified; on the other hand, conservation cannot be achieved through biotechnology because the necessary technologies are either unavailable or uneconomic. In livestock populations with a high degree of genetic variation, molecular markers are being increasingly used to study the distribution and patterns of genetic diversity. Global surveys indicate that 40% of domestic livestock breeds are at risk of extinction. Most of these breeds are found only in developing countries, and often little is known about them or their potential.

Rapid progress is being made in the preparation of dense microsatellite linkage maps to assist in the search for genetic traits of economic importance. These linkage maps can be used to develop strategies for marker assisted selection and marker assisted introgression that will meet the goals of breeding programmes in developing countries. Molecular markers have been widely used in the identification of genotypes and the 'genetic fingerprinting' of organisms. Genotype verification is used intensively to determine the parentage of domestic animals and to trace livestock products in the food chain back to the farm and animal of origin.

REPRODUCTIVE TECHNOLOGIES

The main objectives of using reproductive biotechnologies in livestock are to increase production, reproductive efficiency and rates of genetic improvement. Over the years, many options have become available for managing the reproduction of the major large and small ruminants. Artificial insemination (AI) and preservation of semen are the main technologies that are used extensively. Assessing the fertilisation capacity of sperms, sexing sperms, synchronisation and fixed-time insemination, superovulation, embryo transfer (ET) and *in vitro* embryo production (IVEP) are additional techniques that can improve reproductive efficiency and pregnancy rates. Reproductive technologies can also be used to control reproductive diseases if procedures and protocols are accurately followed.

Artificial Insemination

The conception rate in field AI programmes in developing countries is very low, and therefore the desired effect in terms of animal improvement has not been achieved. Most semen banks still evaluate semen on the basis of sperm motility, even though significant advances have been made in techniques for semen evaluation. Although detailed guidelines are available regarding the processing, storage and thawing of cattle semen and buffalo semen, the processing and handling procedures in laboratories processing semen are often inadequate. Only when farmers have access to considerably better technical and organisational facilities will AI become more effective. At present, the efficiency of the technology is limited by organisational and logistical constraints and by the failure to provide appropriate training for farmers. Several modifications of the technique have been suggested to increase the conception rate.

Synchronisation with different compounds, and the use of gonadotropin-releasing hormone (GnRH) followed seven days later by prostaglandin F2á (PGF2á) can synchronise oestrus and improves the conception rate. In this protocol giving injections of GnRH on day 0, PGF2á on day 7 and GnRH on day 9 is called the 'Ovsynch' programme and synchronises ovulation, permitting timed insemination. The ability to control ovarian follicular and corpus luteum development has allowed insemination in cattle to be timed without having to detect oestrus, and this has increased the net revenue per cow.

Embryo Transfer

One of the major reproductive technologies that can facilitate genetic improvement in cattle is ET. Unfortunately, commercial ET programmes are limited by the high variability in the ovarian follicular response to gonadotropin stimulation. Multiple ovulation and embryo transfer (MOET) takes AI one step further, in terms of both the possible genetic gains and the level of technical expertise and organisation required. In 2001, 450,000 embryos were transferred globally, mainly in dairy cattle, with 62% being transferred in North America and Europe, 16% in South America and 11% in Asia. The main potential advantage of MOET for developing countries is that the elite females of local breeds can be identified, and bulls can be produced from them for use in a field programme of breed improvement.

Zebu cattle and buffaloes in developing countries exhibit less consistent follicular dynamics after superovulation than *Bos taurus* in the developed world. However, over the last 10 to 15 years, the number of transferable embryos produced by zebu donors has increased from 2.4 to 5.8 embryos per flush in the late 1980s to 5.6 to 9.9 embryos per flush in 2000.

The use of ET has been less successful than envisaged for several reasons. The low reproductive efficiency, poor superovulatory responses, very low primordial follicle population and high incidence of atresia all contribute to low embryo production. In buffaloes, embryo recovery was initially less than one, but has subsequently improved to 2.6 with 1.4 transferable embryos per flush. After transferring buffalo embryos to recipients, the conception rate is only 16%.

The poor success rates have limited the use of ET in buffaloes, which are the main dairy animals in developing countries in Asia, South-East Asia and the Mediterranean region.

In Vitro Production of Embryos

Since the birth of the first buffalo calf from an *in vitro* fertilised oocyte, a number of publications have described the effects of different protocols and media on oocyte and embryo development. Two extensive reviews have been published recently.

However, the practical use of IVEP is limited by high production costs and the low overall efficiency under field conditions. High rates of maturation (70% to 90%), fertilisation (60% to 70%) and cleavage (40% to 50%), and moderate to low rates of blastocyst formation (15% to 30%) and calf production (10.5%) have been reported in the literature.

The efficiency of blastocyst production in buffaloes is much poorer than the 30% to 60% reported for cattle. Although viable buffalo blastocysts have been produced from ovaries obtained from abattoirs, the yield of transferable embryos remains low (15% to 39%). Embryos produced *in vitro* have led successfully to pregnancy and calf birth in buffalo, but the success rate is low. Therefore IVEP must be improved before it can be widely used in cattle and buffaloes in developing countries.

IMPROVING HEALTH THROUGH DEVELOPING VACCINES

Most biotechnologies related to health focus on the needs of the developed world, meaning that 90% of health research is devoted to the health problems of 10% of the world's population. Two main approaches are being used to develop vaccines using recombinant DNA technology. The first involves deleting genes that determine the virulence of the pathogen, thus producing attenuated organisms (non-pathogens) that can be used as live vaccines. Currently, this strategy is more effective against viral and bacterial diseases than against parasites. Attenuated live vaccines have been developed against the herpes viruses that cause pseudorabies in pigs and infectious bovine rhinotracheitis in cattle. A number of candidate *Salmonella* vaccines have also been produced. The second approach is to identify protein subunits of pathogens that can stimulate immunity.

The International Livestock Research Institute (ILRI) used this approach to develop a vaccine against *Theileria parva*, the parasite that causes East Coast fever in African cattle. A novel strategy for developing vaccines against bloodsucking parasites involves using components of the gut wall of the parasite that are not usually exposed to the immune system of the host. When the parasite feeds, it ingests antibodies induced by the vaccine, which destroy the gut wall and, consequently, kill the parasite. This strategy has been used successfully to develop a vaccine against the one-host tick *Boophilus microplus*.

Vaccination is one of the most effective and sustainable methods of controlling disease. Vaccines against parasitic diseases in Africa and viral diseases in Asia have been shown to control disease effectively and increase livestock productivity. A recent approach has been to use vaccines based on DNA. The use of DNA in vaccines is based on the discovery that injecting genes in the form of plasmid DNA can stimulate an immune response to the respective gene products. This immune response is a result of the genes being taken up and expressed by cells in the animal after injection. The live-vector and DNA vaccination systems could be manipulated further to enhance the immunity conferred by the gene products. Experimental studies have demonstrated that these vaccines can potentially induce appropriate and enduring immune responses.

This technology is, in principle, one of the simplest and yet most versatile methods of inducing both humeral and cellular immune responses, as well as protecting against a variety of infectious agents. However, although immune responses have been induced in a number of larger species, most of the information on the efficacy of DNA immunisation comes from studies of mice. An exhaustive review of the information available on the use of DNA vaccines in farm animals, including cattle, pigs and poultry, has identified the areas that need specific attention before this technology can be used routinely.

These areas include the delivery, safety and compatibility of plasmids in multivalent vaccines and the potential for using immune stimulants as part of a DNA vaccine. Korean scientists have developed a combined vaccine against pleuropneumonia, pneumonic pasteurellosis and enzootic pneumonia in swine. Molecular biology has been used to produce an improved vaccine against swine fever. In the Philippines, a vaccine has been developed that protects cattle and water buffalo against haemorrhagic septicaemia, which is the leading cause of death in these animals. The new vaccine provides improved protection at a very low cost.

DIAGNOSTICS AND EPIDEMIOLOGY

Advanced diagnostic tests that use biotechnology enable the agents causing disease to be identified and the impact of disease control programmes to be monitored more precisely than was previously possible. Molecular epidemiology characterises pathogens (viruses, bacteria, parasites and fungi) by nucleotide sequencing, enabling their origins to be traced. This is particularly important for epidemic diseases, in which pinpointing the source of the infection can significantly improve disease control. For example, the molecular analysis of rinderpest viruses has been vital in determining the lineages circulating in the world and instrumental in aiding the Global Rinderpest Eradication Programme. Enzyme-linked immunosorbent assays have become the standard means of diagnosing and monitoring many animal and fish diseases worldwide, and the PCR technique is especially useful in diagnosing livestock disease.

Many diagnostic techniques currently used in developing countries are cumbersome and unsuitable for low-resource settings. Molecular diagnostic technologies that are either already in use or being tested in low-income regions include polymerase chain reaction (PCR), monoclonal antibodies and recombinant antigens. These technologies can be modified to facilitate their application in the developing world. Simple hand-held devices that rely on the binding specificity of monoclonal antibodies or recombinant antigens to diagnose infection may be easily adapted for use in settings without running water, refrigeration or electricity. Molecular characterisation of the virus serotypes causing foot and mouth disease has helped in the vaccination and control programmes in Asia.

In Japan and Taiwan, DNA testing is being used to diagnose hereditary weaknesses of livestock. One test looks for the presence of the gene responsible for porcine stress syndrome in pigs. Pigs with this gene tend to produce pale poor-quality meat because of their reaction to the stress of transport and slaughter. Pigs with this gene can now be excluded from breeding programmes, so the gene will become less common.

In addition, DNA testing is being used in Japan to check for the gene that causes leucocyte adhesion deficiency in Holstein cattle. Cattle with this condition suffer from gum disease, tooth loss and stunted growth. They usually die before they are one year old. By using DNA testing, carriers can be identified and eliminated from breeding herds. Bulls used for breeding can also be tested to make sure that they are not carriers. Another DNA test identifies a gene that leads to anaemia and retarded growth in Japanese Black cattle.

NUTRITION AND FEED UTILISATION

The shortage of feed in most developing countries and the increasing cost of feed ingredients mean that there is a need to improve feed utilisation. Aids to animal nutrition, such as enzymes, probiotics, single-cell proteins and antibiotics in feed, are already widely used in intensive production systems worldwide to improve the nutrient availability of feeds and the productivity of livestock. Gene-based technologies are being increasingly used to improve animal nutrition, either through modifying the feeds to make them more digestible or through modifying the digestive and metabolic systems of the animals to enable them to make better use of the available feeds. Feeds derived from GM plants (a quarter of which are now grown in developing countries), such as grain, silage and hay, have contributed to increases in growth rates and milk yield.

Genetically modified crops with improved amino acid profiles can be used to decrease nitrogen excretion in pigs and poultry. Increasing the levels of amino acids in grain means that the essential amino acid requirements of pigs and poultry can be met by diets that are lower in protein. Metabolic modifiers have also been used to increase production efficiency (weight gain or milk yield per feed unit), improve carcass composition (meat-fat ratio), increase milk yield and decrease animal fat.

The use of recombinant bovine somatotropin (rBST) in dairy cows increases both milk yield and production efficiency and decreases animal fat. In the USA, the use of rBST typically increases milk yield by 10% to 15%. Although trials conducted in developing countries have reported a similar percentage increase, this increase is not significant because of the low milk yields and the high cost-benefit ratio. However, rBST is being used commercially in 19 countries where the economic returns make its use worthwhile. A porcine somatotropin has been developed that increases muscle growth and reduces body-fat deposition, resulting in pigs that are leaner and of greater market value.

CONSTRAINTS ON APPLYING THE TECHNOLOGY

The application of new molecular biotechnologies and new breeding strategies to the livestock breeds used in smallholder production systems in developing countries is constrained by a number of factors. In the developing world, poverty, malnutrition, disease, poor hygiene and unemployment are widespread, and biotechnologies must be able to be applied in this context. Over the last few decades, the green revolution has brought comparative prosperity to farmers with land, but the majority of farmers, who are landless or marginal farmers and subsist only on livestock, have been neglected and remain poor.

The major constraints on applying biotechnologies have been enumerated by Madan and include:

- The absence of an accurate and complete database on livestock and animal owners so that programmes can be implemented
- The biodiversity present within species and breeds in agro-ecological systems
- The fact that models of biotechnological intervention differ distinctly between developed and developing economies
- The fact that many animal species and breeds are unique to the developing world; each has its own distinct developmental, production, disease resistance and nutrient utilisation characteristics
- The lack of trained scientists, technicians and fieldworkers to develop and apply the technologies, both in the government and in the private sectors
- The absence of an interface between industry, universities and institutions, which is necessary to translate technologies into products
- The inability to access technologies from the developed world at an affordable price in order to make a rightful, positive and sustainable contribution to livestock production and the economic welfare of farmers
- The high cost of technological inputs such as materials, biologicals and equipment
- The failure to address issues of biosafety and to conduct risk analyses of new biologicals, gene products, transgenics and modified food items, and, above all
- The negligible investment in animal biotechnology.

The critical issues affecting livestock productivity have recently been re-examined. Research that aims to enhance productivity and sustainability should focus on improving livestock feeds and nutrition, improving animal health, managing natural resources relating to the livestock sector, assesing the impact of technological interventions, and strengthening the capacity of the national agricultural research systems of developing countries.

Furthermore, the potential production capacity and contribution of livestock to the economy are still not being achieved in developing countries because the transfer, adaptation and adoption of technology is hampered by the lack of a clear policy for livestock development that is conducive to the introduction of new proven technology and by the lack of information flow from and to decision makers. In developing countries, there is a wealth gap between urban and rural areas, which persists and may even be widening; the rural-urban divide also tends to be reflected in education and health indicators. In addition, women in rural (and urban) areas who are predominantly involved in animal husbandry have higher illiteracy rates than men. A survey of 21 African countries recently highlighted the substantial disparities in primary schooling between urban and rural areas, in favour of urban dwellers.

Special attention must be given to the knowledge and information needed to enable rural people to apply biotechnology. There is a need to identify alternative delivery systems for animal healthcare and to propose new roles for the state and the private sector in service delivery.

BUILDING CAPACITY

Owing to the constraints outlined above, the economic benefits of animal biotechnology cannot be realised without a conscious, sustained, holistic, multi-stakeholder, participatory approach. There is a great need to ensure that capacity is not just created but also is retained and enhanced. Capacity-building activities must be carried out at all levels: the awareness of policy and decision makers must be raised, the necessary legal and regulatory frameworks must be initiated, the technical and regulatory capacities must be enhanced and institutions may need to be overhauled. More importantly, it is necessary to assess and deploy competent operators and institutional capacity continuously so that, as biotechnology advances, the procedures required for its safe use can be constantly evaluated, upgraded and applied. This is a daunting task, but it can be achieved through firm commitment and partnerships.

FUNDING TO IMPLEMENT TECHNOLOGY

Developing and commercialising improved technologies in most developing countries has been the responsibility of the public sector, and technology has been disseminated freely. This situation will have to continue if superior genetics, diagnostics and vaccines are to be delivered. However, research and almost all commercial development of biotechnology in the developed world are being driven largely by the private sector. The global trends in funding for research and development and production do not address the concerns, needs and opportunities of the developing world. Developing countries are finding it increasingly expensive to access and use new technologies. There is limited private-and public-sector investment in animal health and production, particularly in relation to modern biotechnologies that are 'resource hungry'. Although several discoveries have been made in laboratories in the developing world, in most cases these have not been converted into useful technologies or products. The key potential users–resource-poor often illiterate farmers with a limited knowledge base–do not feel that applying these technologies is worth the effort, cost and risk involved. This is mainly because there is no agency or industry that can scale up and package the technology. Also, in the developed world, there is an economic incentive to market biotechnological services and products; this is lacking in the developing world because of the limited purchasing power of resource-poor stakeholders.

Research in biotechnology in recent years has also been motivated by economic considerations, and little research is conducted in the developing world because of the probable lack of returns on the investment. For understandable reasons, current funding policies in developing countries focus on areas that will yield practical benefit in the short term. In determining future policy, policy-makers and funding bodies must not lose sight of the substantial benefits that can be gained in the longer term by investing in strategic research into vaccine development. Adequate multi-institutional (national and international) support through an international donor consortium is needed to develop cost-effective, cheap and easily adaptable biotechnological products.

The World Bank, the Food and Agriculture Organization, the Consultative Group on International Agricultural Research, the United Nations Development Programme, the United States Agency for International Development, the Swedish International Development Cooperation Agency, the International Development Research Centre, the Asian Development Bank and other donor and funding agencies have to designate a higher percentage of funds to the livestock sector. It has been convincingly shown that investing in livestock has a dramatic and far-reaching impact on the human development index. This is a strong argument in favour of investing heavily in animal production and health biotechnologies in order to bring economic prosperity, nutritional security, rural development and health improvements to poor populations in the developing world.

Although animal production is being changed significantly by advances made in thousands of biotechnology laboratories around the world, benefits are reaching the developing world in only a few areas of conservation, animal improvement, healthcare (including diagnosis and control of disease) and the augmentation of feed resources. Adopting biotechnology has resulted in distinct benefits in terms of animal improvement and economic returns to the farmers. Over the past decade, the ILRI has focused on biotechnological applications, especially in Africa, and several developing countries now have multi-institutional programmes to develop and apply biotechnology. The developing world will have to respond to the many genebased technologies now being developed with a sense of commitment, trained manpower, infrastructure and funding.

THE LIVESTOCK ECONOMY IN DEVELOPING COUNTRIES

Livestock is becoming increasingly important in the growth of agriculture in developing economies. The contributions made by livestock to both agriculture and gross domestic product (GDP) have risen, at a time when the contribution of agriculture to GDP has fallen. The demand for livestock products is a function of income, and sustained growth in per capita income, rising urban populations and changes in diet and lifestyle are fuelling growth in livestock production. Livestock production contributes to socioeconomic development in many ways, by augmenting income and employment and reducing the incidence of rural poverty. Though the role of livestock in ensuring nutritional security is recognised in mixed crop-livestock systems, the importance of livestock goes beyond direct food production. Livestock supply draught power and organic manure to the crop sector, and hides, skins, bones, blood and fibre are used in many industries. Thus, livestock are an important source of income and employment, helping to alleviate poverty and smooth the income distribution among small landholders and the landless, who constitute the bulk of the rural population and the majority of livestock owners. In addition, livestock can easily be converted into cash and thus act as a cushion against crop failure, particularly in less favoured environments. By enabling crop residues and by-products to be used as fodder, livestock production contributes positively to the environment.

Animal owners in the developing world are predominantly resource-poor small-scale operators with little or no land and few animals, who must operate within the constraints of the local climate and who have limited purchasing power and little access to resources or opportunity to determine resource allocation for animal production. The situation of the poorest livestock owners is fast deteriorating owing to the fragmentation of limited holdings, exhaustion of land resources and increasing human and animal population pressure. Low livestock productivity in many developing countries is considered to reflect, among other things, the inadequate supply of animal husbandry and veterinary services. Veterinary services have traditionally been provided by the State, but financial constraints have limited the availability and effectiveness of public services.

THE IMPLICATIONS OF TECHNOLOGY

A major benefit of agricultural research and technology is that the purchasing power of the poor increases, because both average incomes and access to staple food products are improved. Studies by economists have provided empirical support for the proposition that growth in the livestock sector affects the whole economy. Rapid growth of livestock production has stimulated demand for and increased the value of land, labour and nonagricultural goods and services, thereby leading to overall economic growth. The poor spend a relatively high proportion of any additional income on food, so increases in livestock production achieved through the use of biotechnology can have major nutritional implications, particularly if the technology is aimed at the poorest producers. However, studies have revealed that the commerci-alisation of agriculture has reduced the nutritional security of the poor. Once production of milk, meat or eggs has been enhanced through the use of technology, it is hoped that it will also make a significant difference in other areas such as nutrition, prevention of diseases, healthcare and other management practices. It is in these areas that biotechnology shows promise and is currently being used. Green Revolution technologies (*i.e.* those technologies designed to improve the efficiency of agricultural processes and increase crop productivity by relying on the extensive use of chemical fertilizers/pesticides and heavy machinery) are intended to be used in package form (*e.g.* new plant varieties supplied with recommendations on fertilizer, pesticide and herbicide rates and water control measures); however, among livestock producers many components of these technologies have been taken up in a piecemeal, often stepwise, manner. The sequence of adoption is determined by availability and by the potential cost savings.

The sequential adoptions of crop management technologies for rice and wheat have been assessed in detail, but few similar studies have focussed on livestock production in developing countries. Evidence from the People's Republic of China, Mexico, South Africa and India suggests that small farmers have had no more difficulty than larger farmers in adopting the new technologies. The question, therefore, is not whether biotechnology can benefit small-scale resource-poor farmers, but rather how biotechnology can address the agricultural problems faced by farmers in developing countries. Biotechnology is a promising new tool in the development of applied agricultural technologies. The challenge is to focus this potential on the problems experienced by developing countries. The introduction of multi-enterprise systems or, more broadly, agricultural diversification is seen as the way forward for agriculture in the developing world; such systems could lift small-scale and marginal farmers out of poverty.

For example, rotating rice and wheat cropping with dairy farming yields higher profits. Introducing multi-enterprise systems involving livestock enhances the purchasing power of farmers and helps them to obtain nutritional security. It also generates rural (both farming and non-farming) employment, thus preventing excessive migration to urban areas, which is a common problem in developing economies. Multi-enterprise systems also support the natural environment and contribute to capital formation, thus leading to higher overall growth in the agricultural economy. The technology, infrastructure and institutions now exist to make the application of biotechnology in the context of a multi-enterprise system involving livestock production economically viable.

Global Advantage from Livestock of Developing Countries

The multiplicity of genes, species, populations and agroecosystems in the developing countries of South and South East Asia, the People's Republic of

China, Africa and Latin America is viewed as a valuable resource for the genetic improvement of livestock on a global scale. The livestock in these countries, which are an integral part of a fragile ecosystem, are a rich source of animal biodiversity. Buffaloes, sheep, goats, camels and zebu cattle have adapted to their regional environments over thousands of years and have provided an important source of sustenance for the population of the region.

Livestock production in the developing world has a number of advantages over production in more developed countries, for example:

- The unique and valuable production traits of buffaloes, cattle, sheep, goats and camels
- The low-input production system
- The low unit cost of production
- The lean meat produced from sheep, goats and buffaloes
- The considerable biodiversity
- Animal breeds that are resistant to stress and to particular diseases
- The ability of the animals to survive on high-roughage feeds
- The potential for biopharmaceutical developments to lead to significant benefits
- The potential for expanding the microbial food, feed and leather industries
- The integrated production system tailored to the local ecology
- The potential for integrating knowledge and industry.

Several genes and desirable traits have been identified in the livestock of developing countries in Asia and Africa, and some of the livestock species and breeds from these countries have become major contributors to the economy of South America.

Examples of breeds from the developing world that are particularly important on a global level are:

- Buffaloes that produce milk with a high fat content or with the protein quality required to produce mozzarella cheese
- Goats from cold dry regions that produce pashmina and toos (the finest wool in the world)
- Black bengal goats that carry a gene for high prolificacy
- Garole sheep that carry genes for twinning
- Andaman goats that are highly tolerant of salt
- The yak and mithun that are adapted to high altitude
- The camels, sheep and goats that are adapted to a tropical arid environment and can tolerate feed with a high lignin content
- The many species that are resistant to stress or to particular diseases.

Economic Impact of Technologies

The genetic resources possessed by animals in developing countries often affect economic development. Traxler has discussed the economic impacts of biotechnological innovations, but the research and policy options need separate consideration. Animal biotechnology is the result of a multistage process, involving research, development, testing and registration, production and marketing. The goal is to develop a technology, process or product that has clear commercial potential and can be commercialised after due testing and regulatory approval. Developing countries find it difficult to develop biotechnology because the facilities or resources needed to complete all of the stages in the process are often lacking. However, several technologies from developed countries have been successfully adopted by developing countries.

The impact of technology can be analysed by estimating the growth of total factor productivity (TFP) in livestock production. Not many TFP studies on livestock have been reported.

However, separate TFP estimates for the aggregate crop and livestock sectors have been made. TFP analysis has shown a shift towards larger, more commercial and more intensive production systems and has further revealed that, as specialisation has developed over the past decade, the importance of backyard livestock production has declined and the importance of specialised household and commercial enterprise has increased. Studies from India have shown that technological input is responsible for about 45% of total output growth and that the TFP growth may be as much as 1.8%.

Bibliography

A. K. Shrivastava: *Advances in Environmental Biotechnology*, Agrotech Press, Delhi, 2013.

A. Sahoo and R.K. Singh: *Animal Health and Production*, Satish Serial Publishing House, Delhi, 2013.

Chiranjib Chakraborty: *Advances in Biochemistry and Biotechnology, Vol. I*, Daya Publication, Delhi, 2005.

D. Swarup and S.K. Dwivedi: *Environmental Pollution and Effects of Lead and Fluoride on Animal Health*, Indian Council of Agricultural Research, 2002.

Garg Sudhi Ranjan: *Environmental Security : Human and Animal Health*, IBDC Publishers, Delhi, 2011.

H.C.L. Gupta: *Household Pests of Human and Animal Health Importance*, Agrotech Publishing Academy, Delhi, 2011.

Irfan Ali Khan and Atiya Khanum: *Advances in Biotechnology*, Ukaaz Publication, Delhi, 2006.

Mahendra Rai: *Advances in Fungal Biotechnology*, I K International, Delhi, 2009.

Manjula K. Saxena and B.B.S. Kapoor: *Advances in Biotechnology*, Madhu Publications, Delhi, 2009.

P R Gilbert: *Advances in Biotechnology*, Anmol Publication, Delhi, 2008.

P.C. Trivedi: *Advances in Biotechnology*, Agrobios Publication, Delhi, 2005.

P.R. Jatkar, Pradeep Parihar and Leena Parihar: *Advances in Applied Biotechnology and Microbiology*, Agrobios Publication, Delhi, 2010.

Pradeep Parihar and Leena Parihar: *Advances in Applied Biotechnology*, Agrobios Publication, Delhi, 2010.

R.K.S. Kushwaha: *Fungi in Human and Animal Health*, Scientific Publication, Delhi, 2004.

R.P. Singh: *Neem in Human and Animal Health*, IBDC Publishers, Delhi, 2013.

Ramanand Sagar: *Advances in Agricultural Biotechnology*, Cyber Tech Publication, Delhi, 2009.

S. R. Garg: *Nanotechnology for Animal Health and Production*, Daya Publication, Delhi, 2014.

S.K. Jindal and M.C. Sharma: *Biotechnology in Animal Health and Production*, New India Publication, Delhi, 2010.

S.N. Mukhopadhyay: *Advanced Process Biotechnology*, Viva Books, Delhi, 2008.

Sohan Sharma: *Advances in Fungal Biotechnology*, Random Publication, Delhi, 2012.

Subodh K. Singh: *Animal Health*, Bio-Green Books, Delhi, 2012.

Sudhi Ranjan Garg: *Human and Animal Health : Environmental Perspectives*, Satish Serial Publishing House, Delhi, 2013.

Susan C. Cork and Roy W. Halliwell: *The Veterinary Laboratory and Field Manual : A Guide For Veterinary Laboratory Technicians and Animal Health Advisors*, Nottingham University Press, 2009.

V.S. Harikumar: *Advances in Agricultural Biotechnology*, Regency Publication, Delhi, 2003.

Zahoor A. Pampori: *Field cum Laboratory Procedures in Animal Health Care*, Daya Publication, Delhi, 2003.

Index